Communications in Computer and Information Science 2301

Series Editors

Gang Li ⓘ, *School of Information Technology, Deakin University, Burwood, VIC, Australia*
Joaquim Filipe ⓘ, *Polytechnic Institute of Setúbal, Setúbal, Portugal*
Zhiwei Xu, *Chinese Academy of Sciences, Beijing, China*

Rationale
The CCIS series is devoted to the publication of proceedings of computer science conferences. Its aim is to efficiently disseminate original research results in informatics in printed and electronic form. While the focus is on publication of peer-reviewed full papers presenting mature work, inclusion of reviewed short papers reporting on work in progress is welcome, too. Besides globally relevant meetings with internationally representative program committees guaranteeing a strict peer-reviewing and paper selection process, conferences run by societies or of high regional or national relevance are also considered for publication.

Topics
The topical scope of CCIS spans the entire spectrum of informatics ranging from foundational topics in the theory of computing to information and communications science and technology and a broad variety of interdisciplinary application fields.

Information for Volume Editors and Authors
Publication in CCIS is free of charge. No royalties are paid, however, we offer registered conference participants temporary free access to the online version of the conference proceedings on SpringerLink (http://link.springer.com) by means of an http referrer from the conference website and/or a number of complimentary printed copies, as specified in the official acceptance email of the event.

CCIS proceedings can be published in time for distribution at conferences or as post-proceedings, and delivered in the form of printed books and/or electronically as USBs and/or e-content licenses for accessing proceedings at SpringerLink. Furthermore, CCIS proceedings are included in the CCIS electronic book series hosted in the SpringerLink digital library at http://link.springer.com/bookseries/7899. Conferences publishing in CCIS are allowed to use Online Conference Service (OCS) for managing the whole proceedings lifecycle (from submission and reviewing to preparing for publication) free of charge.

Publication process
The language of publication is exclusively English. Authors publishing in CCIS have to sign the Springer CCIS copyright transfer form, however, they are free to use their material published in CCIS for substantially changed, more elaborate subsequent publications elsewhere. For the preparation of the camera-ready papers/files, authors have to strictly adhere to the Springer CCIS Authors' Instructions and are strongly encouraged to use the CCIS LaTeX style files or templates.

Abstracting/Indexing
CCIS is abstracted/indexed in DBLP, Google Scholar, EI-Compendex, Mathematical Reviews, SCImago, Scopus. CCIS volumes are also submitted for the inclusion in ISI Proceedings.

How to start
To start the evaluation of your proposal for inclusion in the CCIS series, please send an e-mail to ccis@springer.com.

Wenwu Zhu · Hui Xiong · Xiuzhen Cheng ·
Lizhen Cui · Zhicheng Dou · Junyu Dong ·
Shanchen Pang · Li Wang · Lanju Kong ·
Zhenxiang Chen
Editors

Big Data

12th CCF Conference, BigData 2024
Qingdao, China, August 9–11, 2024
Proceedings

Editors
Wenwu Zhu
Tsinghua University
Beijing, China

Xiuzhen Cheng
Shandong University
Shandong Province, China

Zhicheng Dou
Renmin University of China
Beijing, China

Shanchen Pang
China University of Petroleum
Qingdao, China

Lanju Kong
Shandong University
Jinan, China

Hui Xiong
Hong Kong University of Science
and Technology
Guangzhou, Hong Kong

Lizhen Cui
Shandong University
Shandong Province, China

Junyu Dong
Ocean University of China
Qingdao, China

Li Wang
Taiyuan University of Technology
Shanxi Province, China

Zhenxiang Chen
Jinan University
Jinan, China

ISSN 1865-0929 ISSN 1865-0937 (electronic)
Communications in Computer and Information Science
ISBN 978-981-96-1023-5 ISBN 978-981-96-1024-2 (eBook)
https://doi.org/10.1007/978-981-96-1024-2

© The Editor(s) (if applicable) and The Author(s), under exclusive license
to Springer Nature Singapore Pte Ltd. 2025

This work is subject to copyright. All rights are solely and exclusively licensed by the Publisher, whether the whole or part of the material is concerned, specifically the rights of translation, reprinting, reuse of illustrations, recitation, broadcasting, reproduction on microfilms or in any other physical way, and transmission or information storage and retrieval, electronic adaptation, computer software, or by similar or dissimilar methodology now known or hereafter developed.
The use of general descriptive names, registered names, trademarks, service marks, etc. in this publication does not imply, even in the absence of a specific statement, that such names are exempt from the relevant protective laws and regulations and therefore free for general use.
The publisher, the authors and the editors are safe to assume that the advice and information in this book are believed to be true and accurate at the date of publication. Neither the publisher nor the authors or the editors give a warranty, expressed or implied, with respect to the material contained herein or for any errors or omissions that may have been made. The publisher remains neutral with regard to jurisdictional claims in published maps and institutional affiliations.

This Springer imprint is published by the registered company Springer Nature Singapore Pte Ltd.
The registered company address is: 152 Beach Road, #21-01/04 Gateway East, Singapore 189721, Singapore

If disposing of this product, please recycle the paper.

Preface

The CCF Big Data Conference is one of the most influential and largest academic conferences, and has been successfully held 11 times. It aims to build an academic and industry-oriented high-end platform, promote exchanges and cooperation in academic research, enterprise application, management and governance at home and abroad, and at the same time, give play to the role of national think tanks to provide decision support for national big data development strategy, application and governance.

The 12th CCF BigData Conference (CCF BigData 2024) was held in Qingdao, Shandong Province from August 9–11, 2024, jointly hosted by Big Data Expert Committee and Shandong University. With the theme of "Model-Driven Innovation, Data Generating the Future", the conference will explored the opportunities and challenges facing the field of big data in the era of the digital economy and big models, and carried out academic exchanges and discussions on the processing, management, analysis and governance of big data. It provided a platform for exchanges and cooperation between experts from academia, industry and relevant functional departments, shared the latest research results and practical experience, explored the innovative application of big data in various fields, promoted the development of big data technology and industry, and contributed wisdom and strength to the construction of digital China.

This book constitutes the refereed proceedings of the 12th CCF Conference on BigData 2024. The 26 full papers presented in this volume were carefully reviewed and selected from 219 submissions. The topics of accepted papers include Big Data, Data Science, System Architecture and Infrastructure, Storage Management, Parallel Computing, Analysis Mining and Intelligent Computing, Collection and Preprocessing, and Governance.

Organization

Honorary Presidents

Guojie Li   Institute of Computing Technology, Chinese Academy of Sciences, China
Hong Mei   Peking University, China

Chairs of the Steering Committee

Xueqi Cheng   Institute of Computing Technology, Chinese Academy of Sciences, China
Xiaoyong Du   Renmin University of China, China
Hai Jin   Huazhong University of Science and Technology, China

Presidents of the General Assembly

Yunsheng Hua   Chinese University of Hong Kong, China
Changjun Jiang   Tongji University, China
Zhen Wu   Shandong University, China

Procedure Committee Chairs

Wenwu Zhu   Tsinghua University, China
Hui Xiong   Hong Kong University of Science and Technology (Guangzhou), China
Xiuzhen Cheng   Shandong University, China

Chairs of the Organizing Committee

Lizhen Cui   Shandong University, China
Zhicheng Dou   Renmin University of China, China
Junyu Dong   Ocean University of China, China
Shanchen Pang   China University of Petroleum, China

Publishing Chairs

Li Wang	Taiyuan University of Technology, China
Lanju Kong	Shandong University, China
Zhenxiang Chen	University of Jinan, China

Publicity Chairs

Guoxian Yu	Shandong University, China
Xiang Ao	Institute of Computing Technology, Chinese Academy of Sciences, China
Fei Teng	Southwest Jiaotong University, China

Forum Chairs

Jianxin Li	Beihang University, China
Huawei Shen	Institute of Computing Technology, Chinese Academy of Sciences, China
Yuanjie Zheng	Shandong Normal University, China

Best Paper Selection Chairs

Enhong Chen	University of Science and Technology of China, China
Peng Cui	Tsinghua University, China
Shaoliang Peng	Hunan University, China

Finance Chairs

Lixuan Shen	China Computer Federation, China
Zhongmin Yan	Shandong University, China
Xiaoming Wu	Qilu University of Technology, China

Sponsoring Chairs

Yidong Li	Beijing Jiaotong University, China
Yilei Lu	Beijing Baihai Technology Co., Ltd., China

Zhaohui Peng	Shandong University, China
Shengke Wang	Ocean University of China, China
Jia Yu	Qingdao University, China
Yongquan Liang	Shandong University of Science and Technology, China

Organizer: China Computer Federation (CCF)
Organizer: CCF Big Data Expert Committee, Shandong University

Contents

A Flexible Knowledge Graph Error Detection Framework Combined
with Semantic Information 1
   Yangwu Zhao, Yang Liu, Xiang Ao, and Qing He

Session-Level Normalization and Click-Through Data Enhancement
for Session-Based Evaluation 15
   Haonan Chen, Zhicheng Dou, and Jiaxin Mao

A Multi-dimensional Early Warning Mechanism for Biological Invasions:
A Case Study of Vespa Mandarinia 34
   Shiqi Zhang and Weidong Xiao

DLP-FR: Learning Predictable Degradation for Robust Blind Face
Restoration 47
   Tao Wu, Jie Cao, Huaibo Huang, Yuang Ai, and Ran He

A Joint Relation Extraction Model Utilizing Message Passing Mechanism
for Feature Fusion to Alleviate FN Problem from Distant Supervision 62
   *Jiran Zhu, Hongyun Du, Shengjie Jia, Yanhui Ding, Hui Yu, Weizhi Xu,
and Xiujie Wu*

Urban Traffic Management: A Predictive Approach Using Mobile Phone
Data 78
   *Xiaolong Niu, Huachao Gao, Wentao Hu, Long Yan, Yawei Ren,
Yang Li, Dazhong Li, Hao Ge, Chunxiao Li, Yanyan Xu, and Yulun Song*

FedPSED: Federated Public Safety Emergency Detection Based
on Adaptive Aggregation Strategy 90
   Jiping Fan, Junping Du, Zhe Xue, Ang Li, Zeli Guan, and Meiyu Liang

Evaluation of Retrieval-Augmented Generation: A Survey 102
   Hao Yu, Aoran Gan, Kai Zhang, Shiwei Tong, Qi Liu, and Zhaofeng Liu

CMNet: Fast Time Series Forecasting Based on Hybrid Convolution-MLP
Architecture 121
   *Yikun Yang, Kailiang Chen, Shufen Chen, Jiaen Chen, Renzhong Niu,
Wenbin Chen, and Zhigang Li*

A Domain Adaptive Based Reinforcement Learning Algorithm for Job
Shop Scheduling Problems 134
  Kuan Miao, Lilan Peng, Wuyang Zhang, Jibao Pan, Chongshou Li,
  and Tianrui Li

Large-Scale Data Generation Using SWLBM on the Sunway TaihuLight
Supercomputer and Subsequent Data Mining with Physics-Informed
Neural Networks 147
  Xuesen Chu, Wei Guo, Tianqi Wu, Shengze Cai, and Guangwen Yang

Enhancing Interaction Graph of Data Schema and Syntactic Structure
with Pre-trained Language Model for Text-to-SQL 159
  Wenbin Zhao, Long Zhao, Feng Wu, Zixuan Zheng, Haoxin Jin,
  and Bin Gu

Spatial Evolution Analysis of the Level of Digital Economy Development
in China 174
  Jing Feng, Tongle Han, Ziqi Xia, Xuehua Zhang, and Yunhan Qu

Predicting Calibrated Conversion Rate of Online Advertising Using
a Multi-task Mixture-of-Experts Calibration Model 189
  Xinyue Zhang, Yuyao Guo, and Xiang Ao

A Transformer-Based Spatio-Temporal Graph Neural Network
for Anomaly Detection on Dynamic Graphs 202
  Yuanjun Gao, Quntao Zhu, Xuanhua Shi, and Hai Jin

Morphological Semantic Ensemble Filtering of Massive Sentence Pairs
for Neural Machine Translation 218
  Lin Wang and Wuying Liu

Improving Event-Level Financial Sentiment Analysis
with Retrieval-Augmented Multipath Chain-of-Thought Prompting 232
  Yiming Zhang, Xiang Ao, Guoxin Yu, and Qing He

POSRho: Efficient Spearman's Rho Calculation for Big Data 247
  Xiaofei Zhao and Fanglin Guo

A Fusion Tuning Method for Named Entity Recognition 260
  Jitian Wang, Yanping Chen, Anqi Zou, Yongbin Qin, and Ruizhang Huang

Design of AXI Bus-Based IP Core for Image Processing 275
  Yutong Chen, Zhongchao Yi, Xuqiang Li, Xingyan Chen,
  and Yanjiang Chen

Automated Clinical Summary Generation via Integrating Structured
and Unstructured Data 290
 Jiaojiao Fu, Bowen Yang, Yi Guo, Yangfan Zhou, and Xin Wang

A Review on Deep Learning for Sequential Recommender Systems: Key
Technologies and Directions 305
 Yuchen Liu, Jianpeng Qi, and Yanwei Yu

TrajUT: Intruder Trajectory Recovery on Utility Tunnel via Video
Surveillance Systems 319
 *Wenbin Song, Baijian Yin, Xinwei Li, Shuai Wang, Shuai Wang,
 and Zhao-Dong Xu*

Spatio-Temporal Graph Fusion Network-Based Multivariate Time Series
Forecasting of Environmental Factors in Utility Tunnels 333
 *Wenbin Song, Peiyi Zhao, Xinwei Li, Shuai Wang, Shuai Wang,
 and Zhao-Dong Xu*

Improving Spatial Co-location Pattern Mining with Enhanced Neighbor
Relationship Measures 345
 Liang Xu, Lizhen Wang, Vanha Tran, and Hongmei Chen

PFG: Generation of Paper-Style Handwritten Formulas for Enhancing
Handwritten Mathematical Expression Recognition 361
 Ze Liu, Kai Zhang, Yanghai Zhang, Zhe Yang, Qi Liu, and Enhong Chen

Author Index 373

A Flexible Knowledge Graph Error Detection Framework Combined with Semantic Information

Yangwu Zhao[1,2,3], Yang Liu[2,3], Xiang Ao[1,2,3], and Qing He[1,2,3(✉)]

[1] Henan Institute of Advanced Technology, Zhengzhou University, Zhengzhou, China
yangwu@gs.zzu.edu.cn
[2] Key Lab of AI Safety, Chinese Academy of Sciences (CAS), Beijing, China
[3] Key Lab of Intelligent Information Processing, Institute of Computing Technology, CAS, Beijing, China
heqing@ict.ac.cn

Abstract. Knowledge graphs (KGs) are extensively utilized in numerous applications, including question-answering systems and recommender systems. However, knowledge graphs are often constructed through web crawling or crowdsourcing, leading to errors in the data. The task of knowledge graph error detection aims to identify inaccurate triplets in KGs and has received substantial attention in recent years. However, the majority of current error detection methods overlook the semantic information of the triplets, which can be vital for accurate error detection. In this paper, we introduce a Flexible Knowledge Graph Error Detection Framework that integrates Semantic Information (FKED), which combines both structural and semantic information to detect errors within the knowledge graph. FKED first extracts the structural information of KGs using a graph embedding model. Next, FKED employs a pre-trained language model (PLM) to extract semantic information from the triplets. Then the structural and semantic information are combined to detect errors within the knowledge graph. FKED can be flexibly added to other structure-based error detection models, enhancing their capabilities in downstream tasks of knowledge graphs. We assess FKED using two benchmark datasets: FB15k-237 and WN18RR. Experimental results indicate that our method is both superior and effective, substantially improving the performance of the base models.

Keywords: Knowledge Graphs · Error Detection · Semantic Information

1 Introduction

Knowledge graphs (KGs) provide a structured way to represent knowledge through entities, relations, and attributes. It finds extensive application across various fields, such as recommender systems [5,7], question answering systems [12,17], and search engines [11,21]. However, knowledge graphs are generally constructed through web crawling or crowdsourcing, which inevitably leads to the presence of errors in their data. Unrelated entities may be erroneously linked, or the relationships between related entities may be incorrectly represented. Such errors can impact the performance of downstream tasks, including question-answering and recommender systems.

These errors can be classified into two categories: missing triplets and incorrect triplets. This gives rise to two main tasks: link prediction and error detection. Link prediction involves forecasting the missing connections in the knowledge graph, whereas error detection aims to identify incorrect triplets or relationships within the graph.

Current error detection methods for knowledge graphs are typically classified into two main categories [20]: path-based [6,9,10,18] and embedding-based [1,13,15]. Path-based methods depend on the inter-entity paths to identify errors. Conversely, embedding methods assess the confidence score of each triplet within the KGs. Different from link prediction, error detection seeks to differentiate incorrect triplets from correct ones. However, the majority of current error detection methods overlook the semantic information of the triplets, which can be vital for accurate error detection.

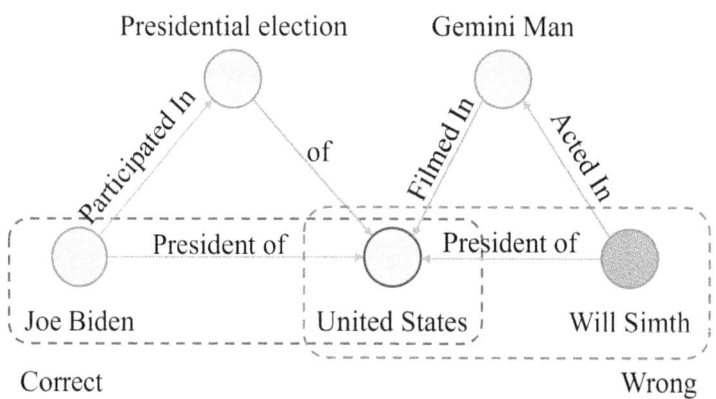

Fig. 1. Example of errors in knowledge graphs. The red border indicates the incorrect triplet. (Color figure online)

The challenges for error detection in knowledge graphs can be identified from the following two aspects. First, the errors in the knowledge graph are often subtle and challenging to detect. The structural information alone may not be sufficient to distinguish between correct and incorrect triplets. Second, the semantic information contained in the knowledge graph can offer valuable insights to improve the effectiveness of error detection. However, the majority of current error detection methods overlook the semantic information of the triplets, which can be vital for accurate error detection. Let us see the example in knowledge graphs (see Fig. 1). In the black border example, the link "President of" correctly associates the head entity "Joe Biden" with the tail entity "United States". Conversely, in the red border example, the link "President of" incorrectly connects the head entity "Will Smith" with the tail entity "United States". The incorrect triplet has similar structural characteristics to the correct triplet, making it challenging to distinguish between the two based solely on structural information. This example shows that distinguishing between correct and incorrect triplets solely based on structural information proves to be a challenging task, as both types contain similar structural characteristics. To overcome these challenges, it is crucial to utilize the semantic information within the KGs to improve the precision of error detection.

To tackle these challenges, we introduce a Flexible Knowledge Graph Error Detection Framework Combined with Semantic Information (FKED). Which combines both structural and semantic information to detect errors within the knowledge graph. We first extract the structural information of KGs using a graph embedding model. Next, we employ a pre-trained language model (PLM) to extract semantic information from the triplets. Then the structural and semantic information are combined to detect errors within the knowledge graph. FKED can be flexibly added to other structure-based error detection models, enhancing their capabilities in downstream tasks of knowledge graphs.

In summary, our contributions can be outlined as follows:

- We introduce a flexible knowledge graph error detection framework that easily integrates with other structure-based models.
- We integrate both structural and semantic information to identify errors within the knowledge graph.
- We have shown the efficacy of our method, which can performance of the base models.

2 Related Work

In this section, we review the existing methods for error detection in KGs and summarize the existing semantic-based knowledge graph representation learning methods.

2.1 KG Error Detection

Error detection in knowledge graphs is a crucial task that has garnered significant attention in recent years. Current approaches for error detection in knowledge graphs can be divided into two main categories: embedding-based and path-based.

The embedding-based model, e.g., TransE [1], ComplEx [15], and RotatE [13], assess the confidence score of each triplet within the knowledge graph. TransE posits that the relation functions as a translation from the head entity to the tail entity within the embedding space, employing the $F = \|h + r - t\|$ formula to calculate the score of the triplet. ComplEx extends TransE to the complex space, which can model symmetric relations. RotatE's use of complex numbers and rotation operations provides a powerful framework for knowledge graph embeddings. Following studies like CKRL [18], CAGED [19], and SCEF [22] incorporate the scoring function from TransE as a vital element. Embedding-based methods rely on negative sampling, but their effectiveness is hindered by the challenge of precisely capturing the true noise distribution. As a result, the negative samples frequently fail to align with the actual noise, which in turn, compromises the error detection capabilities.

Some works extract path information from the knowledge graph to detect errors. PTransE [9] proposed the path-based TransE model for modeling multi-hop relational paths. Previous models only targeted single-hop relationships between entities, but this work can mine multi-hop relationships between entities, which can be used for knowledge inference. Its energy function is $G(h, r, t) = E(h, r, t) + E(h, P, t)$, where P represents the path linking the head entity h and the tail entity t. Other works like KGTtm [6] integrate both embedding information and path information to detect errors. KGTtm uses a tensor-based model to capture the structural information of the KGs. HEAR [20] identifies the two-level hierarchical structure of path information between entity pairs and employs it for error detection, leveraging this hierarchical path structure to identify errors in knowledge graphs. However, path-based methods are constrained by the complexity of the knowledge graph and the challenges associated with extracting path information.

2.2 Semantic-Based Methods

PLM have seen extensive application across a range of NLP tasks. In the domain of knowledge graphs, PLM-based methods encode entities and relations in KGs as text sequences and leverage PLMs to extract semantic information. These approaches utilize PLMs to encode the text and employ the PLM outputs to score the triplets.

The use of semantic information in other tasks has been widely studied. For example, BERT [4] has been employed to derive semantic information from textual data. KG-BERT [8] directly utilizes the BERT model to score triples in knowledge graphs. It converts the triples into corresponding text inputs for the BERT model. By employing tasks such as triple classification, KG-BERT fine-tunes BERT to cater to the specific requirements of knowledge graphs.

To integrate the language model with the structure model, StAR [16] improves the inference speed by encoding entities separately. It proposes an asymmetric twin encoding for triples and suggests interacting the two asymmetric representations for positive and negative sample judgment, achieving better results compared to KG-BERT [8]. CSProm-KG [2] introduces a method that strikes a balance between structural information and textual knowledge, focusing solely on tuning the parameters of *Conditional Soft Prompts* derived from entity and relation representations.

3 Methodology

This section offers an in-depth overview of our approach. Our method consists of two main components: structural information extraction and semantic information extraction.

We begin by extracting the structural information from the knowledge graph using a graph embedding model. Subsequently, we utilize a PLM to derive semantic information from the triplets. Finally, we integrate both structural and semantic information to identify errors within the knowledge graph. The structure of our method is illustrated in Fig. 2.

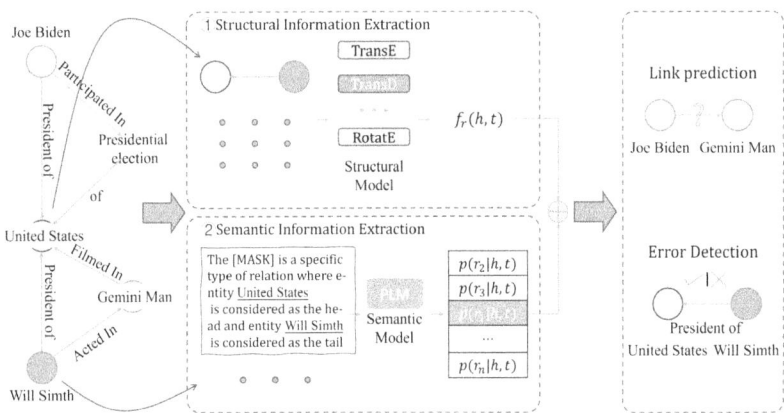

Fig. 2. The structure of FKED integrates the structural model score $f(h, r, t)$ and the semantic model score $p(r|h, t)$ to identify errors within the knowledge graph. This approach can be seamlessly incorporated into other structure-based models.

3.1 Structural Information Extraction

We employ a graph embedding model to derive the structural information of the knowledge graph. We choose TransE [1] as the base model for the structural information extraction. TransE is a widely used graph embedding model that assess the confidence score of each triplet within the KGs. For a given triplet (h, r, t), its score is computed as shown in Eq. (1),

$$f(h, r, t) = ||h + r - t||_2 \qquad (1)$$

where h, r, and t represent the embeddings of the head, relation, and tail, The score $f(h, r, t)$ is employed to assess the confidence level of the triplet. This score is then utilized to identify errors in the KGs. During the training phase, a negative sampling strategy is employed to train the graph embedding model. Given a positive triplet (h, r, t), We randomly choose a head entity h' or tail entity t' from the entity set to generate the negative triplet, thereby simulating noise within the knowledge graph. We employ margin-based loss function to train the graph embedding model. Loss function is defined as shown in Eq. (2):

$$\mathcal{L}_s = \sum_{(h,r,t)\in \mathcal{T}} \sum_{(h',r',t')\in \mathcal{T}'} [\gamma + f(h,r,t) - f(h',r',t')]_+ \qquad (2)$$

where γ represents the margin hyperparameter, $\mathcal{T}$ denotes the set of correct triplets, and $\mathcal{T}'$ signifies the set of incorrect triplets. To train a graph embedding model, it is essential to minimize the loss function. This ensures that correct triplets receive lower scores compared to incorrect triplets.

During the testing phase, we utilize the confidence score $f(h,r,t)$ to identify errors in KGs. A higher score for a triplet signifies an increased probability of the triplet being incorrect.

3.2 Semantic Information Extraction

We utilize a Pretrained-Language-Model (PLM) to derive semantic information from triplets. Given a triplet (h,r,t), we encode the head entity, relation, and tail entity as text sequences using PLM. The design of the prompt is depicted in Fig. 3, where [CLS] is a special token used for classification tasks, and [SEP$_i$] is a special token used to separate the triplet. The text sequence is *"The [MASK] is a specific type of relation where entity **head** is considered as the head and entity **tail** is considered as the tail"*.

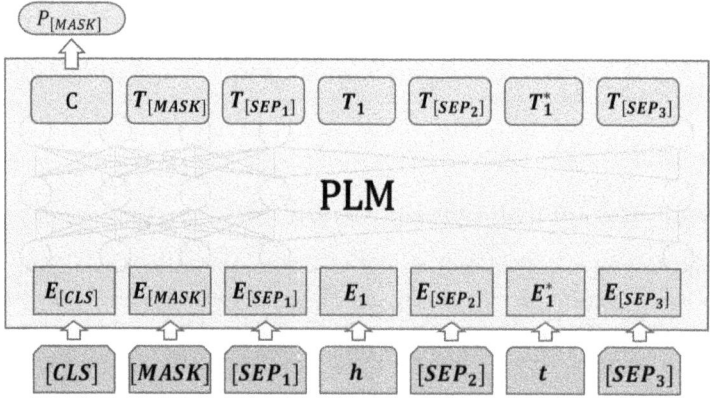

Fig. 3. The text sequence design.

We utilize the [MASK] token to conceal the relation within the text sequence. Subsequently, we input the masked text sequence into PLM, obtaining the logits of the relation. Indeed, we utilize a PLM to execute link prediction, aiming to forecast the relation between the head entity and the tail entity. Consequently,

For a given triplet (h, r, t), we acquire the logits corresponding to the relation as Eq. (3), which we employ as the semantic score for the triplet,

$$g(h, r, t) = P(r|h, t) \tag{3}$$

where $P(r|h, t)$ denotes the conditional probability of relation r given the head entity h and tail entity t. The score $g(h, r, t)$ is used to assess the confidence of the triplet. This score is employed to detect errors in the knowledge graph. The conditional probability $g(h, r, t)$ reflects the likelihood of the relation r given the entities h and t. A higher value of $g(h, r, t)$ suggests a higher probability that the triplet (h, r, t) is correct. In this study, we utilize BERT as the PLM to derive semantic information from the triplets. The score is represented by the logits output by the BERT model.

3.3 Joint Score

Similar to prior research [1,8,15], we assess the certainty of each triple by sorting their scores. We merge the structural and semantic scores to identify inaccuracies in KGs. To ensure the two scores are on a comparable scale, we normalize them prior to combining. The Joint Score is calculated as Eq. (4):

$$s(h, r, t) = \alpha \cdot f(h, r, t) - (1 - \alpha) \cdot g(h, r, t) \tag{4}$$

where α is a hyperparameter governing the relative weighting of the structural and semantic scores.

The semantic score indicates the likelihood of the relation, conditioned on the head and tail entities. Consequently, a low score suggests that the triplet is likely to be incorrect in terms of its probability.

By combining the structural and semantic scores, we can detect errors in the knowledge graph more effectively.

3.4 Pretraining

To fine-tune the BERT model, we employ a negative sampling strategy. We employed the Adam optimizer, setting the learning rate to 5e−5 and using a batch size of 32, and we trained the model for 3 epochs. Loss function is defined as the Eq. (5):

$$\mathcal{L}_t = \sum_{(h,r,t) \in \mathcal{T}} \sum_{(h',r',t') \in \mathcal{T}'} \max(0, g(h', r', t') - g(h, r, t) - \gamma) \tag{5}$$

where $\mathcal{T}$ denotes the set of correct triplets, $\mathcal{T}'$ represents the set of incorrect triplets, and γ serves as the margin hyperparameter. $g(h, r, t)$ indicates the score of a correct triplet, while $g(h', r', t')$ denotes the score of an incorrect triplet. By minimizing the loss function, we fine-tune the BERT model to ensure that correct triplets achieve higher scores compared to incorrect triplets.

The structural information extraction and semantic information extraction are trained separately. The training procedures for both the structural model and the semantic model are outlined in Algorithm 1.

Algorithm 1: FKED training algorithm

Input: Relations and entities sets R and E, training set $S = \{(h, r, t)\}$, margin γ, textual description of entities and relations sets T_e and T_r, structural model and semantic model f and g, train epoch e, batch size B.
Output: Structural model f and semantic model g.
1 initialize f and g;
2      $e \leftarrow$ **uniform** for each $e \in E$;
3      $r \leftarrow$ **uniform** for each $r \in R$;
4 **for** $epoch = 1$ to e **do**
5      $S_{batch} \leftarrow$ **sample**(S, B);   // select a minibatch with a size of B
6      $T_{batch} \leftarrow \emptyset$;   // Initialize the sets of triplets
     // Negative sampling
7   **for** $(h, r, t) \in S_{batch}$ **do**
8          $(h', r, t') \leftarrow$ **sample**(S'_{batch});   // sample a corrupted triplet
9          $T_{batch} \leftarrow T_{batch} \cup \{(h', r, t'), (h, r, t)\}$;
     // Optimize the structural model and semantic model
10   **for** $(h, r, t) \in T_{batch}$ **do**
11   optimize the structural model f using Eq. (2);
12   optimize the semantic model g using Eq. (5);
13   update the parameters of f and g;
14 **return** f and g;
---

4 Experimental Settings

In this section, we will describe the datasets, experimental setup, and evaluation metrics.

4.1 Datasets

We assess the effectiveness of our approach using two benchmark datasets: FB15k-237 [14] and WN18RR [3]. FB15k-237, a subset of the FB15k dataset, comprises 14,541 entities, 237 relations, and 310,116 triplets. WN18RR, a subset of the WN18 dataset, includes 40,943 entities, 11 relations, and 93,003 triplets. The specifics of these datasets are presented in Table 1.

Table 1. Dataset statistics.

Dataset	#Entities	#Relations	#Triplets	Average Degree
FB15k-237	14,541	237	310,116	42.6
WN18RR	40,943	11	93,003	4.5

Most noises and conflicts in the knowledge graph are caused by the incorrect triplets. To evaluate our method's performance, we incorporate 5% incorrect triplets into the test set.

4.2 Baselines

Owing to the flexibility of our approach, it can be effortlessly integrated into other structure-based models. We choose three structure-based models as our baselines: TransE, TransD, and RotatE. We add our method to these models and denote them as TransE+, TransD+, and RotatE+, respectively.

4.3 Evaluation Metrics

We assess our method on two tasks: link prediction and error detection.

In the context of link prediction, we employ the Hit@k metric, which quantifies the fraction of accurate triplets among the top-k predictions. The Hit@k metric is determined as follows:

$$\text{Hit@k} = \frac{\text{The quantity of correct triplets within the top-k predictions}}{\text{The quantity of test triplets}}$$

In the context of error detection, we utilize the Precision@k and Recall@k metrics to evaluate the model's effectiveness. These metrics are derived as follows:

$$\text{Precision@k} = \frac{\text{The quantity of correct triplets within the top-k predictions}}{k}$$

$$\text{Recall@k} = \frac{\text{The quantity of correct triplets within the top-k predictions}}{\text{The quantity of correct triplets}}$$

4.4 Hyperparameter Tuning

We tune the hyperparameters of our method on the validation set. The hyperparameters are tuned using grid search. In the Structural model, the final hyperparameters are determined as follows. Margin hyperparameter γ: 1.0, Learning rate: 0.02, Batch size: 1024, Optimizer: Sgd, Embedding dimension: 100, Negative sampling: 10. In the Semantic model, the final hyperparameters are determined as follows. Learning rate: 1e−5, Negative sampling: 4, Batch size: 32. The weight of the structural score α is 0.5.

5 Experimental Results

We report the outcomes of our approach on the FB15K-237 and WN18RR datasets. Error detection results are detailed in Table 2, while link prediction results are illustrated in Table 3. Our approach significantly enhances the performance of the base models across both tasks.

Table 2. The Results of error detection on the WN18RR and FB15k-237 datasets.

METRIC	MODEL	FB15k-237					WN18RR				
		k = 1%	k = 2%	k = 3%	k = 4%	k = 5%	k = 1%	k = 2%	k = 3%	k = 4%	k = 5%
Precision@k	TransE	0.946	0.774	0.606	0.498	0.423	0.367	0.246	0.217	0.180	0.150
	TransE+	0.956	0.819	0.682	0.591	0.526	0.367	0.279	0.217	0.205	0.176
	TransD	0.962	0.871	0.733	0.636	0.563	0.433	0.295	0.239	0.197	0.163
	TransD+	0.966	0.862	0.751	0.663	0.605	0.400	0.311	0.272	0.238	0.222
	RotatE	0.435	0.413	0.387	0.368	0.346	0.896	0.624	0.474	0.383	0.324
	RotatE+	0.649	0.528	0.452	0.393	0.347	0.941	0.629	0.476	0.390	0.333
Recall@k	TransE	0.189	0.310	0.364	0.399	0.423	0.075	0.103	0.137	0.151	0.158
	TransE+	0.192	0.329	0.412	0.476	0.529	0.076	0.117	0.138	0.172	0.186
	TransD	0.193	0.350	0.442	0.512	0.566	0.086	0.119	0.146	0.159	0.166
	TransD+	0.190	0.340	0.443	0.523	0.595	0.081	0.128	0.168	0.195	0.228
	RotatE	0.087	0.166	0.233	0.296	0.348	0.177	0.240	0.275	0.283	0.302
	RotatE+	0.130	0.212	0.273	0.316	0.349	0.188	0.251	0.285	0.311	0.333

5.1 Error Detection

Our approach attains superior performance on both datasets. The Precision@k and Recall@k metrics show substantial improvement over the base models. Specifically, in TransE+, the Precision@5% metric exhibits a 10.3% increase on the FB15k-237 dataset and a 2.6% increase on the WN18RR dataset.

5.2 Link Prediction

We also assess our method on the task of link prediction, employing the same settings as outlined in the original TransE paper [1].

Our approach improves the performance of the base models on both datasets. Specifically, in TransE+, the Hit@10 metric increases by 3.3% on the FB15k-237 dataset and by 2.0% on the WN18RR dataset.

5.3 Analysis

Our method significantly enhances the performance of base models across both tasks. However, performance variations are observed between different datasets. Specifically, we noted superior results on the FB15k-237 dataset compared to the WN18RR dataset, likely due to the differing complexities between the two. The FB15k-237 dataset is more complex, with an average degree of 42.6, which surpasses that of the WN18RR dataset. This higher average degree signifies a more intricate knowledge graph, facilitating easier error detection in structural information compared to WN18RR. Conversely, the WN18RR dataset, with an average degree of 4.5, provides less structural information for error detection, necessitating greater reliance on semantic information for improved performance.

Table 3. Link prediction results on FB15k-237 and WN18RR datasets.

MODEL	FB15k-237			WN18RR		
	Hit@1	Hit@3	Hit@10	Hit@1	Hit@3	Hit@10
TransE	0.166	0.257	0.392	0.056	0.126	0.288
TransE+	0.196	0.302	0.425	0.153	0.213	0.308
TransD	0.161	0.254	0.391	0.050	0.120	0.282
TransD+	0.194	0.304	0.428	0.150	0.211	0.307
RotatE	0.078	0.153	0.313	0.027	0.073	0.245
RotatE+	0.116	0.201	0.368	0.082	0.169	0.321

5.4 Ablation Study

We carry out an ablation study to assess the contribution of semantic information. The findings are displayed in Table 4.

Table 4. Relying solely on semantic information for error detection.

METRIC	FB15k-237					WN18RR				
	k=1%	k=2%	k=3%	k=4%	k=5%	k=1%	k=2%	k=3%	k=4%	k=5%
Precision@k	0.014	0.016	0.018	0.017	0.018	0.028	0.029	0.031	0.031	0.029
Recall@k	0.003	0.006	0.011	0.014	0.018	0.006	0.012	0.019	0.025	0.029

As shown in Table 4, only using BERT for error detection achieves poor performance on both datasets. The Precision@k and Recall@k metrics are significantly lower compared to those combined with the structure-based model. This demonstrates the importance of combining the structural and semantic information for error detection in knowledge graphs. Structural information is necessary for KG error detection. Figure 4 shows an example of the failure case when using semantics only.

In this example, "Jackie Chan was born in China" and "Bruce Lee acted in the movie Hero" can be deemed acceptable when scrutinized solely through a semantic lens, given their plausibility. However, when considering the structural information, it becomes evident that the latter fact is erroneous. The movie "Hero" was released in 2002, while Bruce Lee passed away in 1973. Consequently, the fact that Bruce Lee acted in the movie "Hero" is incorrect. Detecting errors in a knowledge graph solely based on semantic information can be challenging and impractical due to the sheer volume of such triples. This example validates the necessity of combining structural and semantic information for error detection in KGs.

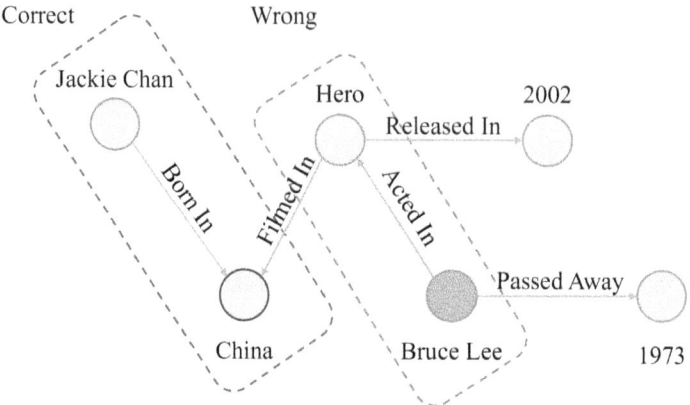

Fig. 4. An example demonstrating the failure case when using semantics.

6 Conclusion

In this paper, we introduce a flexible knowledge graph error detection framework that leverages the semantic information within KGs. By integrating semantic and structural information, our method enhances the effectiveness of structural models. Additionally, it can be easily and quickly incorporated into other structure-based models. The experimental results demonstrate the validity of this approach.

Acknowledgments. The research work is supported by National Key Research and Development Program of China (No. 2022YFC3303302), the National Natural Science Foundation of China under Grant (No. 61976204, U2436209). Xiang Ao is also supported by the Project of Youth Innovation Promotion Association CAS and the Beijing Nova Program. Yang Liu is also supported by the China Postdoctoral Science Foundation (No. 2023M743567).

References

1. Bordes, A., Usunier, N., Garcia-Duran, A., Weston, J., Yakhnenko, O.: Translating embeddings for modeling multi-relational data. In: Burges, C., Bottou, L., Welling, M., Ghahramani, Z., Weinberger, K. (eds.) Advances in Neural Information Processing Systems, vol. 26. Curran Associates, Inc. (2013). https://proceedings.neurips.cc/paper_files/paper/2013/file/1cecc7a77928ca8133fa24680a88d2f9-Paper.pdf
2. Chen, C., Wang, Y., Sun, A., Li, B., Lam, K.Y.: Dipping PLMs sauce: bridging structure and text for effective knowledge graph completion via conditional soft prompting. arXiv preprint arXiv:2307.01709 (2023)
3. Dettmers, T., Minervini, P., Stenetorp, P., Riedel, S.: Convolutional 2D knowledge graph embeddings. In: Proceedings of the AAAI Conference on Artificial Intelligence (2018)

4. Devlin, J., Chang, M.W., Lee, K., Toutanova, K.: BERT: pre-training of deep bidirectional transformers for language understanding. arXiv preprint arXiv:1810.04805 (2018)
5. Guo, Q., et al.: A survey on knowledge graph-based recommender systems. IEEE Trans. Knowl. Data Eng. **34**(8), 3549–3568 (2020)
6. Jia, S., Xiang, Y., Chen, X., Wang, K.: Triple trustworthiness measurement for knowledge graph. In: The World Wide Web Conference (2019). https://doi.org/10.1145/3308558.3313586
7. Li, D., Qu, H., Wang, J.: A survey on knowledge graph-based recommender systems. In: 2023 China Automation Congress (CAC), pp. 2925–2930 (2023). https://doi.org/10.1109/CAC59555.2023.10450693
8. Liang, Y., Mao, C., Luo, Y.: KG-BERT: BERT for knowledge graph completion. arXiv Computation and Language (2019)
9. Lin, Y., Liu, Z., Luan, H., Sun, M., Rao, S., Liu, S.: Modeling relation paths for representation learning of knowledge bases. In: Proceedings of the 2015 Conference on Empirical Methods in Natural Language Processing (2015). https://doi.org/10.18653/v1/d15-1082
10. Lin, Y., Liu, Z., Sun, M., Liu, Y., Zhu, X.: Learning entity and relation embeddings for knowledge graph completion. In: Proceedings of the AAAI Conference on Artificial Intelligence (2015). https://doi.org/10.1609/aaai.v29i1.9491
11. Peng, C., Xia, F., Naseriparsa, M., Osborne, F.: Knowledge graphs: opportunities and challenges (2023)
12. Saxena, A., Tripathi, A., Talukdar, P.: Improving multi-hop question answering over knowledge graphs using knowledge base embeddings. In: Proceedings of the 58th Annual Meeting of the Association for Computational Linguistics (2020). https://doi.org/10.18653/v1/2020.acl-main.412
13. Sun, Z., Deng, Z.H., Nie, J.Y., Tang, J.: RotatE: knowledge graph embedding by relational rotation in complex space. arXiv preprint arXiv:1902.10197 (2019)
14. Toutanova, K., Chen, D., Pantel, P., Poon, H., Choudhury, P., Gamon, M.: Representing text for joint embedding of text and knowledge bases. In: Proceedings of the 2015 Conference on Empirical Methods in Natural Language Processing (2015). https://doi.org/10.18653/v1/d15-1174
15. Trouillon, T., Welbl, J., Riedel, S., Gaussier, E., Bouchard, G.: Complex embeddings for simple link prediction. In: International Conference on Machine Learning (2016)
16. Wang, B., Shen, T., Long, G., Zhou, T., Wang, Y., Chang, Y.: Structure-augmented text representation learning for efficient knowledge graph completion. In: Proceedings of the Web Conference 2021 (2021). https://doi.org/10.1145/3442381.3450043
17. Wang, Y., Lipka, N., Rossi, R.A., Siu, A., Zhang, R., Derr, T.: Knowledge graph prompting for multi-document question answering. In: Proceedings of the AAAI Conference on Artificial Intelligence **38**(17), 19206–19214 (2024). https://doi.org/10.1609/aaai.v38i17.29889, https://ojs.aaai.org/index.php/AAAI/article/view/29889
18. Xie, R., Liu, Z., Lin, F., Lin, L.: Does William Shakespeare really write hamlet? Knowledge representation learning with confidence. In: Proceedings of the AAAI Conference on Artificial Intelligence (2018)
19. Zhang, Q., Dong, J., Duan, K., Huang, X., Liu, Y., Xu, L.: Contrastive knowledge graph error detection. In: Proceedings of the 31st ACM International Conference on Information & Knowledge Management, pp. 2590–2599 (2022)

20. Zhang, Z., Zhang, F., Zhuang, F., Xu, Y.: Knowledge graph error detection with hierarchical path structure. In: Proceedings of the 32nd ACM International Conference on Information and Knowledge Management, pp. 4430–4434 (2023)
21. Zhao, X., Chen, H., Xing, Z., Miao, C.: Brain-inspired search engine assistant based on knowledge graph. IEEE Trans. Neural Netw. Learn. Syst. **34**(8), 4386–4400 (2023). https://doi.org/10.1109/TNNLS.2021.3113026
22. Zhao, Y., Liu, J.: SCEF: a support-confidence-aware embedding framework for knowledge graph refinement. arXiv preprint arXiv:1902.06377 (2019)

Session-Level Normalization and Click-Through Data Enhancement for Session-Based Evaluation

Haonan Chen, Zhicheng Dou[✉], and Jiaxin Mao

Gaoling School of Artificial Intelligence, Renmin University of China, Beijing, China
{hnchen,dou,maojiaxin}@ruc.edu.cn

Abstract. Since a user usually has to issue a sequence of queries and examine multiple documents to resolve a complex information need in a search session, researchers have paid much attention to evaluating search systems at the session level rather than the single-query level. Most existing session-level metrics evaluate each query separately and then aggregate the query-level scores using a session-level weighting function. The assumptions behind these metrics are that all queries in the session should be involved, and their orders are fixed. However, if a search system could make the user satisfied with her first few queries, she may not need any subsequent queries. Besides, in most real-world search scenarios, due to a lack of explicit feedback from real users, we can only leverage some implicit feedback, such as users' clicks, as relevance labels for offline evaluation. Such implicit feedback might be different from the real relevance in a search session as some documents may be omitted in the previous query but identified in the later reformulations. To address the above issues, we make two assumptions about session-based evaluation, which explicitly describe an ideal session-search system and how to enhance click-through data in computing session-level evaluation metrics. Based on our assumptions, we design a session-level metric called Normalized U-Measure (NUM). NUM evaluates a session as a whole and utilizes an ideal session to normalize the result of the actual session. Besides, it infers session-level relevance labels based on implicit feedback. Experiments on two public datasets demonstrate the effectiveness of NUM by comparing it with existing session-based metrics in terms of correlation with user satisfaction and intuitiveness. We also conduct ablation studies to explore whether these assumptions hold. (Student paper).

Keywords: Session-level Normalization · Click-through Data Enhancement · Session Search · Evaluation Metrics

1 Introduction

With the development of search engines, researchers increasingly focus on building better evaluation methods. In the early years, the Cranfield paradigm [1] was the dominant approach in evaluating the search results of a single query.

However, when a user is trying to complete a complex search task, she may issue multiple queries and browse a series of documents to obtain sufficient information in a *search session* [2–4]. Many works have emerged to design session-based evaluation metrics. Some of them have already been used in some evaluation tasks, *e.g.*, Session-based DCG [5] in the TREC Session Track [6] and Recency-aware Session-based Metric [7] in the recent NTCIR Session Search (SS) Task [8]. However, there are still some remaining challenges for session-based evaluation. In this work, we identify two major challenges of session-based evaluation as some existing session-based metrics are based on oversimplified or problematic assumptions.

The first challenge is that most existing session-level metrics (*e.g.*, the metrics used in these tasks) evaluate each query in a session based on an existing query-level metric, and then aggregate those query-level scores with some session-level discount factors or weighting schema to evaluate systems at the session level. For example, sDCG [5] is based on cascade hypothesis, which assigns smaller weights to search results that ranked lower and queries that issued later, and RSMs [7] gives larger weights to the recently issued queries. While previous studies show that these aggregation metrics correlate well with users' session-level satisfaction feedback [7], these metrics all implicitly assume that a user's query sequence in a search session will not be altered by the systems. However, it would be expected that if a user is satisfied with the information retrieved by the present and past queries, she may not need any subsequent queries. (Following [9,10], we assume that in the same session, the queries that may represent different sub-topics serve the same primary information need.) To put it another way, we presume that an ideal search system would return all relevant documents in a session before all irrelevant documents, so the user can spend the least effort in completing the search task. For example, a user issues a query "Java" and clicks "What is Java Language". After a minute, she issues another query "Java Project", and clicks "Java Projects for Beginners". We assume the user would prefer a system if it could predict that she is seeking the second clicked document (*e.g.*, using a personalized search model) and place it at the beginning of this session. However, none of the existing session-level evaluation metrics can fully take this reduction in effort into consideration and give a maximum score to such an ideal search system. Note that it is true that a user may learn something in the session which triggers her to seek new pieces of information. However, we believe that an ideal system can predict the change of her interest, lead her to discover all information needs as soon as possible, and rank all the documents that serve her needs high in the session.

Another challenge that may hinder the computation of session-level evaluation metrics is that in real-world search scenarios, search engines can only record implicit user feedback to the documents, *i.e.*, click-through data. Therefore, most metrics have to assume that the clicked documents are "relevant" to the query in offline evaluation. However, users may skip some relevant documents during the search because of position bias. This problem is more common in a long search session as Price et al. [11] found out that users may omit documents in previous queries yet recognize and click them in later reformulations, *i.e.*, the session's subsequent queries. For example, a user issues a query "MacBook" then clicks "MacBook on Amazon". A minute before, the search system had already ranked this document among the top ten results when she searched "Apple", but she omitted it. We assume that when evaluating this system, we should also mark "MacBook on Amazon" relevant to "Apple" in order to reward this system for successfully predicting the user's search intent and saving her efforts on issuing another query. Thus, we need to consider the "relevance" of a document at the session level but not at the level of each separate query. However, existing metrics do not account for these omitted documents and simply assume the unclicked documents are irrelevant (when only implicit feedback is available), because these metrics only take clicks as per-query relevance judgments, rather than considering clicks in the entire session.

To resolve these issues, in this paper, we make two **assumptions** about a user's search behaviors in a search session:

Assumption 1: An ideal search system should rank all the documents that the user requires, *i.e.*, all the relevant documents, above all the irrelevant ones in the entire session. By doing so, it can save the user's effort because she may not even have to reformulate her queries. In other words, we need to remove the boundary of queries in the session-level evaluation.

Assumption 2: A document that is clicked in a subsequent query but omitted in a preceding query in the same session is relevant to that preceding query. Consequently, the first occurrence of this document is assumed to be relevant to the information need of the current session.

We will refer to Assumption 1 as A1 and Assumption2 as A2.

Most offline evaluation in the industry can only use implicit user feedback (click-through data) to infer relevance labels because it is costly to get human relevance labels. Besides, most human-labeled relevance is query-level relevance, not the session-level relevance that is preferred when evaluating session search systems. Thus, our assumptions in this paper are based on actual click-through data (implicit feedback) and user behaviors rather than manual labels. Under this condition, we describe an ideal session search system (A1) and try to mine session-level relevance labels from click-through data (A2).

The common idea implied in these assumptions is that we need to reduce the impact of query boundaries in session search evaluation. Based on these assumptions, we design a session-level metric called **Normalized session level U-Measure (NUM)** based upon the original U-Measure [12]. This metric evaluates a session as a single "virtual query" and employs an ideal session to normalize the evaluation result according to A1. NUM also converts the click-through feedback into session-level relevance labels based on A2. Experiments on two public datasets (TianGong-SS-FSD [7] and NTCIR-16 Session Search Task [8]) show that NUM is an intuitive session-level metric that correlates well with user satisfaction. Furthermore, ablation studies confirm that A1 and A2 are **valid**.

To summarize, the contributions of the paper are as follows:

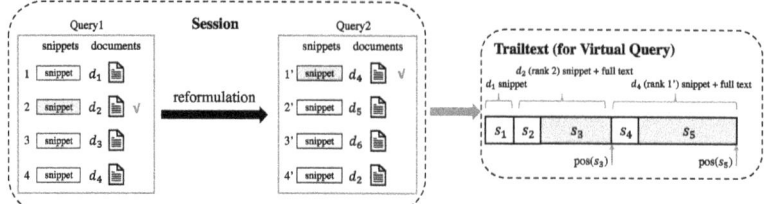

Fig. 1. The illustration of how U-measure constructs trailtext from a two-query session. The results clicked by the user are marked with red checkmarks and the results marked as relevant are filled with color gray. The right part is the constructed trailtext, where "s_i" is the i-th string of it. (Color figure online)

(1) We make two assumptions about what we should do in session-based evaluation. With these assumptions, we describe what an ideal session should be and discuss how to use the click-through data to derive session-level relevance labels.
(2) We design a session-level metric called NUM based on U-measure [12]. It treats a session as a virtual query and uses the evaluation score of the ideal session for normalization (**session-level normalization**).
(3) We show that NUM correlates well with user satisfaction. Further studies also demonstrate its intuitiveness. In addition, the ablation studies explore the reasonability of our two assumptions.

2 Related Work

2.1 Session-Based Evaluation Metrics

Because in real-world search scenarios, users usually issue a sequence of queries to complete a complex search task, researchers have gone beyond the traditional query-level metrics (*e.g.*, MAP, nDCG, and MRR) and designed some session-level metrics.

Session-based DCG (sDCG) [5] is a multi-query metric based on Discounted Cumulative Gain (DCG) [13]. According to the cascade hypothesis, sDCG discounts the weights of the results that ranked lower and queries issued later. Similarly, Lipani et al. [14] proposed a session-based version of Rank-Biased Precision (RBP) [15], which added a new parameter over RBP to balance between query reformulation behaviors or continuing to examine documents. Yang and Lad [16] proposed a utility-based evaluation framework. They evaluated the Expected Utility of the search system over all possible interaction patterns. Van Dijk et al. [17] leverages a Markovian Chain of users' behaviors in a session to substitute the fixed discount of documents. Moffat et al. [18], expanded the C/W/L framework to evaluate effectiveness in a session-level context.

Liu et al. [19,20] showed how the recency effect can affect users' session-level satisfaction with user studies. They proposed that the later-issued queries should receive higher weights, which is in contrast to the cascade hypothesis. Based on these findings, Zhang et al. [7] proposed Recency-aware Session-based Metrics (RSMs) integrate the recency effect into session-based evaluations. Their results demonstrated that RSMs exhibit the highest correlation with user satisfaction among these metrics and achieve state-of-the-art performance in estimating user satisfaction.

Although various session-based evaluation metrics have been proposed, and some perform quite well in estimating user satisfaction and/or measuring system effectiveness, most of these session-based metrics evaluate each query separately and aggregate the evaluation scores using some discount factors or weighting schema. Therefore, they all implicitly assume that all queries in the session should be considered in the session-based evaluation and the orders of the queries are definite. However, as we stated in **A1**, if a search system could make the user satisfied with the first few queries, she may not need any subsequent queries. Few existing session-based metrics have taken this into consideration.

Besides, most existing metrics are based on the Cranfield/test collection approach (test collections with explicit relevance judgments). However, as stated in Sect. 1, most offline scenarios in the industry can only use implicit user feedback (click-through data). Therefore, these metrics would have to directly treat click-based labels as relevance labels. Our work attempts to bridge this gap by enhancing click-through data based on **A2**. By this, we infer session-level relevance labels with implicit feedback.

2.2 U-Measure

In this part, we will briefly review U-measure [12], based on which we design a session-level metric NUM. U-measure is a framework for evaluating information access that can be used to evaluate various IR tasks.

Figure 1 is an example in its original paper of how U-measure constructs a trailtext from a two-query session. When evaluating a session, U-measure treats the session as a single "virtual query" (as a whole) by building a **trailtext** for this session. A trailtext tt is made up of n strings concatenated together: $tt = s_1 s_2 \cdots s_n$. Each string s_i could be a snippet or the entire content. Sakai and Dou assume that trailtext accurately represents the sequence in which a user reads information during a search. Besides, they assume that a user reads $F\%$ (20% in [12]) of the content of the document, i.e., only $F\%$ of the document's length is counted in the trailtext. They define $pos(s_i) = \sum_{j=1}^{i} |s_j|$ as the offset position of s_i. Specifically, trailtexts are derived from session data that follows the algorithm in [12] (Fig. 5). The general computation of U-measure is:

$$\mathrm{U} = \sum_{pos=1}^{|tt|} \mathrm{gain}(pos)\mathrm{D}(pos), \qquad (1)$$

where $\mathrm{D}(pos)$ is a decay function based on position and $\mathrm{gain}(pos)$ is the corresponding gain. Specifically, if s_i is not relevant, $\mathrm{gain}(pos(s_i)) = 0$, and if s_i

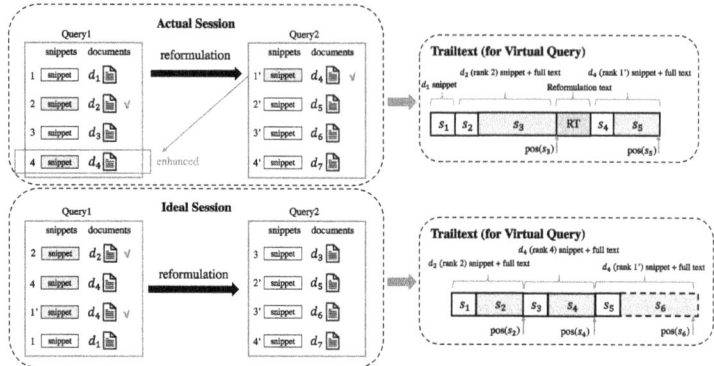

Fig. 2. The illustration of NUM. The upper part is the actual session, and the lower part is the ideal session. The results clicked by the user are marked with red checkmarks, and the results marked as relevant are filled with color gray. Rank 4 of the first query is marked relevant even though it is not clicked here because it is clicked in the subsequent query. We construct the trailtext based on actual user clicks for the actual session and based on the enhanced session-level labels for the ideal session. (Color figure online)

is regarded l-relevant, $\text{gain}(pos(s_i)) = gv_l$, where gv_l is relevance level l's gain value for . $D(pos)$ is a decay function:

$$D(pos) = \max(0, 1 - \frac{pos}{L}), \tag{2}$$

We choose U-measure as our metric's backbone because: (1) Rather than aggregating the evaluation results of all queries in a session, which may contain queries that the user does not need, U-measure evaluates a session as a single "virtual query" by building a trailtext. (2) U-measure accounts for document length and incorporates the diminishing return property, rendering it more realistic compared to rank-based metrics.

Note that the original paper of U-measure [12] primarily introduces it as a general evaluation framework for a variety of IR tasks, rather than specifically for session search. Therefore, many improvements are required to bring it to session-level evaluation (based on our two assumptions). Experiments conducted in Sect. 5.3 demonstrate the effectiveness and necessity of these modifications.

3 The Proposed Methods

In this section, we will introduce the proposed metric **Normalized U-Measure (NUM)**. When evaluating a session, instead of aggregating the scores of each query in the session, it evaluates the session as a "virtual query" by building a comprehensive trailtext for the whole session. Besides, it uses the ideal session to normalize the evaluation result in accordance with **A1**. Based on **A2**, it uses click-through data to infer session-level relevance labels.

3.1 Normalizing U-Measure

3.1.1 Ideal Session

As we stated in A1, we believe that an ideal search system should rank all relevant documents before all irrelevant documents so that the user may not even need to reformulate her queries, saving her time and effort. As illustrated in Fig. 2, we place all the relevant documents before all irrelevant documents to build the ideal trailtext because it allows the user to get all of the information she needs without reformulation. By doing so, we manage to explicitly tell the search system what an optimal session-level ranking is and lead it to limit the number of reformulations, which we believe will make the user more satisfied. Note that the relevance label we refer to here is the session-level labels derived from click-through data based on A2.

We notice that there could be different re-ranking strategies of the documents with the same relevance scores in the ideal session. If the user scans the ranked documents from top to bottom, she will learn by reading the preceding document that subsequent documents are relevant. Thus, we maintain the same order of clicks in the trailtext of the ideal session as the actual session.

3.1.2 Session-Level Normalization

The value of NUM represents the proximity of an actual session to the ideal session. Note that when building trailtext for the actual session, based on [12,21], the user is assumed to read all the documents she clicks and all the passages before the last-clicked document in the session. And for the ideal session, we simply presume the user reads the documents that are marked session-level relevant (inferred by click-through data and A2) since all of them are presented at the beginning of the session and they all support the user's main information need. The session-level normalization technique is defined as follows:

$$\text{NUM} = \frac{\text{U}(S_{\text{actual}})}{\text{U}(S_{\text{ideal}})}, \tag{3}$$

as illustrated in Eq. (1) and Eq. (2), U(S) can be computed as follows:

$$\text{U}(S) = \sum_{pos=1}^{|tt|} \text{gain}(pos)\max(1 - \frac{pos}{L}), \tag{4}$$

where L is the longest Maximal Trailtext Length (MTL) across all conceivable search sessions. The MTL for a session is calculated by summing the text lengths of: (1) all passages before the last click for each query, (2) all passages clicked during the session, and (3) all reformulation texts (illustrated in Sect. 3.1.3). As a result, L reflects the most text the user has to read in a single session.

3.1.3 Reformulation Text

As shown in the upper right part of Fig. 2, we add an empty text named **reformulation text** in the trailtext between two queries. We believe it can penalize

Table 1. The statistics of the pre-processed datasets.

	TianGong-SS-FSD	NTCIR-16 SS
# Sessions	994	2,000
# Queries	3,411	6,420
Avg. # Query per Session	3.43	3.21
# Results per Query	10	10
Avg. # Clicks per Session	3.49	3.08

query reformulation behaviors to an adjustable degree, which the original U-measure does not account for. The intuition here is that U-measure uses text length to simulate reading time, thus we use empty texts to represent the user's reformulation time.

We determine the length of the reformulation text by exploring a Chinese field study dataset TianGong-SS-FSD [7] to find out how much time users spend between the exit of a query and the issuance of the following query. We compute the query reformulation time for each query. Besides, we eliminate the queries with negative reformulation time (due to multi-tabbing or logging errors) and 4% of queries with largest reformulation time values. More details are in Sect. 5.1.

3.2 Click-Through Data Enhancement

We use the click-through data to infer session-level relevance labels based on **A2**. We believe the same document should be marked relevant to this session if it was clicked in a subsequent query but skipped in a preceding one. As illustrated in the left part of Fig. 2, the document d_4 at rank 4 is marked relevant because it is clicked in the subsequent query.

Note that there will be duplicate documents that are considered relevant in the ideal session after the enhancement, e.g., d_4 at rank 4 and rank 1'. Among the existing works, some do not provide any special treatment to duplicate relevant documents [12,22] and some choose to discount these documents [16]. We identify three possible choices of dealing with d_4: (1) We include it in the trailtext because we believe that both rank 4 and rank 1' are informative. Besides, users may hope to find their clicked documents remain in the top results, which can facilitate the re-finding behaviors and the trust of search systems [22]; (2) We include it in the trailtext but give a discount to its relevance score; (3) We exclude it in the trailtext because it is redundant.

4 Experimental Setup

4.1 Datasets

We conduct our experiments on two public datasets: TianGong-SS-FSD [7] and NTCIR-16 Session Search (SS) Task [8]. For simplicity, this two datasets are denoted as *FSD* and *NST* in the following discussions.

4.1.1 *FSD*

FSD is a dataset collected from field studies. It records the users' session-level satisfaction rating, which we believe is a good standard for evaluating the metrics because it reflects the user's actual feelings about the search system. We explore the correlation of metrics (NUM and the baseline metrics) with user satisfaction ratings on this dataset. Following [7], we filter out the sessions containing more than one SERP. Besides, to facilitate NUM usage, we filtered out the sessions that do not contain any clicks and only have one query. We believe this kind of session can be considered **good abandonment** because in each of these sessions, the user's information need is resolved only by the results pages (*e.g.*, the information of snippets), without having to click on a result or do any query reformulations [23].

4.1.2 *NST*

NST is collected from a Chinese search engine. Its full data has 147,154 sessions, and there are 2,000 of them that have human-labeled relevance. We test the intuitiveness of the metrics on these 2,000 sessions because we may use these manual labels in future work and the scale of 2,000 sessions are large enough to draw conclusions (already larger than the dataset used in many works that are based on Cranfield/test collection [5,7,14]). We use the remaining 145,154 sessions to estimate L and the length of the reformulation text when testing intuitiveness on *NST*. These sessions do not contain any good abandonment.

The statistics of the pre-processed datasets are shown in Table 1.

4.2 Meta-evaluation Approaches

We utilize two meta-evaluation approaches to evaluate and compare NUM with existing session-based metrics:

4.2.1 Correlation with User Satisfaction

Since user satisfaction in information retrieval can be defined as the fulfillment of a specific objective [24] and it assesses users' actual feelings about a system, it can be considered as the ground truth to evaluate the evaluation metrics [25,26]. With the session-level user satisfaction feedback in *FSD*, we can evaluate the performance of various session-based metrics by calculating their correlation with user satisfaction using this dataset.

4.2.2 Intuitiveness

The Concordance Test [27] is proposed to quantify the intuitiveness of diversity metrics. We believe it can also predict the intuitiveness of session-based metrics. We will first choose some golden standard measures and presume them to actually represent intuitiveness. Given a pair of metrics (M1, M2), the relative intuitiveness of M1(or M2) is computed in terms of preference agreement with the golden standard measures.

Table 2. Spearman's ρ and Kendall's τ between session-based metrics and user satisfaction on FSD. The best performance is in bold, and the second-best performance is underlined. "Improv." reflects the improvements of NUM over RS-RBP.

Metric	sDCG	sRBP	sDCG/q	sRBP/q	U-measure
Spearman's ρ	0.0335	0.0508	0.3136	0.3150	−0.2282
Kendall's τ	0.0242	0.0382	0.2492	0.2507	−0.1800
Metric	U-measure/q	RS-DCG	RS-RBP	NUM	Improv.
Spearman's ρ	0.1008	0.3473	0.3508	**0.3611**	2.94%
Kendall's τ	0.0789	0.2771	0.2795	**0.2884**	3.18%

Table 3. Performance of ablated metrics on FSD.

	Spearman's ρ		Kendall's τ	
NUM w/o. SN	−0.1960	−154.28%	−0.1561	−154.13%
NUM w/o. RT	0.3276	−9.28%	0.2639	−8.50%
NUM w/o. SE	0.3522	−2.46%	0.2789	−3.29%
NUM (Full)	**0.3611**	–	**0.2884**	–

4.3 Generating Runs for NST

Since the organizers of NTCIR-16 Session Search (SS) Task have not released the run data of the participants, we use some ranking models to re-rank NST to generate some runs for the experiments of intuitiveness. The models are comprised of: (1) **ad-hoc ranking models**, including KNRM [28], ARC-II [29], Conv-KNRM [30], and DUET [31]; (2) **session-based ranking models**, including HBA-Transformers [32], COCA [2], and RICR [4]. The settings of these models are all the same as in their original papers.

Moreover, since the current session-based ranking models are not advanced enough to consider the two assumptions we put forward, we artificially construct two types of runs based on these ranking models: (1) **Ideal runs.** For each query of the session, we first add the candidate documents of the subsequent queries in the same session into its pool. Then we re-rank the candidates using these models and keep the top ten results. We believe this can make the re-ranked session closer to the ideal session that we defined. (2) **Diversified runs.** For each query of the session, based on the extended candidate pool described above, we discard the candidate documents that are already presented (included in the top ten results) in the preceding queries. We believe this can make the session more "diversified".

We generate 7 original runs, 7 ideal runs, and 7 diversified runs based on 7 ranking models aforementioned for NST.

4.4 Baselines

To demonstrate the effectiveness of NUM and verify our two assumptions, we compare NUM with existing DCG-based and RBP-based metrics sDCG [5], sRBP [14], sDCG/q, sRBP/q, RS-DCG [7] and RS-RBP [7]. In addition, to verify the effectiveness of the session-level normalization introduced in Sect. 3.1.2, we also compare our metric with U-measure [12] and U-measure/q.

Following the settings of [7], supposing each query has N documents and each session S has M queries, the computation of sDCG can be described as follows:

$$\text{sDCG}(S) = \sum_{m=1}^{M} \sum_{n=1}^{N} \frac{g(d_{m,n})}{(1+\log_{b_q} m)(1+\log_{b_r} n)}, \tag{5}$$

where $g(d_{m,n})$ maps the score of the n-th document in the m-th query of the session. And sRBP is computed as follows:

$$\text{sRBP}(S) = (1-p) \sum_{m=1}^{M} \left(\frac{p-bp}{1-bp}\right)^{m-1} \sum_{n=1}^{N} (bp)^{n-1} g(d_{m,n}). \tag{6}$$

"/q" in sDCG/q and sRBP/q is a way of normalizing metrics by simply the number of queries in the session. It is described as: Metric/q = Metric$(S)/M$.

For the computations of RS-DCG and RS-RBP, we apply the settings of their original paper [7].

Following [12], we calculate the *gain* value of a l-relevant document as: $g(d) = (2^l - 1)/2^H$, where H is the highest relevance level. Since this paper mainly discusses the scenarios where we only have implicit feedback, the *gain* value of a session-level relevant document is $(2^1 - 1)/2^1 = 0.5$. For NUM, the session-level labels are inferred from click-through data, according to A2. For the other metrics, we treat the clicked document directly as relevant ones, and **do not** use A2 to enhance the corresponding relevance labels.

For U-measure, we build a trailtext from user clicks and compute U-measure on it as illustrated in Eq. (4). The value of L in [12] is estimated using Microsoft's Bing (September 7, 2012, US market) data. However, we have to re-estimate L due to the inconsistency between the language of Microsoft's Bing (English) and the datasets we utilize in this study (Chinese). Furthermore, we believe that we should take the search engine that the dataset uses into consideration when we estimate L. This is because we believe that a user's tolerance for "the largest amount of text that the user may have to read in one session" [12] can be different from one search engine to another. Thus we estimated L independently for TianGong-SS-FSD [7] and NTCIR-16 Session Search (SS) Task [8]. We estimated the MTL for each session by assuming that each snippet is 80 characters long (which is a reasonable assumption for Chinese search engines), and discarded 1% of the sessions with the highest MTL values. Note that we set F as 20 following the original U-measure paper [12].

For the instantiations of the baseline metrics and NUM, we adopt different approaches with respect to different meta-evaluation techniques and the corresponding datasets. More details can be found in Sect. 5.1 and Sect. 5.2.

5 Results and Analysis

5.1 Correlation with User Satisfaction

We first compare the performance of NUM to the baseline metrics by computing Spearman's ρ [33] and Kendall's τ [34] with user satisfaction on FSD.

For the instantiations of the baseline metrics and NUM in this experiment, we adopt a 5-fold cross-validation method following [7]. We repeat this approach ten times. For each time, we use one fold of data to test the metrics' correlation with user satisfaction and use the other four folds to tune the parameters. We tune the parameters of DCG-based and RBP-based baselines to fit user satisfaction based on Spearman's ρ. For DCG-based metrics, b_r and b_q are searched in range $(1.0, 5.0]$ with step of 0.1. For RBP-based metrics, b and p are searched in range $(0, 1)$. For U-measure-based metrics (including NUM), we only estimate (not tune) their parameters (L and the length of the reformulation text) because these parameters should be consistent with users' real reading behavior. Note that we estimate them on the same folds of data as other baselines for fair comparisons.

The average estimated L is 19,336 for U-measure-based metrics. Besides, We found that a query reformulation takes an average of 206 s, or 3.43 min, which is a relatively high cost. Additionally, native Chinese speakers can usually read at an average speed of 255 ± 29 words per minute [35]. As a result, the average estimated length of the reformulation text is ($255 \times 3.43 = 875.5$) words (Chinese characters).

We report the average Spearman's ρ and Kendall's τ between each metric and user satisfaction across all ten times of 5-fold cross-validation on FSD. The results are shown in Table 2, which demonstrate the effectiveness of our method Furthermore, we can make the following observations:

(1) NUM achieves the best results among all metrics, demonstrating its effectiveness of estimating user satisfaction. For example, when compared to the state-of-the-art baseline RS-RBP, our metric has improved Kendall's τ by 3.18%.

(2) Compared to the original U-measure and the simply-normalized U-measure/q, NUM has a stronger correlation with user satisfaction, which demonstrates the effectiveness of the proposed session-level normalization. We can observe that NUM outperforms the original U-measure, indicating that it is necessary to apply the session-level normalization and the click-through data enhancement for session-based evaluation. Moreover, NUM performs better than U-measure/q, which further demonstrates the effectiveness of session-level normalization.

5.2 Intuitiveness

Since there are few works trying to evaluate the intuitiveness of session-based metrics, the golden standard measures that represent the intuitiveness of session search have yet to be discovered. In this work, we suggest two metrics for the intuitiveness test of session-level metrics:

(1) Average Precision (**AP**). We believe this measure is simple but intuitive, by which we simply compute the precision of each query and average these values. Note that because MAP depends on the recall base, which can not be estimated with click data, we thus use average precision instead.
(2) The position of the Last Clicked Document (**LCD**). This measure records the position of the last clicked document in the whole session and takes the reciprocal of this position as the score: $\text{LCD}(S) = 1/Index_{lc}$, where $Index_{lc}$ is the position of the last clicked document in the session S. This value depicts the number of snippets a user has to examine in order to obtain all the information she needs. For example, if each query has 10 candidate documents and the last clicked document of the session is the fourth document of the second query, then $Index_{lc} = 10+4 = 14$, $\text{LCD}(S) = 1/14$. We believe that the session with a higher LCD value should be preferred because the user can scan fewer snippets to complete her search task, saving her time and effort. Thus, we believe LCD also represents the intuitiveness of session search.

For the instantiations of the DCG-based and RBP-based metrics in this experiment, since there are no golden standard labels like user satisfaction in NST, we use the mean of the parameters tuned on each fold in Sect. 5.1. For U-measure-based metrics, we only need to estimate (not tune) L as we explained in Sect. 5.1. Thus, we use the remaining 145,154 sessions of NST to estimate them and test the intuitiveness on the other 2,000 sessions. The estimated L is 12,792. For the length of the reformulation text, since there is no start and end timestamp of a query in NST, we simply use the estimated one in Sect. 5.1 (362).

We exclude sDCG/q, sRBP/q, and U-measure/q from this experiment because the "/q" normalization does not affect the concordance test (two runs have the same number of queries). The concordance test is performed on all 21 runs (7 original runs + 7 ideal runs + 7 diversified runs, i.e., $(21 \times 20/2 \times 2000 =)$ 420,000 session pairs.

The results are presented in Table 4. For example, the result at the top left represents that sDCG and sRBP disagree in 75,655 pairs of sessions. Among these disagreed pairs, sDCG agrees with AP on around 89% of them, while sRBP agrees on about 83%, which implies sDCG is more intuitive than sRBP in terms of AP. Furthermore, we can observe that:

(1) In terms of all golden standard measurements, NUM is more intuitive than all baselines. For example, NUM agrees with AP on about 94% of the 182,192 disagreement session pairs, whereas RS-DCG is only consistent with AP on around 56%.
(2) Incorporating the recency effect makes the metric less intuitive. After incorporating the recency effect (RS-DCG and RS-RBP), we can

Table 4. Intuitiveness based on preference agreement with the proposed golden standard measures(AP and LCD). For each metric combination, the higher score is in bold, and the number of disagreements between these two metrics is stated in the parentheses below. The abbreviation "UM" stands for "U-measure".

AP	sRBP	UM	RS-DCG	RS-RBP	NUM
sDCG	**0.89**/0.83 (75,655)	**0.89**/0.70 (114,588)	**0.89**/0.65 (117,044)	**0.84**/0.67 (130,035)	0.64/**0.91** (154,371)
sRBP	– –	**0.85**/0.71 (125,447)	**0.86**/0.67 (122,497)	**0.84**/0.69 (125,116)	0.62/**0.92** (154,137)
UM	– –	– –	**0.86**/0.79 (96,944)	**0.79**/0.78 (115,498)	0.58/**0.94** (172,019)
RS-DCG	– –	– –	– –	0.82/**0.90** (72,863)	0.56/**0.94** (182,192)
RS-RBP	– –	– –	– –	– –	0.57/**0.92** (179,640)

LCD	sRBP	UM	RS-DCG	RS-RBP	NUM
sDCG	0.81/**0.86** (75,655)	**0.93**/0.59 (114,588)	**0.84**/0.62 (117,044)	**0.75**/0.68 (130,035)	0.56/**0.85** (154,371)
sRBP	– –	**0.92**/0.58 (125,447)	**0.84**/0.61 (122,497)	**0.77**/0.67 (125,116)	0.57/**0.84** (154,137)
UM	– –	– –	0.72/**0.86** (96,944)	0.62/**0.88** (115,498)	0.45/**0.94** (172,019)
RS-DCG	– –	– –	– –	0.73/**0.95** (72,863)	0.48/**0.87** (182,192)
RS-RBP	– –	– –	– –	– –	0.53/**0.83** (179,640)

Table 5. The ablation experiments of intuitiveness based on preference agreement with LCD.

Metric	NUM (Full)	# disagreements
NUM w/o. SN	0.46/**0.93**	(157,785)
NUM w/o. RT	0.78/**0.99**	(89,495)
NUM w/o. SE	0.75/**0.91**	(96,245)

observe that the intuitiveness of metrics decreases in terms of all golden standard measures. These findings reveal that, while the recency effect is beneficial for estimating user satisfaction, it degrades the metrics' intuitiveness.

5.3 Ablation Study

To further explore the reasonability of our two assumptions and the effectiveness of the improvements we make over the original U-measure, we design several variants of NUM. Specifically, we conduct the ablation experiments on FSD and NST as follows:

- **NUM w/o. SN.** We remove the session-level normalization part (SN, illustrated in Sect. 3.1.2). In another word, we only evaluate the actual session without considering the ideal session.
- **NUM w/o. RT.** We discard the reformulation text (RT, introduced in Sect. 3.1.3). We do not add extra empty texts into the trailtext to penalize query reformulations.
- **NUM w/o. SE.** We eliminate the click-through data enhancement, which states that the same document should be tagged relevant to the preceding query if it is clicked in a subsequent query but skipped in the previous one, *i.e.*, the supplemental enhancement (SE) based on **A2**.

The results of the ablation experiments are shown in Table 3 and Table 5. From which we can draw the following conclusions:

(1) Normalizing the evaluation result at the session level is effective. In **A1**, we presume that in an ideal session, all relevant documents should be ranked before all irrelevant documents. We can evaluate the similarity between this session and the ideal session by a session-level normalization. After removing this technique, our metric's intuitiveness and correlation with user satisfaction both drop. For example, Spearman's ρ on FSD decreases by about 154.28%. Furthermore, NUM without session-level normalization agrees with LCD on 46% of the disagreement pairs, whereas NUM agrees with LCD on 93%. These declines show that normalizing the actual session's evaluation score with the ideal session's score is effective. It also supports **A1** that an ideal session should rank all relevant documents before all irrelevant documents in the session.

(2) It is useful to include a reformulation text in the trailtext for each query reformulation. We propose to add a reformulation text in the trailtext between every two queries (as illustrated in Sect. 3.1.3). We believe it can penalize the query reformulation behavior to an adjustable degree, which the original U-measure does not take into account. After eliminating the reformulation texts, the performance of our metric drops. For example, the performance of estimating user satisfaction decreases by about 8.50% in terms of Kendall's τ. Furthermore, NUM agrees with LCD on about 78% of the disagreement pairs without the reformulation text, whereas with the reformulation text, NUM agrees with around 99%. This indicates that penalizing query reformulations by adding empty texts into the trailtext is effective.

(3) The supplemental enhancement (SE) makes our metric more intuitive and correlate better with user satisfaction. To verify **A2**, we propose that if a document is clicked in a subsequent query but skipped in a preceding one, it should be tagged relevant to the preceding query. After removing SE, the performance decreases. For example, it causes a decrease of 3.29% in terms of Kendall's τ. Moreover, NUM agrees with LCD on about 75% of disagreement pairs without SE, but the full NUM agrees with LCD on around 91%. These results demonstrate the effectiveness of SE and verify **A2**. The reason these reductions are smaller than those of the preceding two removals is that the number of documents that require SE is small (approximately 1% of FSD and 3% of NST).

6 Conclusions

In this work, we identify two challenges in session-based evaluation and make two assumptions about evaluating a session. **A1** states that an ideal search system should rank all relevant documents before all irrelevant documents in the session. **A2** believes that the documents clicked in a subsequent query but omitted in a preceding query are also relevant to that preceding query. To verify these assumptions, we design a session-level metric called **Normalized U-Measure (NUM)**. NUM evaluates a session as a virtual query, uses the score of an ideal session to normalize the evaluation result (**A1**) and enhances the click-through data (**A2**). Experiments on two public datasets demonstrate that NUM is intuitive and able to estimate user satisfaction well. In addition, ablation studies demonstrate the effectiveness of **A1** and **A2**.

Acknowledgement. This work was supported by the National Natural Science Foundation of China No. 62272467, the fund for building world-class universities (disciplines) of Renmin University of China, and Public Computing Cloud, Renmin University of China. The work was partially done at the Engineering Research Center of Next-Generation Intelligent Search and Recommendation, MOE.

References

1. Cleverdon, C.W., Mills, J., Keen, E.M.: Factors determining the performance of indexing systems (volume 1: Design), vol. 28. College of Aeronautics, Cranfield (1966)
2. Zhu, Y., et al.: Contrastive learning of user behavior sequence for context-aware document ranking. In: CIKM 2021: The 30th ACM International Conference on Information and Knowledge Management, Virtual Event, Queensland, Australia, 1–5 November 2021, pp. 2780–791. ACM (2021). https://doi.org/10.1145/3459637.3482243
3. Zuo, X., Dou, Z.,Wen, J.: Improving session search by modeling multi-granularity historical query change. In: WSDM 2022, The Fifteenth ACM International Conference on Web Search and Data Mining, 21–25 February 2022, Tempe, AZ, USA. ACM (2022). https://doi.org/10.1145/3488560.3498415
4. Chen, H., Dou, Z., Zhu, Q., Zuo, X., Wen, J.-R.: Integrating representation and interaction for context-aware document ranking. ACM Trans. Inf. Syst. (2022). https://doi.org/10.1145/3529955

5. Järvelin, K., Price, S.L., Delcambre, L.M.L., Nielsen, M.L.: Discounted cumulated gain based evaluation of multiple-query IR sessions. In: Macdonald, C., Ounis, I., Plachouras, V., Ruthven, I., White, R.W. (eds.) ECIR 2008. LNCS, vol. 4956, pp. 4–15. Springer, Heidelberg (2008). https://doi.org/10.1007/978-3-540-78646-7_4
6. Kanoulas, E., Clough, P.D., Carterette, B., Sanderson, M.: Overview of the TREC 2010 session track. In: Proceedings of The Nineteenth Text REtrieval Conference, TREC 2010, Gaithersburg, Maryland, USA, 16–19 November 2010. NIST Special Publication, vol. 500-294. National Institute of Standards and Technology (NIST) (2010)
7. Zhang, F., Mao, J., Liu, Y., Ma, W., Zhang, M., Ma, S.: Cascade or recency: constructing better evaluation metrics for session search. In: Proceedings of the 43rd International ACM SIGIR Conference on Research and Development in Information Retrieval, SIGIR 2020, Virtual Event, China, 25–30 July 2020, pp. 389–398. ACM (2020). https://doi.org/10.1145/3397271.3401163
8. Chen, J., Wu, W., Mao, J., Wang, B., Zhang, F., Liu, Y.: Overview of the NTCIR-16 session search (SS) task. In: Proceedings of NTCIR-16 (2022)
9. Jones, R., Klinkner, K.L.: Beyond the session timeout: automatic hierarchical segmentation of search topics in query logs. In: Proceedings of the 17th ACM Conference on Information and Knowledge Management, CIKM 2008, Napa Valley, California, USA, 26–30 October 2008, pp. 699–708. ACM (2008). https://doi.org/10.1145/1458082.1458176
10. Wang, H., Song, Y., Chang, M.W., He, X., White, R.W., Chu, W.: Learning to extract cross-session search tasks. In: WWW 2013 - Proceedings of the 22nd International Conference on World Wide Web, 1353–1363 (2013). https://doi.org/10.1145/2488388.2488507
11. Price, S., Nielsen, M.L., Delcambre, L.M.L., Vedsted, P.: Semantic components enhance retrieval of domain-specific documents. In: Proceedings of the Sixteenth ACM Conference on Information and Knowledge Management, CIKM 2007, Lisbon, Portugal, 6–10 November 2007, pp. 429–438. ACM (2007). https://doi.org/10.1145/1321440.1321502
12. Sakai, T., Dou, Z.: Summaries, ranked retrieval and sessions: a unified framework for information access evaluation. In: The 36th International ACM SIGIR Conference on Research and Development in Information Retrieval, SIGIR 2013, Dublin, Ireland, 28 July–01 August 2013, pp. 473–482. ACM (2013). https://doi.org/10.1145/2484028.2484031
13. Järvelin, K., Kekäläinen, J.: Cumulated gain-based evaluation of IR techniques. ACM Trans. Inf. Syst. **20**(4), 422–446 (2002). https://doi.org/10.1145/582415.582418
14. Lipani, A., Carterette, B., Yilmaz, E.: From a user model for query sessions to session rank biased precision (SRBP). In: Proceedings of the 2019 ACM SIGIR International Conference on Theory of Information Retrieval, ICTIR 2019, Santa Clara, CA, USA, 2–5 October 2019, pp. 109–116. ACM (2019). https://doi.org/10.1145/3341981.3344216
15. Moffat, A., Zobel, J.: Rank-biased precision for measurement of retrieval effectiveness. ACM Trans. Inf. Syst. **27**(1), 2–1227 (2008). https://doi.org/10.1145/1416950.1416952
16. Yang, Y., Lad, A.: Modeling expected utility of multi-session information distillation. In: Azzopardi, L., et al. (eds.) ICTIR 2009. LNCS, vol. 5766, pp. 164–175. Springer, Heidelberg (2009). https://doi.org/10.1007/978-3-642-04417-5_15

17. van Dijk, D., Ferrante, M., Ferro, N., Kanoulas, E.: A Markovian approach to evaluate session-based IR systems. In: Azzopardi, L., Stein, B., Fuhr, N., Mayr, P., Hauff, C., Hiemstra, D. (eds.) ECIR 2019, Part I. LNCS, vol. 11437, pp. 621–635. Springer, Cham (2019). https://doi.org/10.1007/978-3-030-15712-8_40
18. Wicaksono, A.F., Moffat, A.: Modeling search and session effectiveness. Inf. Process. Manag. **58**(4), 102601 (2021). https://doi.org/10.1016/j.ipm.2021.102601
19. Liu, M., Liu, Y., Mao, J., Luo, C., Ma, S.: Towards designing better session search evaluation metrics. In: The 41st International ACM SIGIR Conference on Research & Development in Information Retrieval, SIGIR 2018, Ann Arbor, MI, USA, 08–12 July 2018, pp. 1121–1124. ACM (2018). https://doi.org/10.1145/3209978.3210097
20. Liu, M., Mao, J., Liu, Y., Zhang, M., Ma, S.: Investigating cognitive effects in session-level search user satisfaction. In: Proceedings of the 25th ACM SIGKDD International Conference on Knowledge Discovery & Data Mining, KDD 2019, Anchorage, AK, USA, 4–8 August 2019, pp. 923–931. ACM (2019). https://doi.org/10.1145/3292500.3330981
21. Joachims, T., Granka, L.A., Pan, B., Hembrooke, H., Radlinski, F., Gay, G.: Evaluating the accuracy of implicit feedback from clicks and query reformulations in web search. ACM Trans. Inf. Syst. **25**(2), 7 (2007). https://doi.org/10.1145/1229179.1229181
22. Kanoulas, E., Carterette, B., Clough, P.D., Sanderson, M.: Evaluating multiquery sessions. In: Proceeding of the 34th International ACM SIGIR Conference on Research and Development in Information Retrieval, SIGIR 2011, Beijing, China, 25–29 July 2011, pp. 1053–1062. ACM (2011). https://doi.org/10.1145/2009916.2010056
23. Li, J., Huffman, S.B., Tokuda, A.: Good abandonment in mobile and PC internet search. In: Proceedings of the 32nd Annual International ACM SIGIR Conference on Research and Development in Information Retrieval, SIGIR 2009, Boston, MA, USA, 19–23 July 2009, pp. 43-50. ACM (2009). https://doi.org/10.1145/1571941.1571951
24. Kelly, D.: Methods for evaluating interactive information retrieval systems with users. Found. Trends Inf. Retr. **3**(1–2), 1–224 (2009). https://doi.org/10.1561/1500000012
25. Al-Maskari, A., Sanderson, M., Clough, P.D.: The relationship between IR effectiveness measures and user satisfaction. In: SIGIR 2007: Proceedings of the 30th Annual International ACM SIGIR Conference on Research and Development in Information Retrieval, Amsterdam, The Netherlands, 23–27 July 2007, pp. 773–774. ACM (2007). https://doi.org/10.1145/1277741.1277902
26. Huffman, S.B., Hochster, M.: How well does result relevance predict session satisfaction? In: SIGIR 2007: Proceedings of the 30th Annual International ACM SIGIR Conference on Research and Development in Information Retrieval, Amsterdam, The Netherlands, 23–27 July 2007, pp. 567–574. ACM (2007). https://doi.org/10.1145/1277741.1277839
27. Sakai, T.: Evaluation with informational and navigational intents. In: Proceedings of the 21st World Wide Web Conference 2012, WWW 2012, Lyon, France, 16–20 April 2012, pp. 499–508. ACM (2012). https://doi.org/10.1145/2187836.2187904
28. Xiong, C., Dai, Z., Callan, J., Liu, Z., Power, R.: End-to-end neural ad-hoc ranking with kernel pooling. In: Proceedings of the 40th International ACM SIGIR Conference on Research and Development in Information Retrieval, Shinjuku, Tokyo, Japan, 7–11 August 2017, pp. 55–64. ACM (2017). https://doi.org/10.1145/3077136.3080809

29. Hu, B., Lu, Z., Li, H., Chen, Q.: Convolutional neural network architectures for matching natural language sentences. In: Advances in Neural Information Processing Systems 27: Annual Conference on Neural Information Processing Systems 2014, 8–13 December 2014, Montreal, Quebec, Canada, pp. 2042–2050 (2014)
30. Dai, Z., Xiong, C., Callan, J., Liu, Z.: Convolutional neural networks for softmatching n-grams in ad-hoc search. In: Chang, Y., Zhai, C., Liu, Y., Maarek, Y. (eds.) Proceedings of the Eleventh ACM International Conference on Web Search and Data Mining, WSDM 2018, Marina Del Rey, CA, USA, 5–9 February 2018, pp. 126–134. ACM (2018). https://doi.org/10.1145/3159652.3159659
31. Mitra, B., Diaz, F., Craswell, N.: Learning to match using local and distributed representations of text for web search. In: Proceedings of the 26th International Conference on World Wide Web, WWW 2017, Perth, Australia, 3–7 April 2017, pp. 1291–1299. ACM (2017). https://doi.org/10.1145/3038912.3052579
32. Qu, C., Xiong, C., Zhang, Y., Rosset, C., Croft, W.B., Bennett, P.: Contextual re-ranking with behavior aware transformers. In: Proceedings of the 43rd International ACM SIGIR Conference on Research and Development in Information Retrieval, SIGIR 2020, Virtual Event, China, 25–30 July 2020, pp. 1589–1592. ACM (2020). https://doi.org/10.1145/3397271.3401276
33. Spearman, C.: The proof and measurement of association between two things (1961)
34. Kendall, M.G.: A new measure of rank correlation. Biometrika **30**(1/2), 81–93 (1938)
35. Trauzettel-Klosinski, S., Dietz, K., Group, I.S., et al.: Standardized assessment of reading performance: the new international reading speed texts IReST. Invest. Ophthalmol. Vis. Sci. **53**(9), 5452–5461 (2012)

A Multi-dimensional Early Warning Mechanism for Biological Invasions: A Case Study of Vespa Mandarinia

Shiqi Zhang and Weidong Xiao

National Key Laboratory of Information Systems Engineering,
National University of Defense Technology, Changsha, China
zhangshiqi@nudt.edu.cn

Abstract. The prevention and control of biological invasions have always been a focal issue for governments worldwide. However, due to the large number, wide distribution, and varied morphologies of invasive species, existing target detection methods struggle to effectively identify and distinguish these organisms. Moreover, traditional prediction methods find it challenging to capture the habits of different invasive species, which poses difficulties in predicting their locations and probabilities of occurrence. To address these challenges, we innovatively combine cubic polynomial regression with logistic functions to develop the "Tick-Tock" strategy, and integrate it with YOLOv5 and Geo-fencing technology. This allows us to identify and predict the growth trends, external characteristics, and geographical locations of invasive species from multiple dimensions. We integrate these methods into a priority recommendation system (PRS) to effectively monitor and provide early warnings for biological invasions. We focus on the case of Vespa Mandarinia (VM) in Washington State to demonstrate the superiority of our approach.

Keywords: Biological invasion · "Tick-Tock" strategy · YOLOv5 · Geo-fencing · Monitoring and Early warning

1 Introduction

Invasive species have severely impacted agricultural, forestry, fishery, and livestock production in various countries, posing a threat to the stability of ecosystems. However, monitoring and controlling biological invasions is becoming increasingly challenging for the following reasons. First, invasive species often have large populations, diverse forms, and extensive distributions, making it difficult for traditional target detection models to effectively distinguish them from other species, especially for insects with external characteristics very similar to those of their kind. Second, different invasive species usually have distinct growth patterns and habits, and existing prediction models are unable to dynamically predict based on the various growth stages of invasive species, which may lead to incorrect assessments. Therefore, constructing accurate and comprehensive models for predicting and identifying invasive species is very important, as it helps address the control and prevention of biological invasions.

Existing work [1], often utilizes conventional time series models such as ARIMA to predict the growth trend of invasive species. However, due to their complex growth habits and unstable trends, it is difficult to accurately capture and predict their growth using a single method. Some studies also use black-box models like CNN to predict the growth trends of invasive species, but this approach has drawbacks in terms of poor interpretability, making it challenging to adapt the prediction strategies to the variable and complex growth environments of invasive species [2]. Additionally, since the morphological features of invasive species are mostly similar to those of their counterparts, especially in insects, traditional target detection methods find it difficult to distinguish them, leaving room for improvement in the accuracy of species identification. Therefore, there are a series of challenges in predicting, identifying, and monitoring early warnings of invasive species. So we construct an integrated monitoring system that encompasses prediction, identification, and early warning in order to address these challenges.

According to a report by the Washington State Department of Agriculture (WSDA), Vespa Mandarinia (VM) has recently arrived in North America, drawing significant attention from the U.S. government [3]. VM is the world's largest hornet species and is known for its aggressive attacks on honeybee hives, posing a devastating threat to the beekeeping industry. Additionally, VM reproduces quickly and can cause fatal injuries to other insects and humans, which could have severe negative impacts on local environments, economies, and public health. Therefore, we take the VM in Washington State as an example to demonstrate the advanced and robustness of our system.

In summary, our contribution is three-fold:

- We innovatively propose a "Tick-Tock" strategy that dynamically predicts the growth trend of invasive species based on their different growth stages.
- We build a target detection model using YOLOv5 and establish a geographical coordinate model using Geo-fencing, which effectively recognizes the external features and geographic locations of invasive species. Experimental results demonstrate that our approach achieves higher accuracy.
- We establish an integrated system for prediction, identification, and early warning (PRS) that helps various levels of government and departments monitor the spread of invasive species in real-time.

In Sect. 2, we review and organize the related work. In Sect. 3, we provide a detailed discussion of our model and methods, and explain the steps for building an invasive species monitoring and early warning system. In Sect. 4, we conduct experiments with our model on specific datasets and discuss the results. Finally, we draw conclusions in Sect. 5.

2 Related Work

2.1 Control of Biological Invasions

Biological invasions not only disrupt the stability of local ecosystems but also threaten biodiversity. Existing work often utilizes traditional machine learning models such as ARIMA, random forest (RF), and gradient boosting machine (GBM) to predict the growth trends of invasive species [4]. However, these models require a large amount of observational data and labels, and a single prediction model cannot adapt to the variable growth phases of species. Therefore, some research has adopted pseudo-labels for predicting the growth of invasive species, yet this heavily relies on the quality of the pseudo-labels [5]. Recent studies [5,6] have used species distribution models (SDMs) and Sentinel-2 time series models to monitor the spatial distribution and habitats of invasive species, achieving relatively good results. However, they only consider how to more accurately monitor the occurrence of invasive species and neglect the characteristics of their population reproduction and growth, failing to serve as an early warning.

2.2 Invasive Species Target Detection

The traditional target detection algorithm is based on the general location of the target object to extract the features to identify the method. For instance, DPM [7] achieves object detection by sliding a window at uniformly spaced locations across the image and running classifiers on the components. R-CNN [8] uses a segmentation algorithm to divide the complete image into small segments and runs classifiers on these segments to identify objects. Although various algorithms have their own advantages, due to the inconsistency and uncertainty of the target, these methods have unsolvable difficulties in terms of speed, optimization and accuracy.

Recent work has introduced the combination of YOLOv5 and attention mechanisms for multi-scale training to enhance the accuracy of invasive species identification [9]. There have also been efforts to automatically construct features to achieve more precise classification [10]. While these methods have achieved good recognition accuracy, they often overlook geographic location factors. Existing work [11] utilizes Geographic Information Systems (GIS) and Remote Sensing (RS) to monitor the geographic locations of invasive species, yet there has been no research integrating target detection with geographic monitoring systems to achieve better classification effects and generalization.

3 Methodology

In this section, we provide a detailed introduction to our proposed "Tick-Tock" growth trend prediction strategy and explain how to utilize YOLOv5 and Geofencing for object recognition and geographic location prediction. Finally, we outline the steps for constructing the PRS system.

3.1 Tick-Tock: A Novel Growth Prediction Strategy

Polynomial Regression. First, we use the polynomial regression[1] method to explore the quantitative relationship between the number of days until the initial discovery of an invasive species and the total number of discoveries.

$$y = w_0 + w_1x + w_2x^2 + \cdots + w_nx^n = \sum_{i=1}^{n} w_i x^i, \quad (1)$$

where x represents the number of days since the first discovery of the invasive species, y is the total number of discovered invasive species, n is the order of the polynomial, $w_0, \cdots, w_1$ are the coefficients of the polynomial, denoted as W.

Secondly, the purpose of fitting data is to minimize the error function. Since the error function is a quadratic function of the polynomial coefficients W, it is possible to determine W through matrix operations and to use mean squared error and goodness-of-fit test as methods for evaluating the polynomial.

Growth Model Based on Logistic Function. Logistic function is a common sigmoid function. It was originally named by Pierre-François Verhulst in 1838 when he studied its relationship with population growth. Therefore, its typical application is the population growth model. Due to the analysis of the phenomenon of suspension of reproduction during hibernation, we also consider environmental factors such as survival pressure and inter-species competition that may exist in nature, so we consider using an improved Malthusian population growth model over a period of time to predict the spread behavior of the invasive species during that interval [13].

"Tick-Tock" Strategy. From the two models, based on experimental results (see Sect. 4), we choose to use a third-order polynomial regression to predict the reproduction of the invasive species during hibernation, and a logistic function (an improved Malthusian bee colony growth model) to predict the reproduction of the invasive species during the normal spawning period. By combining these two models, we derive our "Tick-Tock" strategy, which is formulated as follows:

$$TTS = TT_x - (TT_{i*(x-180)}) = \begin{cases} y_{x+74} & x \in \{December, \cdots, May\} \\ 0.34y_{x+57} + 0.8P_{x-170} & x \in \{June, \cdots, November\} \end{cases}, \quad (2)$$

$$TTS_{unit} = Unit \cdot Hive, \quad i \in \{0, 1, 2, 3 \cdots\} x_{unit} = Day. \quad (3)$$

where i represents the number of years since the first discovery of the invasive species and TTS represents our "Tick-Tock" strategy. In addition, since the data of x is the time interval from the first discovery of the invasive species, it also starts from the beginning of December of the previous year as the starting date and ends at the end of November of the current year, so the time is added and subtracted in this formula, to match the above model Match.

---

[1] Referencing [12], in the experiments of this paper, we set the mortality rate of VM during hibernation at 20% to ensure that the conclusion has practical significance.

3.2 Image Classification and Position Coordinates Prediction

We use the YOLOv5 [14] target detection algorithm to locate and recognize the image information of the invasive species. Subsequently, we use the GIS system of the observers (sighting reports) to locate and determine whether the geographical location falls within our designated target range using the Geo-fencing system.

Geo-fencing technology is a new application of LBS, designed to use virtual barriers to define virtual geographic boundaries, thereby establishing spatial limits for specific areas, as well as distances to nearby physical obstacles [15]. When an observer enters, leaves, or moves within a specific geographic area, devices such as smartphones can receive automatic notifications and other information [16]. This helps us to automatically register the locations where sighting reports occur. We train the Geo-fencing model on a specific dataset, and each additional sighting report effectively increases the sample size. According to the law of large numbers, this continuously enhances the reliability of the model. If a sighting is false, it confirms the model's reliability; if the sighting is accurate, it adds a new sample, thereby providing better assistance in determining the range of invasive species.

3.3 PRS: A Comprehensive Monitoring System

Now, we integrate the three technologies previously mentioned—prediction of invasive species growth trends, external feature identification, and geographical location early warning—into a Priority Recommendation System (PRS), as shown in Fig. 1.

Specifically, we integrate our "Tick-Tock" strategy, target detection model, and the life habits of various invasive species to establish a comprehensive evaluation index system. We employ the Analytic Hierarchy Process (AHP) to assign weights to different evaluation indicators, and combine Geo-fencing provided geographical locations to determine if sighting reports are within a reasonable Geo-fencing. Furthermore, we establish an updated propagation model, which uses the PRS system to detect sighting reports in real-time, thereby achieving the purpose of early warning. Once there is a potential for positive sighting reports, our monitoring system will issue an alert, which helps us to timely sense the possible presence of invasive species. If the system alarms frequently, we update our model daily; if the interval between system alarms is long, we update the model in real-time according to the number of alarms. This way, we can promptly update our model based on the appearance of invasive species, and load the new report locations into the database to promptly alert the public within the area to be more vigilant.

4 Experiment

In this section, we apply our model to specific datasets and calculate the model's specific values using VM in Washington State as an example. Then we analyze the results and conclusions.

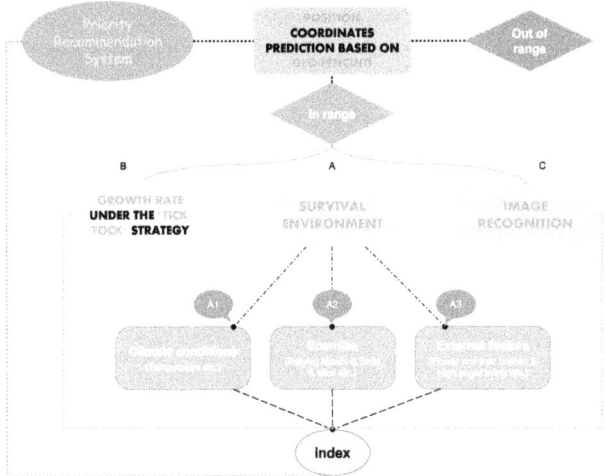

Fig. 1. Priority recommendation system flow chart.

4.1 Experiment Settings

We conduct experiments on the dataset provided by the Asian Giant Hornet Public Dashboard on the official website of the Washington State Department of Agriculture in the United States[2]. The dataset includes 4440 sighting reports along with the corresponding image information for 3305 of these reports. Each sighting report dataset includes a Global ID, Detection Date, Notes, Lab Status, Lab Comments, local latitude and longitude, and Submission Date.

For analysis, all images are standardized to a resolution of 1280 × 720 pixels, with the training, testing, and validation samples split in a ratio of 8:1:1. In the network architecture, the batch size is set to 8, and the Adam optimizer is used. Each phase of training consists of 200 epochs with a learning rate of 1e-4. As the number of iterations increases, the loss function exhibits a consistent downward trend without oscillations, indicating good convergence.

4.2 Results Analysis

"Tick-Tock" Strategy. Before applying the "Tick-Tock" strategy, we observe the distribution of the sample data. Considering the large time intervals in the sample data, which affect the accuracy of the final model, we use interpolation methods to fill in the sample data. Through experiments, we find that the quadratic curve is an approximation of the curve with the least curvature passing through the nodes. The curve is smooth, and the points only affect locally, resulting in a relatively small error. Therefore, we use spline interpolation for processing.

---

[2] https://agr.wa.gov/departments/insects-pests-and-weeds/insects/hornets/data
https://www.comap.com/contests/mcm-icm.

Through experimentation, we find that after interpolation, the cubic polynomial regression has a smaller mean squared error and higher goodness of fit. Subsequently, we compare the goodness of fit of quartic polynomial regression with cubic polynomial regression longitudinally, as shown in Table 1.

Table 1. Polynomial regression mean square error and goodness of fit.

Result	Degree	RMSE	r^2 score
Fourth order uninterpolated	4	1.01	0.92
Third order interpolated	3	0.98	0.94

It is clear that the cubic polynomial has a higher goodness of fit and the smallest mean squared error. Therefore, we use interpolation to process the data and ultimately employ a cubic polynomial regression model to study the quantitative relationship between the time intervals from the first discovery of the VM and the total number of sightings, confirming the robustness of this technique through the Bisquare method.

$$y = -0.1805x + 2.6x^2 + 3.087x^3 + 4.824. \tag{4}$$

Furthermore, we present and analyze the results of the logistic regression. The Sigmoid function curve and the final result of the model we use are shown in Fig. 2 and Fig. 3.

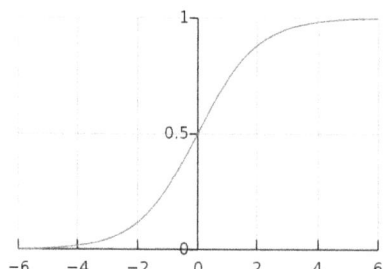

Fig. 2. Sigmoid function.   **Fig. 3.** Growth model fitting results.

From Fig. 3, the red line represents the potential growth trend of the VM. Since the fitting of our model is a comprehensive consideration of the existing data and the actual situation, there is a time lag, so the red line in the figure does not completely correspond to the positive report.

1) In the first ten days of March of the year, the temperature rose to above 10 °C and the wintering bees began to disperse, and no Positive Report appeared at this time.

2) When the temperature is constant above 17 °C, the bees begin to enter the hive and spawn stage, which is closer to mid-April, and there will be more VM going out for food thereafter, so the positive reports will begin to appear in May of the following year (2020). The growth model we constructed began to operate at this time.
3) After VM have been breeding in the hive for a period of time, they began to emerge into adults in June and July. The larvae in the hive did not increase the probability of people seeing them. Therefore, a small number of Positive Reports at this stage are expected.
4) When the model runs for the 100th day (mid-August), a large number of larvae have hatched and reached the peak of K value. At the same time, VM has filled the space with large-scale quantities in a short time, so it is not surprising that people frequently submit Positive Reports since then (the model's prediction and time are lagging, due to the small sample size, it is impossible to discuss in more detail)
5) However, under the assumptions, the model does not exceed the estimated environmental carrying capacity when it runs to the end until the hibernation begins, and the model has a good interpretability for the actual sample data.

We can get the fitting accuracy of the third-order polynomial regression and the improved Malthusian colony growth model. Experimental results demonstrate that the fitting accuracy of the third-order polynomial regression is 0.94, and the fitting accuracy of the improved Malthusian colony growth model is 0.64. According to our "Tick-Tock" strategy, in the comprehensive model, we

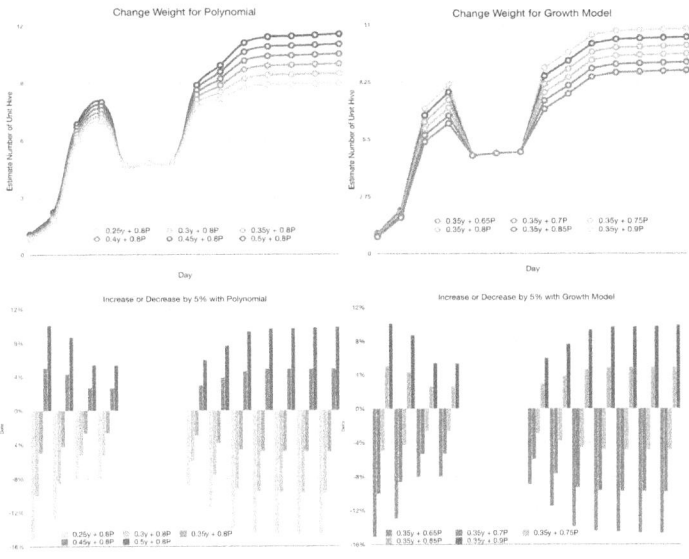

Fig. 4. Plots of adjusted coefficient.

calculate the goodness of fit to be 87.9% based on the real fit results. After considering VM habits and environmental resistance, the model we constructed has a prediction accuracy of 88%. Under the premise of small samples and large time span, a relatively good model is constructed with scalability.

Sensitivity Analysis. In our "Tick-Tock" Strategy model, we give each model 0.4 and 0.8 weight respectively. We think our Growth Model is more important than the previous one, which stands for forecasting the total colony number instead of just fitting a model to predict the data rigidly. We assume our combining has enough rationality. Indeed, we pass the Bisquare method. We try to adjust our coefficient (±0.05) respectively, and the results are shown in Fig. 4.

Our relative errors trending to change approximately at 5% for each recursive term, only a minority of them are merely above the line. Thus, the change of weight has no significant impact on the results of the model, so the weight of 0.4 and 0.8 we set has certain rationality.

Image Classification and Position Coordinates Prediction. Due to the large number of parameters in YOLOv5, having too few images in the dataset can easily lead to model overfitting. Therefore, we choose to perform web scraping from Google to obtain more image data. We eventually collect a total of 17,540 images related to VM for network training, and set up negative examples for contrastive learning to achieve better classification results. Furthermore, we put unprocessed reports into the model for testing, and the results are shown in Fig. 5. We find that given any VM image can automatically recognize its authenticity, and the model has an accuracy of 98.81.

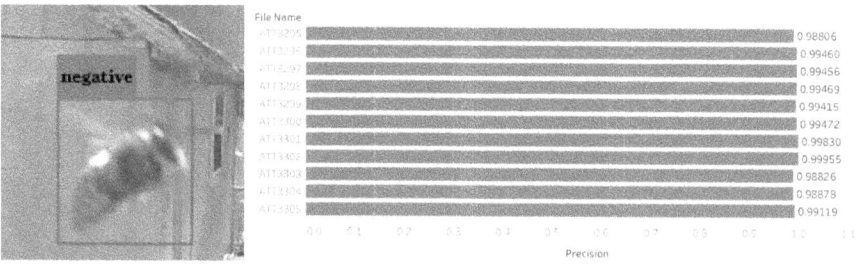

Fig. 5. Image recognition results and accuracy.

Subsequently, we use VM in Washington State as an example to set up the Geo-fencing model. As far as we know, after the winter, a new queen will build a hive and lay eggs within 30 km. Thus, we visualize the geographic location corresponding to the Positive ID of two years with a radius of 30 km, as shown in Fig. 6. Among them, the blue circle is based on Positive ID data in 2019, and the yellow circle is based on Positive ID data in 2020. In order to test the

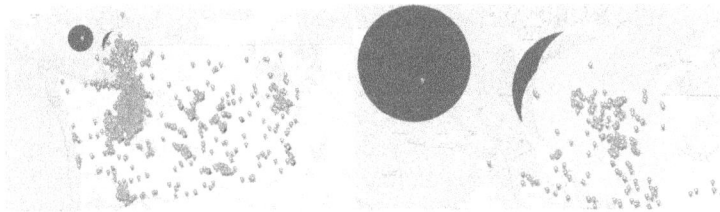

Fig. 6. Positive report range.

Fig. 7. Positive report classification results.

accuracy, we will conduct a classified survey based on all the sample data in the circle, and the result is shown in Fig. 7. It can be seen from Fig. 7 that in the classification within the circle, there are still a large number of reports that have not been confirmed as positive reports, and only a few have been confirmed. Therefore, we found that even if the report falls within the circle, we cannot be completely sure of its accuracy, and further investigation is needed.

Construction of PRS. Referring to the research conclusions of [17], we find that VM have stringent requirements for their survival environment. They are not only sensitive to temperature but also have natural predators such as praying mantises, birds, and rats, and they tend to gather in villages and small spaces rich in flowering plants, shrubs, succulents, and with high sweetness. VM typically start building beehives at temperatures between 16 °C and 18 °C, prepare for winter at 6 °C to 10 °C, and cease all activities when winds are at 44 °C to 46 °C or above wind force level 3. In other words, if a region's local temperature falls below 0 °C or the wind force exceeds level 3, the likelihood of VM occurrences in that area is lower. Also, if the temperature at a VM location is around 50 °C, the probability of VM occurrences is also very low.

Furthermore, under our "Tick-Tock" strategy, VM growth rate and VM image recognition can help verify the authenticity of reports.

Based on this, we construct a recommendation system evaluation index system aimed at predicting the authenticity of reports from three dimensions: survival environment, growth rate under the "Tick-Tock" strategy, and image recognition. Within the survival environment of the VM, we have set three sub-indicators: climate temperature, predator competition, and external factors.

We use the AHP method to assign weights to different indicators. The calculated CR for the primary indicators is 0.0088 < 0.1, and the CR for the secondary indicators is 0.0176 < 0.1, thus both judgment matrices are consistent. The evaluation index system and corresponding weights are shown in Table 2. Based on the above steps, we try to combine the geographic coordinate model with the evaluation index system to establish a comprehensive recommendation system. The formula is as follows:

$$y = G * [(A * (A_1 * E_1 + A_2 * E_2 + A_3 * E_3) + B * R + C * P)], \qquad (5)$$

Among them, G represents the judgment result according to the geographic coordinate model. If the geographic coordinate is within the above-mentioned limited range, 1 is output, otherwise it is 0. E, R, and P respectively represent the primary indicators of Survival environment, Growth rate under the "Tick-Tock" strategy, and Image recognition. A, B, and C respectively represent the weights of the three primary indicators E, R, and P. A_1, A_2, and A_3 respectively represent the weights of E_1, E_2, and E_3, which are three secondary indicators.

Table 2. Index weight.

Target layer	First level indicator	Weights	Secondary indicators	Weights
Priority recommendation system	Survival environment (E)	0.5396	Climate temperature E_1	0.5584
			Natural enemy competition E_2	0.1220
			External factors E_3	0.3196
	Growth rate under the "Tick-Tock" strategy (R)	0.1634	–	–
	Image recognition (P)	0.2970	–	–

According to the above-mentioned positive report priority recommendation system, we can gradually screen a new report, and then go through the screening of geographical location, local environmental factors, natural enemies, and VM image recognition, and finally automatically identify the most likely positive priority recommendations for witness reports, effectively improving accuracy.

5 Conclusion

To establish an automatic system that accurately identifies and verifies sighting reports, effectively predicting and preventing the spread of VM, we devise the "Tick-Tock" strategy to dynamically predict VM growth trends during different periods. We extract and mine VM's external features and growth environments through object detection and Geo-fencing technologies. By establishing a geographic coordinate system and an evaluation index system, we effectively filter sighting reports across latitudes and automatically recommend the most likely positive sighting reports to be addressed first. The model is updated based on alert frequency and intervals, achieving real-time monitoring and early warning effects. Future work can focus on expanding the dataset and enhancing the effects of object detection through iterative training.

References

1. Liu, F., Zhu, H., Cheng, X.: Vespa mandarinia recognition and prediction strategy based on geographic location and image recognition. In: IOP Conference Series: Earth and Environmental Science, vol. 769, no. 3, p. 032063. IOP Publishing (2021)
2. Qi, Y.: Research on Vespa mandarinia's invasion in the state of Washington. In: Proceedings of the 2021 5th International Conference on Intelligent Systems, Metaheuristics & Swarm Intelligence, pp. 84–87 (2021). https://doi.org/10.1145/3461598.3461612
3. Washington State Department of Agriculture, 2020 Asian Giant Hornet Public Dashboard. https://agr.wa.gov/departments/insects-pests-and-weeds/insects/hornets/data. Accessed 5 Nov 2020
4. Wang, H., Yi, T., Ling, L.: Statistics is much more powerful than murder hornet:—a statistical method to predict the spread of Vespa Mandarinia. In: 2021 2nd International Conference on Artificial Intelligence and Information Systems, pp. 1–8 (2021). https://doi.org/10.1145/3469213.3470380
5. Kim, E., Moon, J., Shim, J., et al.: Predicting invasive species distributions using incremental ensemble-based pseudo-labeling. Eco. Inform. **79**, 102407 (2024). https://doi.org/10.1016/J.ECOINF.2023.102407
6. Mouta, N., Silva, R., Pinto, E.M., et al.: Sentinel-2 time series and classifier fusion to map an aquatic invasive plant species along a river-the case of water-hyacinth. Remote Sens. **15**(13), 3248 (2023). https://doi.org/10.3390/RS15133248
7. Lu, C., Zhou, Y., Bao, F., et al.: DPM-solver: a fast ode solver for diffusion probabilistic model sampling in around 10 steps. Adv. Neural. Inf. Process. Syst. **35**, 5775–5787 (2022)
8. Bharati, P., Pramanik, A.: Deep learning techniques—R-CNN to mask R-CNN: a survey. In: Computational Intelligence in Pattern Recognition: Proceedings of CIPR, pp. 657–668 (2020)
9. Wang, Q., Cheng, M., Huang, S., et al.: A deep learning approach incorporating YOLO v5 and attention mechanisms for field real-time detection of the invasive weed Solanum rostratum Dunal seedlings. Comput. Electron. Agric. **199**, 107194 (2022). https://doi.org/10.1016/J.COMPAG.2022.107194
10. Zhang, D., Lee, D.J., Zhang, M., et al.: Object recognition algorithm for the automatic identification and removal of invasive fish. Biosyst. Eng. **145**, 65–75 (2016)
11. Duarte, L., Castro, J P., Sousa, J J., et al.: GIS application to detect invasive species in aquatic ecosystems. In: IGARSS 2022-2022 IEEE International Geoscience and Remote Sensing Symposium, pp. 6013–6018 (2022). https://doi.org/10.1109/IGARSS46834.2022.9884895
12. Keeling, M.J., Franklin, D.N., Datta, S., et al.: Predicting the spread of the Asian hornet (Vespa velutina) following its incursion into Great Britain. Sci. Rep. **7**(1), 6240 (2017)

13. Michel, P.: Model of neo-Malthusian population anticipating future changes in resources. Theor. Popul. Biol. **140**, 16–31 (2021)
14. Zhu, X., Lyu, S., Wang, X., et al.: TPH-YOLOv5: improved YOLOv5 based on transformer prediction head for object detection on drone-captured scenarios. In: Proceedings of the IEEE/CVF international conference on computer vision, pp. 2778–2788 (2021). https://doi.org/10.1109/ICCVW54120.2021.00312
15. Ullah, F., Haq, H.U., Khan, J., et al.: Wearable IoTs and geo-fencing based framework for COVID-19 remote patient health monitoring and quarantine management to control the pandemic. Electronics **10**(16), 2035 (2021)
16. Koshti, D., Kamoji, S., Cheruthuruthy, K., et al.: A detection, tracking and alerting system for COVID-19 using geo-fencing and machine learning. In: 2021 5th International Conference on Intelligent Computing and Control Systems (ICICCS), pp. 1499-1506 (2021)
17. Looney, C., Carman, B., Cena, J., et al.: Detection and description of four Vespa mandarinia (Hymenoptera, Vespidae) nests in western North America. J. Hymenopt. Res. **96**, 1–20 (2023)

DLP-FR: Learning Predictable Degradation for Robust Blind Face Restoration

Tao Wu[1,2], Jie Cao[2], Huaibo Huang[2], Yuang Ai[2], and Ran He[2(✉)]

[1] School of Information Science and Technology, ShanghaiTech University, Shanghai, China
wutao2022@shanghaitech.edu.cn
[2] New Laboratory of Pattern Recognition (NLPR), Institute of Automation, Chinese Academy of Sciences, Beijing, China
{jie.cao,huaibo.huang}@cripac.ia.ac.cn, yuang.ai@ia.ac.cn,
rhe@nlpr.ia.ac.cn

Abstract. Blind Face Restoration (BFR) focuses on the intrinsic challenge of transforming low-quality facial images with unknown and varied degradation into high-quality counterparts. To rectify the issue of varying degradation levels may lead to sub-optimal restoration or over-correction, this paper introduces a novel approach, Degradation Level Predictable Face Restoration (DLP-FR), to leverage a fusion of a degradation assessment framework and Stable Diffusion. The core of DLP-FR lies in its two primary components: a degradation probability predictor that quantifies the degradation severity of the input image and a multi-modal prompt-guided Stable Diffusion process to dynamically adapt the restoration efforts based on the predicted degradation level. Degraded datasets derived from the CelebA-Test are specifically crafted for the model to encompass a broad spectrum of degradation severity. Abundant experiments indicate that DLP-FR significantly outperforms existing state-of-the-art methods, allowing us to comprehensively demonstrate DLP-FR's superior performance in handling various levels of image degradation.

Keywords: Image Restoration · Blind Face Restoration · Diffusion Model

1 Introduction

Blind Face Restoration (BFR) primarily aims to restore high-quality (HQ) face images based on the corresponding low-quality (LQ) counterparts that have suffered from various degradation processes, including but not limited to low-resolution [1,2], noise addition [3], blurriness [4,5], JPEG compression [6], *etc*.

The enhancing ability to generate high-quality facial images is useful across various fields ranging from face recognition in unconstrained environments to high-definition digital avatars.

Existing research on BFR is diversified through several categories. Earlier approaches introduce several utilization of facial priors to leverage deep facial information [1,7–13]. These methods utilize pre-trained models, mostly GANs-based, to capture generative priors, investigated as guidance to help improve the color enhancement performance for the facial details and photo-realistic textures generation [14–20]. However, despite the remarkable advancements in visual quality with the adoption of deep learning methodologies, face images generated by these pre-trained models often highly rely on pre-defined degradation to synthesize training data.When applied to real-world images with more severe degradation, these models further suffer from the lack of detailed textures, manifested by excessively smooth skin and hair texture [21].

Other models [22,23] then advance in crafting restorations by scaling up with larger datasets to bridge the gap between synthetic scenarios and real-world facial images. This design offers a more robust solution to the BFR challenge caused by the pre-defined degradation. Though such design provides the models with augmented access to wilder input information, these models do not possess any inherent understanding of the degradation in low-quality images. Such nature causes difficulties for models to deal with varying degradation levels, leading to either sub-optimal restoration or over-correction. This issue adversely affects the model's efficacy on datasets containing different degrees of degradation (Fig. 1).

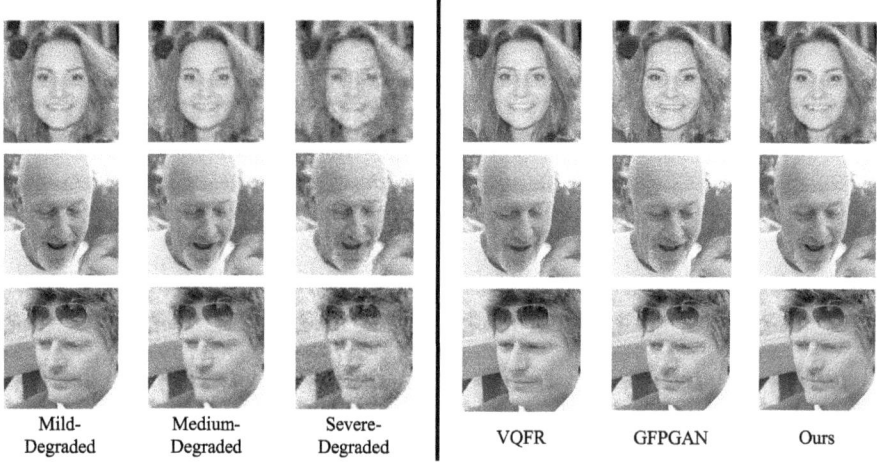

Fig. 1. Left: Comparison of the degradation severity of different synthetic degradation levels. Right: Comparison of the restoration between our method and other baselines on the medium synthetic degradation.

To address the above issue, we propose a diffusion-based blind face restoration pipeline, dubbed **D**egradation **L**evel **P**redictable **F**ace **R**estoration (DLP-FR).

Our motivation lies in endowing the model with the capability to make predictions for the degradation severity in the LQ images. We leverage the diffusion model's superiority in aspects of generating long-tail distribution to alleviate the dependency on synthetic degradation.

Specifically, we intricately design two components to address the challenges posed by the varying degrees of degradation in BFR. (1) The first component is a degradation probability predictor that evaluates the severity of degradation in input facial images. Diverse combinations of degradation types are merged with three levels of adjusted weights to simulate degradation intensity. This degradation synthetic method mimics real-world data and improves the predictor's ability to handle extreme degradation. (2) The second is a Stable-Diffusion (SD) based restoration network that utilizes dynamic from the predictor to craft textual and visual prompts essential for the diffusion process. We generate visual and textual prompts from LQ input image embeddings respectively with CLIP [24–26] and VAE [27,28]. Then, we dynamically integrate them into HQ embeddings based on degradation probabilities estimated by the lightweight predictor. Finally, these prompts are fed to the Stable Diffusion to restore the HQ output from the LQ images with a denoising U-Net. Experimental results demonstrate our model surpasses existing state-of-the-art methods in fidelity and detail restoration, manifested as a 0.5 increase in PSNR and a 0.03 increase in SSIM metric compared with other baseline methods.

In summary, the contributions of this work are as follows:

– We propose DLP-FR that utilizes a Stable Diffusion model [29] to remove degradation that applies both text and visual prompts, achieving robustness restoration against severe and unknown degradation conditions.
– We trained a degradation probability predictor based on a three-level degradation severity dataset to simulate real-world data. This design enhances the model's flexibility in mitigating degradation.
– We carried out abundant experiments to demonstrate DLP-FR's superiority in achieving adaptability and robust restoration when addressing intricate degradation.

2 Related Work

2.1 Image Restoration

Image Restoration generally focuses on super-resolution, denoising, deblurring, removing compression, *etc.* [2–6,30]. Recent advancements in Image Restoration highlight a shift towards complex degradation models to approximate real-world degradation like BSRGAN [31] and Real-ESRGAN [32]. They aim to synthesize practical degradation by employing GANs-based methods and give corresponding restoration based on the known degradation model, such as Deblur-GAN [33,34]. However, the real-world degradation for LQ images varies and often remains unknown. Challenges remain in generating realistic details with indeterminate degradation scales, thereby limiting the performance of Image Restoration.

2.2 Blind Face Restoration

The research of Blind Face Restoration, aimed at regeneration realism and high-fidelity face images from LQ counterparts, has evolved through several stages. Earlier attempts utilized diverse facial priors to alleviate the dependency on degraded inputs and enhance the robustness. Geometry priors including facial landmarks [1,7,8], parsing map [1,9,35] and component heatmaps [10], though achieve remarkable performance in recovering shape accurately, contain no detailed information. Reference priors [36–43] are later proposed as information guidance to reconstruct facial details to retain identity features. However, since degraded images contain only scarce prior information, using only real-world LQ images still fails to ensure the realism of the generated faces.

To further improve the restoration quality, deep generative models gained popularity due to their ability to utilize the rich prior information encapsulated in them. StyleGAN [14,15] first leverages generative priors to recover detailed textures. GPEN [16], GFP-GAN [17], GLEAN [19] utilize pre-trained GANs to capture the generative face priors and feed the priors into the encoder-decoder structure which significantly boost the restoration performance.

Besides the GAN-based category, VQFR [22] and CodeFormer [23] apply pre-trained Vector-Quantize dictionaries on entire faces to acquire abundant expressiveness.

2.3 Diffusion Models

In recent years, Denoising Diffusion Probabilistic Models (DDPM) [44,45] and Denoising Diffusion Implicit Models (DDIM) [46] create a new mainstream in the field of computer vision, due to diffusion model's superiority of avoiding training collapse and generating long-tail distribution compare to GAN-based methods. The research utilizes a diffusion model from earlier super-resolution tasks to later BFR scenarios. SR3 [47] conditions DDPM on low-resolution images to carry a channel-wise super-resolution, GLIDE [48] takes text conditions for class-free HQ realism image synthesis. DR2 [49] and DifFace [50] employ the diffusion model's strength on generative priors to address the issue of scarce information, which serves the LQ images as a guiding constraint for more realism restoration.

3 Methodology

3.1 Overview

We propose DLP-FR for blind face restoration in complex real-world scenarios. The proposed framework is depicted in Fig. 2. The rest of this section is organized as follows. In Sect. 3.2, we first apply a pre-trained CLIP image encoder [24] to obtain the input's image embedding and feeds it to the degradation probability predictor to predict its degradation severity level performed as a dynamic weight. Then in Sect. 3.3, we utilize this dynamic weight to guide the multi-modal adaptor to generate correspond high-quality textual and visual prompts [51,52] for Stable Diffusion [29]. Finally in Sect. 3.4, we introduce a post-processing method to solve the inherent color shift issue of the restoration network.

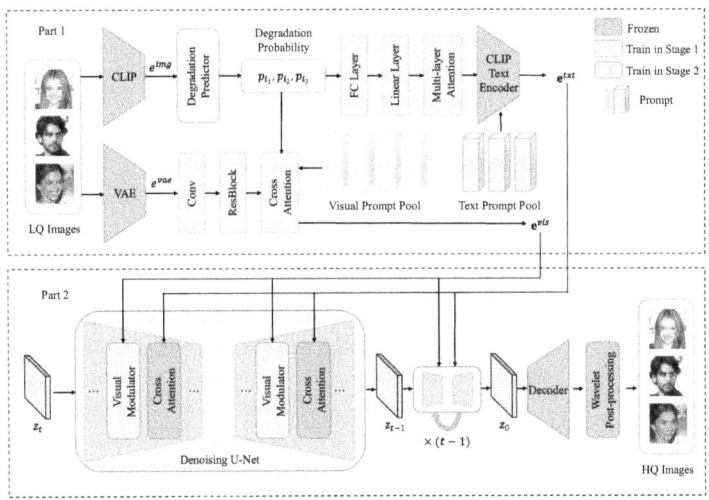

Fig. 2. Illustration of DLP-FR's pipeline with a two-stage training. The first part takes LQ images as input, trains a degradation severity level predictor to predict the CLIP image feature embedding into dynamic weight, and then generates both textual and visual prompts. The second part utilizes these prompts to train an restoration U-Net and applies a wavelet transformer-based post-processing for image color fix to get HQ images.

3.2 Degradation Probability Predictor

This module aims to address the real-world challenges underlying the fact of the inherent ill-posed nature of BFR, where a single LQ image could correspond to multiple high-quality restorations. One practical solution requires the models to accurately assess the severity of the degradation for appropriate restoration to prevent over-smoothing and artifact generation. Hence, we introduce a degradation probability predictor to quantify the degradation severity level, formulating it as dynamic weights to guide the subsequent restoration model.

To train the degradation probability predictor, we first synthesize the degraded LQ images then use them to guide the prediction. High-definition facial images first undergo combinations of diverse degradation through blurriness, noise addition, down-sampling, and JPEG compression to generate LQ images **I**. By this approach we create three distinct LQ datasets, namely representing mild, medium, and severe degradation.

Motivated by the CLIP model's impressive capability [24–26] on image representation, we apply a pre-trained CLIP model $\mathcal{E}^{img}$ to extract features $\mathbf{e}^{img}$ for the predictor to learn about estimating the degradation severity of unclassified image embeddings. This procedure can be formulated as:

$$p(i) = \mathrm{P}(\mathbf{e}_i^{img}), \quad \text{where} \quad \mathbf{e}_i^{img} = \mathcal{E}^{img}(\mathbf{I}), \tag{1}$$

where $p(i) = [p_{i_1}, p_{i_2}, p_{i_3}]$ represents the degradation probability for the input image, signifying the three distinction among different levels of degradation. The predictor P() is structured with several fully connected layer. It is optimized with cross-entropy loss:

$$\mathcal{L}_{\text{DPP}} = -\sum_{n=1}^{N} y_{o,c} \log(p_{o,c}), \qquad (2)$$

where $y_{o,c}$ is a binary indicator of whether class label c presents the correct classification for observation o and $p_{o,c}$ is the predicted probability.

3.3 Multi-modal Prompt Guided Stable Diffusion

Preliminary: Stable Diffusion. The second part is a restoration model based on the Stable Diffusion [29] backbone, a text-to-image denoising diffusion probabilistic model that learns to generate data samples through a series of denoising sequences and to estimate the data distribution. For stable and efficient training, SD takes a pre-trained encoder $\mathcal{E}$ to deliver the input image $\mathbf{I}$ to a latent space embedding $z = \mathcal{E}(\mathbf{I})$ and reconstruct it with decoder $\mathcal{D}$, where the diffusion and denoising are conducted:

$$z_t = \sqrt{\hat{\alpha}_t} z + \sqrt{(1 - \hat{\alpha}_t)} \epsilon, \qquad (3)$$

here $\epsilon \sim \mathcal{N}(0, \mathbf{I})$, $\alpha_t = 1 - \beta_t$ and $\hat{\alpha}_t = \prod_{s=1}^{t} \alpha_s$, where $\beta_t \in (0, 1)$ and time t is added for producing the noisy latent. z_t will nearly be standard Gaussian distribution as t is getting large enough. The denoising U-Net ϵ_θ is then trained in the latent space, which can be formulated as

$$\mathcal{L}_{\text{LDM}} = \mathbb{E}_{z,c,t,\epsilon}[||\epsilon - \epsilon_\theta(z_t, c, t)||_2^2], \qquad (4)$$

where $\mathbf{I}, c$ represent the input image and its condition information sampled from the dataset, a randomly picked time-step t is uniformly sampled, and $\epsilon \in \mathcal{N}(0, \mathbf{I})$ is the ground truth noise map at t.

Multi-modal Prompt Guidance. As shown in Fig. 2, to leverage the full potential of Stable Diffusion in text-to-image generation, we propose the concurrent use of a multi-model prompt, integrated by both text and visual prompts.

For the creation of the text prompt, the objective is to capture textual descriptions for the desired corresponding HQ facial images. Ways to achieve this goal involve employing a CLIP text encoder $\mathcal{E}^{txt}$, which transforms the input LQ image's embedded features into textual features, guided by dynamic degradation probability weights obtained from the probability predictor. A textual prompt pool $\mathbf{T} = \{\mathbf{T}_1, \mathbf{T}_2, \mathbf{T}_3\} \in \mathbb{R}^{3 \times L \times C}$ is established to record three severity levels of degradation information, where L represents the number of tokens and C represents CLIP's embedding dimension. The whole text prompt creation could be formulated as:

$$\mathbf{e}^{txt} = \sum_{i=1}^{N} p_i \mathcal{E}^{txt}(\text{AD}(\mathbf{e}_i^{img}), \mathbf{T}_i), i \in \{1, 2, 3\}, \qquad (5)$$

where AD is the image restoration adaptor with multi self-attention layers corresponding to each degradation severity level.

For the visual prompt part, we project the input LQ images through the VAE encoder of SD into latent embeddings and then decompose it into multi-scale features $\mathbf{e}^{vae} = \{\mathbf{e}_1^{vae}, \mathbf{e}_2^{vae}, ..., \mathbf{e}_N^{vae}\}$. Similar to the textual prompt, we construct $N = 4$ visual prompt pools $\mathbf{V} = \{\mathbf{V}_1, \mathbf{V}_2, ..., \mathbf{V}_N\}$ for each scale. The visual prompt is then dynamically integrated with the degradation severity probability with:

$$\mathbf{e}^{vis} = \mathrm{RB}_k(\mathbf{e}_k^{vae} + \sum_{i=1}^{N} p_i \mathrm{MHCA}_k(\mathbf{e}_k^{vae}, \mathbf{V}_k^i, \mathbf{V}_k^i)), i \in \{1, 2, 3\}, k \in \{1, ..., N\}, \tag{6}$$

where RB represents the residual block and $\mathrm{MHCA}_k(q, k, v)$ is a multi-head cross-attention layer of the k-th scale.

With the above degradation severity-aware probability's guidance, the model can distinguish the input from different degradation types and appropriately select the corresponding textual and visual prompt for the restoration network.

3.4 Post-processing

During the training, we observed the color shift phenomenon that the generated facial color images created by the diffusion model exhibit coherent shifts in color and brightness, where the shift direction varied as the training continued.

This phenomenon stems from inaccuracies in the prediction of spacial means within generated imagery, suggesting the model may face struggles with tasks requiring global features understanding and underscoring potential limitations during the face restoration.

To address this issue, we incorporate a wavelet-based color modification detection [53, 54] as a post-processing module. It is grounded on variance ratio analysis to adjust to the color shift in output images. This module employs a mathematical framework to detect and correct color discrepancies, formulated by:

$$\mathbf{I}_{\mathrm{OUTPUT}'} = \mathrm{WT}^{-1}(\frac{\mathrm{Var}(\mathbf{I}_{\mathrm{LQ}})}{\mathrm{Var}(\mathbf{I}_{\mathrm{OUTPUT}})} \times \mathrm{WT}(\mathbf{I}_{\mathrm{OUTPUT}})), \tag{7}$$

in this formulation, $\mathbf{I}_{\mathrm{OUTPUT}'}$ represents the image after the wavelet-based color correction. Such correction is achieved by applying a wavelet transform WT to the generated color shift image $\mathbf{I}_{\mathrm{OUTPUT}}$ and scaling this transformed data by the variance ratio of the reference low-quality data $\mathbf{I}_{\mathrm{LQ}}$. Finally, the inverse wavelet transform WT^{-1} is applied to produce the color-corrected image.

This process aligns the output image's color characteristics to be closer to those of the LQ image, thereby mitigating the color shift effects.

4 Experiment

4.1 Dataset Preparation

Training Dataset. We train our models on the FFHQ [14] dataset, which consists of 70,000 high-quality facial images with a resolution of 1024^2. We crop the images into 512^2 resolution and then degrade them into low-quality training data pairs with the following formula:

$$y = [(x \otimes k_\sigma) \downarrow_r + n_\delta]_{\text{JPEG}_q} \qquad (8)$$

We convolve the HQ image x with the Gaussian blur kernel k_σ, down-sample it with the scale factor r, apply a randomly chosen additive noise n_δ from Gaussian, and finally utilize JPEG compression with quality q. To generate different severity levels for the degradation, we choose three value splits for the above hyper-parameter for each we generate 1,000 images to train the degradation probability predictor. The *mild* split randomly samples σ, r, δ and q from {0.1, 3}, {0.1, 4}, {0, 10}, {80, 100}. The *medium* from {1, 5}, {1, 8}, {5, 15}, {70, 90}. And the *severe* from {3, 7}, {4, 16}, {10, 25}, {60, 80}.

All three splits will be applied for training the degradation probabilistic predictor, while during the training of the restoration network, a given HQ image will randomly take one of the three degradations to form the LQ pair.

Testing Dataset. We evaluate our model on CelebA-Test, a synthetic dataset consisting of 3000 HQ facial images sourced from the CelebA-HQ dataset [55]. None of the test images overlap with our training data. To further determine our model's capability while dealing with extremely severe degraded images, we apply the above degradation method to generate two splits of test data. The *mild* from {0.1, 5}, {0.1, 8}, {0, 15}, {70, 100}. And the *severe* from {3, 7}, {4, 16}, {10, 25}, {60, 80}.

4.2 Implementation Details

During this work, the LQ facial images are cropped into $512 \times 512 \times 3$ dimensions. We applied Adam optimizer [56] with a 1e−4 learning rate to train the degradation probability predictor for 10 epochs. For the training of the Stable Diffusion model, we applied a 5e−5 base learning rate with a scale factor of 0.18215. The in, out, and model channels for the diffusion U-net are respectively 4, 4, and 320, with a {4, 2, 1} attention resolution and a {1, 2, 4, 4} channel multiplication factor. The number of residual blocks is set to 2 while the number of channel heads is set to 64. The training process is conducted with 3 NVIDIA RTX 4090 GPUs for 72 h consisting of 300,000 iterations. For inference, we adopt DDPM sampling with 200 timesteps/image with sides <512, we first up-sample and enlarge them to 512 then carry out the inference.

4.3 Evaluation Metrics

Regarding the evaluation with the ground truth, we employ the traditional metrics, PSNR and SSIM, to measure image restoration fidelity, reflecting deviation and structural similarity respectively. LPIPS [57] evaluates perceptual similarity through deep feature comparison. FID [58] assesses distributional divergence between restored and original HQ images to gauge perceptual quality. NIQE [59] as a non-reference metric evaluates naturalness without ground truth.

4.4 Analysis of Results

To validate the effectiveness of our model, we compared its performance with several other state-of-the-art BFR methods, including GFPGAN [17], GPEN [16], VQFR [22], Codeformer [23] and DiffBIR [60], to cover diverse base generative methods including GANs, transformer and diffusion-based models.

Table 1. Quantitative comparison with blind face restoration baselines on the mild degraded dataset. The best and second best of each metric will be highlighted in **boldface** and underline format, respectively. ↓ indicates the lower is better, and ↑ higher is better.

Dataset	CelebA-Test Mild				
Methods	PSNR↑	SSIM↑	LPIPS↓	NIQE↓	FID↓
VQFR [22]	25.9830	**0.6948**	0.4048	4.4777	49.4098
GPEN [16]	25.1402	0.6664	0.3689	4.6536	47.8691
GFP-GAN [17]	27.0200	0.6743	0.3701	<u>4.3829</u>	48.4983
CodeFormer [23]	26.6040	<u>0.6917</u>	0.3708	4.5321	**43.0551**
DiffBIR [60]	26.7648	0.6741	<u>0.3330</u>	4.9192	56.2783
DLP-FR (ours)	**27.1605**	0.6850	**0.3296**	**4.2805**	<u>44.7989</u>

Table 2. Quantitative comparison with blind face restoration baselines on the severe degraded dataset. The best and second best of each metric will be highlighted in **boldface** and underline format, respectively. ↓ indicates the lower is better, and ↑ higher is better.

Dataset	CelebA-Test Severe				
Methods	PSNR↑	SSIM↑	LPIPS↓	NIQE↓	FID↓
VQFR [22]	23.6347	0.6284	0.4410	<u>4.5760</u>	75.2051
GPEN [16]	23.9538	<u>0.6369</u>	0.4256	4.7124	86.1075
GFP-GAN [17]	24.0178	**0.6400**	0.4436	4.6219	76.8136
CodeFormer [23]	23.7750	0.6146	0.4284	4.8768	**61.6245**
DiffBIR [60]	<u>24.3442</u>	0.6305	**0.4196**	5.9073	75.9340
DLP-FR (ours)	**25.0348**	0.6334	<u>0.4251</u>	**4.5448**	<u>64.8424</u>

Quantitative Comparison. To demonstrate the superiority of our model when dealing with images suffering from extremely severe degradation, we calculate evaluation metrics on two splits based on the synthetic dataset CelebA-Test, namely medium degradation and severe degradation. As shown in Tables 1 and 2, our proposed DLP-FR notably surpasses other SOTA methods for the PSNR, NIQE, and LPIPS assessment, highlighting its effectiveness in image fidelity and perceptual similarity. The slightly lower score in SSIM and FID metrics indicates that there may be room for improvement in maintaining structural similarity and natural image quality. One possible explanation for this scoring tendency can result in the model's prioritization of reducing high-frequency errors while ignoring the exact structural fidelity and perceived naturalness.

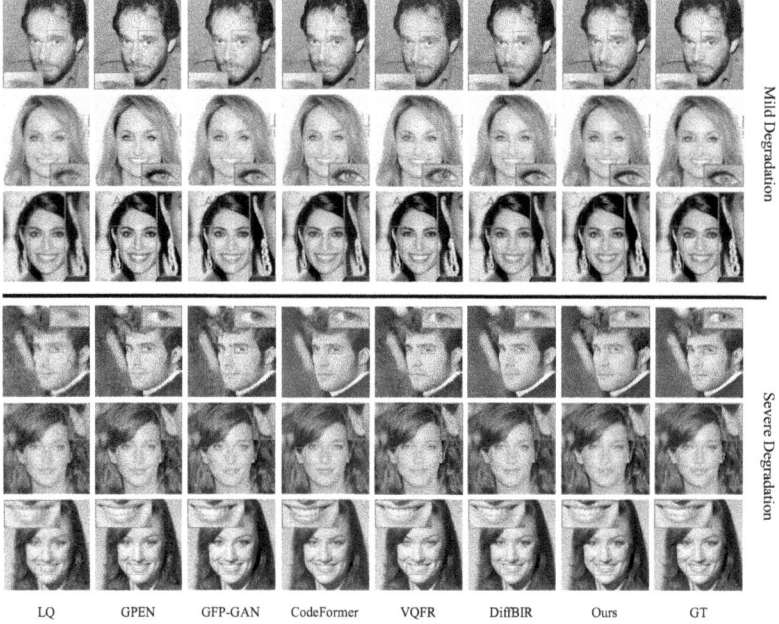

Fig. 3. Qualitative comparisons on CelebA-Test dataset with different degradation severity. Our method generates a more natural detailed appearance compared to other baselines.

Qualitative Comparison. Qualitative comparisons are presented in Fig. 3 with the same settings as the quantitative comparison. The visualized results indicate that our proposed method can enhance the output images' realism and provide richer details, especially in critical areas such as mouth, eyes, hair, and accessories like earrings. In a detailed explanation, in rows 3 and 5, our proposed DLP-FR outperforms other models in restoring the correct shape and textural for the earring accessories. For rows 2 and 4, previous methods fail to complement adequate eye and pupil shapes, while our outputs are visually more pleasant.

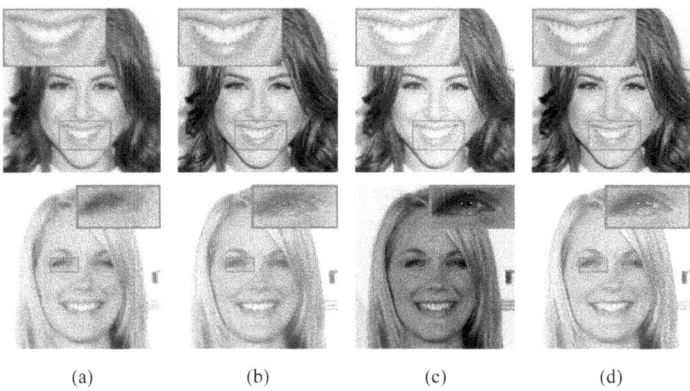

Fig. 4. Ablation study on CelebA-Test dataset variant different structure. (a) LQ input; (b) w/o post-processing; (c) w/o degradation probability predictor; (d) Our proposed method.

Ablation Study. We also make an ablation study to further address the role of the degradation probability predictor and the post-processing design in our approach. To make it clear, we designed two variant pipelines: (a) w/o post-processing, indicating the performance of only using probability prompt generator for the restoration network and does not apply the wavelet color fix module. (b) w/o probability predictor, denoting the absence of the pre-trained CLIP feature classifier which refers to a model that generates prompts for the restoration network randomly. As shown in Table 3, the ablation study conducted on the synthesized CelebA-Test strongly persuades that our model outperforms the other variant pipelines. The visualization is also provided in Fig. 4. The result reveals the fact that the post-processing helps the model to correctly restore the color information while the degraded probability predictor is effective in building detailed facial information, especially for severely degraded images.

Table 3. Ablation study for the DLP-FR framework. The best of each metric will be highlighted in **boldface**. ↓ indicates the lower is better, and ↑ higher is better.

Configurations	PSNR↑	SSIM↑	LPIPS↓	NIQE↓	FID↓
a) w/o post-processing	26.0514	0.6561	0.3532	4.9025	50.4569
b) w/o DPP	26.8321	0.6669	0.3631	4.7791	47.0319
Ours	**27.1605**	**0.6850**	**0.3296**	**4.2805**	**44.7989**

5 Conclusion

Our proposed method, DLP-FR is to tackle the real challenge of Blind Face Restoration, addressing the inherent one-to-many mapping from LQ facial images to the corresponding HQ restorations. The two specially designed modules allow our model to discern the appropriate level of restoration required. Extensive experiments have validated the superiority of DLP-FR over existing SOTA methods especially for severe degradation inputs. Further development in combining the model with other model light-weighting techniques is encouraged.

Acknowledgement. This work is partially funded by Beijing Municipal Science and Technology Project (No. Z231100010323005), National Natural Science Foundation of China (Grant No. 62206277), Youth Innovation Promotion Association CAS (Grant No. 2022132), and Beijing Nova Program (20230484276).

References

1. Chen, Y., Tai, Y., Liu, X., Shen, C., Yang, J.: FSRNet: end-to-end learning face super-resolution with facial priors. In: Proceedings of the IEEE Conference on Computer Vision and Pattern Recognition, pp. 2492–2501 (2018)
2. Zhang, Y., Li, K., Li, K., Wang, L., Zhong, B., Fu, Y.: Image super-resolution using very deep residual channel attention networks. In: Proceedings of the European Conference on Computer Vision (ECCV), pp. 286–301 (2018)
3. Liu, A., Liu, Y., Gu, J., Qiao, Yu., Dong, C.: Blind image super-resolution: a survey and beyond. IEEE Trans. Pattern Anal. Mach. Intell. **45**(5), 5461–5480 (2022)
4. Zhang, K., Zuo, W., Chen, Y., Meng, D., Zhang, L.: Beyond a gaussian denoiser: residual learning of deep CNN for image denoising. IEEE Trans. Image Process. **26**(7), 3142–3155 (2017)
5. Tao, X., Gao, H., Shen, X., Wang, J., Jia J.: Scale-recurrent network for deep image deblurring. In: Proceedings of the IEEE Conference on Computer Vision and Pattern Recognition, pp. 8174–8182 (2018)
6. Dong, C., Deng, Y., Loy, C.C., Tang, X.: Compression artifacts reduction by a deep convolutional network. In: Proceedings of the IEEE International Conference on Computer Vision, pp. 576–584 (2015)
7. Kim, D., Kim, M., Kwon, G., Kim, D.-S.: Progressive face super-resolution via attention to facial landmark. arXiv preprint arXiv:1908.08239 (2019)
8. Zhu, S., Liu, S., Loy, C.C., Tang, X.: Deep cascaded bi-network for face hallucination. In: Computer Vision–ECCV 2016: 14th European Conference, Amsterdam, The Netherlands, 11–14 October 2016, Part V, pp. 614–630. Springer (2016)
9. Shen, Z., Lai, W.-S., Xu, T., Kautz, J., Yang, M.-H.: Deep semantic face deblurring. In: Proceedings of the IEEE Conference on Computer Vision and Pattern Recognition, pp. 8260–8269 (2018)
10. Yu, X., Fernando, B., Ghanem, B., Porikli, F., Hartley, R.: Face super-resolution guided by facial component heatmaps. In: Proceedings of the European Conference on Computer Vision (ECCV), pp. 217–233 (2018)
11. Huang, H., Luo, M., He, R.: Memory uncertainty learning for real-world single image deraining. IEEE Trans. Pattern Anal. Mach. Intell. **45**(3), 3446–3460 (2022)
12. Cao, J., Luo, M., Yu, J., Yang, M.-H., He, R.: ScoreMix: a scalable augmentation strategy for training GANs with limited data. IEEE Trans. Pattern Anal. Mach. Intell. **45**(7), 8920–8935 (2022)
13. He, R., Hu, B., Yuan, X., Wang, L., et al.: Robust recognition via information theoretic learning. Springer (2014)

14. Karras, T., Laine, S., Aila, T.: A style-based generator architecture for generative adversarial networks. In: Proceedings of the IEEE/CVF Conference on Computer Vision and Pattern Recognition, pp. 4401–4410 (2019)
15. Karras, T., Laine, S., Aittala, M., Hellsten, J., Lehtinen, J., Aila, T.: Analyzing and improving the image quality of styleGAN. In: Proceedings of the IEEE/CVF Conference on Computer Vision and Pattern Recognition, pp. 8110–8119 (2020)
16. Yang, T., Ren, P., Xie, X., Zhang, L.: GAN prior embedded network for blind face restoration in the wild. In: Proceedings of the IEEE/CVF Conference on Computer Vision and Pattern Recognition, pp. 672–681 (2021)
17. Wang, X., Li, Y., Zhang, H., Shan, Y.: Towards real-world blind face restoration with generative facial prior. In: Proceedings of the IEEE/CVF Conference on Computer Vision and Pattern Recognition, pp. 9168–9178 (2021)
18. Huang, H., He, R., Sun, Z., Tan, T.: Wavelet domain generative adversarial network for multi-scale face hallucination. Int. J. Comput. Vision **127**(6), 763–784 (2019)
19. Chan, K.C.K., Wang, X., Xu, X., Gu, J., Loy, C.C.: GLEAN: generative latent bank for large-factor image super-resolution. In: Proceedings of the IEEE/CVF Conference on Computer Vision and Pattern Recognition, pp. 14245–14254 (2021)
20. He, R., Cao, J., Song, L., Sun, Z., Tan, T.: Adversarial cross-spectral face completion for NIR-VIS face recognition. IEEE Trans. Pattern Anal. Mach. Intell. **42**(5), 1025–1037 (2019)
21. Qiu, X., Han, C., Zhang, Z., Li, B., Guo, T., Nie, X.: DiffBFR: bootstrapping diffusion model for blind face restoration. In: Proceedings of the 31st ACM International Conference on Multimedia, pp. 7785–7795 (2023)
22. Gu, Y., et al.: VQFR: blind face restoration with vector-quantized dictionary and parallel decoder. In: European Conference on Computer Vision, pp. 126–143. Springer (2022)
23. Liu, G., Zhou, X., Pang, J., Yue, F., Liu, W., Wang, J.: CodeFormer: a GNN-nested transformer model for binary code similarity detection. Electronics **12**(7), 1722 (2023)
24. Radford, A., et al.: Learning transferable visual models from natural language supervision. In: International Conference on Machine Learning, pp. 8748–8763. PMLR (2021)
25. Goh, G., et al.: Multimodal neurons in artificial neural networks. Distill **6**(3), e30 (2021)
26. Materzyńska, J., Torralba, A., Bau, D.: Disentangling visual and written concepts in CLIP. In: Proceedings of the IEEE/CVF Conference on Computer Vision and Pattern Recognition, pp. 16410–16419 (2022)
27. Kingma, D.P., Welling, M.: Auto-encoding variational bayes. arXiv preprint arXiv:1312.6114 (2013)
28. Dai, X., et al.: Emu: enhancing image generation models using photogenic needles in a haystack. arXiv preprint arXiv:2309.15807 (2023)
29. Rombach, R., Blattmann, A., Lorenz, D., Esser, P., Ommer, B.: High-resolution image synthesis with latent diffusion models. In: Proceedings of the IEEE/CVF Conference on Computer Vision and Pattern Recognition, pp. 10684–10695 (2022)
30. Zhou, X., Huang, H., Wang, Z., He, R.: RISTRA: recursive image super-resolution transformer with relativistic assessment. IEEE Trans. Multimed. (2024)
31. Zhang, K., Liang, J., Van Gool, L., Timofte, R.: Designing a practical degradation model for deep blind image super-resolution. In: Proceedings of the IEEE/CVF International Conference on Computer Vision, pp. 4791–4800 (2021)

32. Wang, X., Xie, L., Dong, C., Shan, Y.: Real-ESRGAN: training real-world blind super-resolution with pure synthetic data. In: Proceedings of the IEEE/CVF International Conference on Computer Vision, pp. 1905–1914 (2021)
33. Kupyn, O., Budzan, V., Mykhailych, M., Mishkin, D., Matas, J.: DeblurGAN: blind motion deblurring using conditional adversarial networks. In: Proceedings of the IEEE Conference on Computer Vision and Pattern Recognition, pp. 8183–8192 (2018)
34. Kupyn, O., Martyniuk, T., Wu, J., Wang, Z.: DeblurGAN-v2: deblurring (orders-of-magnitude) faster and better. In: Proceedings of the IEEE/CVF International Conference on Computer Vision, pp. 8878–8887 (2019)
35. Chen, C., Li, X., Yang, L., Lin, X., Zhang, L., Wong, K.-Y.K.: Progressive semantic-aware style transformation for blind face restoration. In: Proceedings of the IEEE/CVF Conference on Computer Vision and Pattern Recognition, pp. 11896–11905 (2021)
36. Dogan, B., Gu, S., Timofte, R.: Exemplar guided face image super-resolution without facial landmarks. In: Proceedings of the IEEE/CVF Conference on Computer Vision and Pattern Recognition Workshops (2019)
37. Li, X., Liu, M., Ye, Y., Zuo, W., Lin, L., Yang, R.: Learning warped guidance for blind face restoration. In: Proceedings of the European Conference on Computer Vision (ECCV), pp. 272–289 (2018)
38. Li, X., Chen, C., Zhou, S., Lin, X., Zuo, W., Zhang, L.: Blind face restoration via deep multi-scale component dictionaries. In: European Conference on Computer Vision, pp. 399–415. Springer (2020)
39. Li, X., Li, W., Ren, D., Zhang, H., Wang, M., Zuo, W.: Enhanced blind face restoration with multi-exemplar images and adaptive spatial feature fusion. In: Proceedings of the IEEE/CVF Conference on Computer Vision and Pattern Recognition, pp. 2706–2715 (2020)
40. Zhou, X., Fu, C., Huang, H., He, R.: Dynamic graph memory bank for video inpainting. IEEE Trans. Circ. Syst. Video Technol. (2024)
41. Zhao, H., Zhou, X., Cao, J., Huang, H., Zheng, A., He, R.: Semantic-aware detail enhancement for blind face restoration (2024)
42. Huang, H., He, R., Sun, Z., Tan, T.: Wavelet-SRNet: a wavelet-based CNN for multi-scale face super resolution. In: Proceedings of the IEEE International Conference on Computer Vision, pp. 1689–1697 (2017)
43. Huang, H., Yu, A., He, R.: Memory oriented transfer learning for semi-supervised image deraining. In: Proceedings of the IEEE/CVF Conference on Computer Vision and Pattern Recognition, pp. 7732–7741 (2021)
44. Ho, J., Jain, A., Abbeel, P.: Denoising diffusion probabilistic models. Adv. Neural. Inf. Process. Syst. **33**, 6840–6851 (2020)
45. Nichol, A.Q., Dhariwal, P.: Improved denoising diffusion probabilistic models. In: International Conference on Machine Learning, pp. 8162–8171. PMLR (2021)
46. Song, J., Meng, C., Ermon, S.: Denoising diffusion implicit models. arXiv preprint arXiv:2010.02502 (2020)
47. Saharia, C., Ho, J., Chan, W., Salimans, T., Fleet, D.J., Norouzi, M.: Image super-resolution via iterative refinement. IEEE Trans. Pattern Anal. Mach. Intell. **45**(4), 4713–4726 (2022)
48. Nichol, A., et al.: GLIDE: towards photorealistic image generation and editing with text-guided diffusion models. arXiv preprint arXiv:2112.10741 (2021)
49. Wang, Z., et al.: DR2: diffusion-based robust degradation remover for blind face restoration. In: Proceedings of the IEEE/CVF Conference on Computer Vision and Pattern Recognition, pp. 1704–1713 (2023)

50. Yue, Z., Loy, C.C.: DiFace: blind face restoration with diffused error contraction. arXiv preprint arXiv:2212.06512 (2022)
51. He, R., Zhang, M., Wang, L., Ji, Y., Yin, Q.: Cross-modal subspace learning via pairwise constraints. IEEE Trans. Image Process. **24**(12), 5543–5556 (2015)
52. Ai, Y., Huang, H., Zhou, X., Wang, J., He, R.: Multimodal prompt perceiver: empower adaptiveness generalizability and fidelity for all-in-one image restoration. In: Proceedings of the IEEE/CVF Conference on Computer Vision and Pattern Recognition, pp. 25432–25444 (2024)
53. Jeon, J.J., Eom, I.K.: Wavelet-based color modification detection based on variance ratio. EURASIP J. Image Video Process. **2018**(1), 1–12 (2018). https://doi.org/10.1186/s13640-018-0286-6
54. Huang, H., Yu, A., Chai, Z., He, R., Tan, T.: Selective wavelet attention learning for single image deraining. Int. J. Comput. Vision **129**(4), 1282–1300 (2021)
55. Liu, Z., Luo, P., Wang, X., Tang, X.: Deep learning face attributes in the wild. In: Proceedings of the IEEE International Conference on Computer Vision, pp. 3730–3738 (2015)
56. Kingma, D.P., Ba, J.: Adam: a method for stochastic optimization. arXiv preprint arXiv:1412.6980 (2014)
57. Zhang, R., Isola, P., Efros, A.A., Shechtman, E., Wang, O.: The unreasonable effectiveness of deep features as a perceptual metric. In: Proceedings of the IEEE Conference on Computer Vision and Pattern Recognition, pp. 586–595 (2018)
58. Heusel, M., Ramsauer, H., Unterthiner, T., Nessler, B., Hochreiter, S.: GANs trained by a two time-scale update rule converge to a local nash equilibrium. In: Advances in Neural Information Processing Systems, vol. 30 (2017)
59. Mittal, A., Soundararajan, R., Bovik, A.C.: Making a "completely blind" image quality analyzer. IEEE Signal Process. Lett. **20**(3), 209–212 (2012)
60. Lin, X., et al.:. DiffBIR: towards blind image restoration with generative diffusion prior. arXiv preprint arXiv:2308.15070 (2023)

A Joint Relation Extraction Model Utilizing Message Passing Mechanism for Feature Fusion to Alleviate FN Problem from Distant Supervision

Jiran Zhu[1,2], Hongyun Du[2], Shengjie Jia[2], Yanhui Ding[2], Hui Yu[3], Weizhi Xu[2], and Xiujie Wu[1(✉)]

[1] Shandong Academy of Innovation and Development, Jinan, China
123127449@qq.com
[2] School of Information Science and Engineegring, Shandong Normal University, Jinan, China
[3] Business School, Shandong Normal University, Jinan, China
huiyu0117@sdnu.edu.cn

Abstract. While distant supervision enables automatic labeling of large-scale data, mislabeling is inevitable. Previous work has focused more on the relationship reduction (False Positive, FP) of mislabeling, but little work has explored the missing relationship (False Negative, FN) caused by an incomplete knowledge base. It is statistically proven that the dataset composed by distant supervision contains a large number of FN cases. One approach proposes a two-stage Pipeline model of "relationship first, entity later", which can effectively alleviate the FN problem. However, this model cannot effectively utilize the intrinsic connections and dependencies between the two subtasks. Utilizing the relationship information may reduce the extraction of semantically irrelevant entities in the later entity extraction module, thus alleviating the extraction of redundant triples and improving the performance of the model. In order to solve these problems and utilize good relationship information, we present a joint relation extraction model that uses the message passing mechanism to feature fusion. First, the subject-object entity correspondence module generates a two-dimensional matrix containing the correspondence scores between all entities. At the same time, the model predicts all relationships that may exist in the sentence. Then, entity extraction is performed on these relationships, with the special feature that the input vectors to the entity extraction module are sentence vectors that have been fused with relationship information features through a message passing mechanism. This allows the model to focus on selecting entities with high relevance to this relationship in the entity extraction module. Finally, the correctness of the extraction triples is verified by the results of the subject-object entity correspondence module. Experiments show that our model achieves a state-of-the-art (SOTA) F1 score on the distantly supervised dataset NYT10-HRL and provides comparable results on dataset NYT11-HRL.

Keywords: Deep neural network · Natural language processing · Distant supervision · Joint relation extraction

1 Introduction

Information extraction can obtain structured text information and knowledge from unstructured text data. Relation extraction is a subfield of information extraction. It is an essential technique in natural language processing (NLP), the purpose of which is to extract triples (head entity, relation, tail entity) information from sentences.

With the rapid expansion of data resources and the significant improvement of computing power in the Internet era, deep learning has greatly promoted the development of various fields of natural language processing [1]. The relationship extraction method based on deep learning can automatically learn semantic features and relational patterns from massive data. However, the model is extremely dependent on large-scale training data, and it takes a lot of manpower and time to manually label the entity pairs and relational data in sentences accurately. In order to realize the automation of relation extraction, Mintz et al. proposed the distant supervision hypothesis [2]. Distant supervision aligns entity mentions in text with entity mentions in the knowledge base to automatically generate relationship instances, enabling automatic annotation of large-scale data.

Despite the rich training data obtained by distant supervision, there are also non-negligible errors in labeling. There are two types of errors [3]. For the first type, the marked relationship does not match the original meaning of the sentence, and this type of error is called a false positive (FP). For the second type, relationships are missing in a large number of sentences due to the incompleteness of the knowledge base, which is called false negative (FN).

Previous work has focused more on the relationship reduction (FP) of mislabeling, but few have explored the missing relationship (FN) caused by an incomplete knowledge base. However, the datasets constituted by the distant supervision approach contain a large number of FN cases. For example, NYT and SKE datasets frequently used on distant supervised relation extraction tasks have at least 33% and 35% of FN cases, respectively [4]. In addition, the number of negative samples in the data generated using distant supervision overwhelmingly exceeded the number of positive samples in the previous question formulation. Xie et al. [2] proposed a "relationship first, entity second" Pipeline model, proving that this paradigm can effectively alleviate the FN problem arising from distant supervision. However, this Pipeline model does not take advantage of the interaction information between the two subtasks of relation extraction and entity recognition to reduce the extraction of redundant triples. If relationship information can be introduced before entity extraction, the extraction of semantically irrelevant entities can be greatly reduced [5], thus effectively alleviating the redundant extraction problem.

In order to reduce the noise problem caused by these negative samples, effectively utilize the intrinsic connection between the two subtasks, and at the same time reduce the extraction of redundant triples, this paper presents a joint relation extraction model using the message passing mechanism to feature fusion. Firstly, the subject-object entity correspondence module generates a two-dimensional matrix containing the correspondence scores between all entities. At the same time, the model predicts all the relationships that may exist in a sentence. Then, entity extraction is performed on these potential relationships, with the special feature that the vectors input to the entity extraction task is vectors of sentences that introduce the message passing mechanism to fuse relationship information features. Finally, the correctness of the extracted triples is verified by the results of the module corresponding to the subject-object entity. The contributions of this paper can be summarized as follows:

- We use a joint relation extraction model, which first extracts the relationships that may exist in the sentences and then extracts the entities of these relationships. In dealing with the negative data of distant supervision, we can not only effectively alleviate the FN problem caused by distant supervision but also effectively utilize the internal relations and dependencies of the two subtasks through joint relation extraction.
- In the entity extraction module, we introduce the message passing mechanism to fuse the relationship information features into the sentence vector features as inputs to enhance the interaction between the relationship information and the sentence information, and focus on selecting entities with high relevance to this relationship in the entity extraction module to effectively reduce the extraction of semantically irrelevant entities, alleviate the redundant extraction of the triples, and improve the performance of the model.
- Experimental results on the NYT10-HRL dataset show that our model outperforms existing methods and achieves SOTA performance.

2 Related Work

Much of the current work on distantly supervised relation extraction tasks is directed toward how to solve the FP problem. Lin et al. [6] combined multi-instance learning with neural network models to build relation extractors based on distantly supervised data to mitigate annotation errors. Feng et al. [7] converted the entity selection problem into a reinforcement learning problem. They selected high-quality sentences through reinforcement learning to solve the noise problem in distantly supervised relation extraction. Jia et al. [8] used the attention mechanism to let the model focus on the relationship indicator to identify the noisy data, and select the high quality labeled data gradually by Bootstrapping method. Li et al. [9] proposed a hierarchical comparative learning framework that makes full use of semantic interactions within and across specific levels, reduces the impact of noisy data in distant supervision, and so on. Tasks aimed at solving the FP problem have gradually matured.

In fact, due to the imperfect knowledge base, some entity pairs are not in the knowledge base, but some sentences can express certain relationships, which will lead to the relationship in this part of the sentence being mislabeled as NA, and many FN samples will be generated in the extraction process [10]. But there is little work being done to solve this problem. Some of the earlier methods based on feature engineering, for example, Min et al. [11] proposed optimization models that learn only from positive examples and unlabeled samples for situations where the knowledge base is incomplete. Xu et al. [12] dealt with FN training samples by using pseudo correlation feedback information retrieval technology. Ritter et al. [13] proposed a new latent variable method to simulate missing data and solve the problem of missing labels in training data. Although these methods can reduce the noise problems caused by negative samples, feature engineering-based methods are prone to errors when extracting features, which will be transmitted to the relational model and reduce the accuracy of the model.

Due to the widespread research and adoption of deep learning techniques, it is possible to use deep neural networks to solve FN problems. Based on BERT, Wang et al. [14] improved the task performance in two stages. First, they determined whether there was a relationship between the entity pairs. Then, they extracted the specific relationship between the entity pairs in which the relationship existed to solve the NA relationship problem. Xie et al. [4] proved that paradigms that first detect the relationship of sentences and then extract entity pairs suffer less overwhelming negative labels. Therefore, based on this advantage, Xie et al. [4], for the first time, used a model related to deep neural networks to solve the problem of missing relationship (FN) caused by an incomplete knowledge base in distantly supervised relation extraction, and proposed a novel two-stage Pipeline model, RERE, which firstly uses relational labels to perform sentence classification. The head/tail entity is then extracted, formulating the relationship extraction problem as an active unlabeled learning task to mitigate the FN problem [15]. However, the two subtasks in the Pipeline method are relatively independent, and the internal relationship and dependency between the two subtasks cannot be effectively utilized.

The joint relation extraction model can alleviate the problems of the Pipeline method. The general joint relation extraction is based on entity identification to extract the relationship between known entities, such as CasRel [16], TPLinker [17], etc. However, this model is prone to entity redundancy, causing subsequent error accumulation. In order to enhance the interaction between entity recognition models and relationship classification models, Takanobu et al. [18] proposed a new joint extraction paradigm in the framework of hierarchical reinforcement learning, which first detects the relationships and then extracts the corresponding entities as the parameters of the relationships. The PRGC model proposed by Zheng et al. [19] was also based on this paradigm, but effectively alleviates the CasRel, TPLinker et al. joint decoding-based relation extraction models in terms of extraction efficiency.

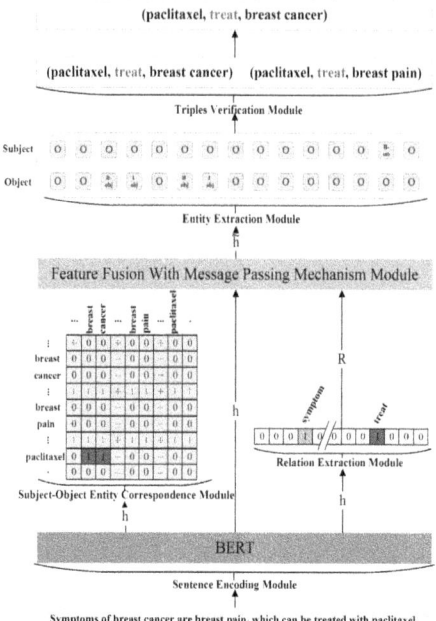

Fig. 1. The overall architecture of our model.

3 Model

3.1 Overall Architecture

In this paper, we present a joint relation extraction model using the message passing mechanism to feature fusion. Assuming that a sentence corresponds to multiple relation labels, the model splits a sentence sample into multiple sentence samples in the training phase, each of which contains only one relation label. The model takes the treat relationship as an example. As shown in Fig. 1, the overall framework of the model consists of six modules. Sentence Encoding Module uses BERT as a sentence encoder for the model to transform sentences into vector form. Subject-Object Entity Correspondence Module generates a matrix containing the scores of the correspondences between all the entities. Relation Extraction Module is modeled as a multi-labeled binary classification task that predicts all the potential relationships that may be present in a sentence. Feature Fusion With Message Passing Mechanism Module introduces a message passing mechanism to fuse relationship information features into sentence vector features as input to the entity extraction module to reduce the extraction of semantically irrelevant entities. Entity Extraction Module restricts the entity extraction to the relationships predicted by the relationship extraction module, and performs sequence labeling operations for the subject entity extraction and the object entity extraction, respectively. Triples Verification Module verifies the correctness of the extracted triples by means of the two-dimensional matrices generated by the module corresponding to the subject and object entity.

3.2 Sentence Encoding Module

As shown in Fig. 1, we use BERT as the sentence encoder for the model. The main framework of the BERT model is built with layers of Transformer. It not only solves the long-term dependency problem, but also makes the vector representation of each word in the output represent all the information of the input sentence as comprehensively and accurately as possible. Take the input as the sentence $S = [x_1, x_2, \ldots, x_n]$, a vector representation of the sentence can be obtained by BERT:

$$h = [h_1, h_2, \ldots, h_n] = BERT(x_1, x_2, \ldots, x_n) \qquad (1)$$

where h is the vector representation of the encoded sentence. h_n is a token vector representation that contains context information. x_n is the token of the input sentence. n is the number of tokens for the input sentence.

3.3 Subject-Object Entity Correspondence Module

This module computes the correspondence between subject and object in order to determine the correct triples. As shown in Fig. 1, it is a two-dimensional matrix of $n \times n$, where n is the number of tokens. Each element in the matrix represents the confidence that two entities are subject and object relationships. The calculation method is to concatenate the token vector representation of the two entities directly after a full connection and sigmoid activation function. Each element of the matrix is calculated as follows:

$$C_{i,j} = \sigma(W_a[h_i, h_j] + b_a) \qquad (2)$$

where $C_{i,j}$ represents the confidence level of the relationship that the entity of the i-th starting coordinate is subject and the entity of the j-th starting coordinate is object, respectively. The higher the $C_{i,j}$, the higher the confidence that the subject and object belongs to the true triples. If the confidence is greater than the threshold value, the $C_{i,j}$ is 1. Otherwise it is 0. h_i and h_j represent the encoded representation of the i-th token and j-th token in the input sentence. W_a is a trainable weight. σ is the sigmoid activation function. b_a is the bias vector.

3.4 Relation Extraction Module

For the distantly supervised relation extraction task, most of the work is to extract the relation on the basis of the entity pair already given. However, this introduces too many FN samples and makes the model difficult to learn. Therefore, our model first models the relation extraction at the sentence level and transforms it into a multi-label binary classification task. In this part, the model can capture the relationship expressed by the sentence. It runs in parallel with the Subject-Object Entity Correspondence Module. In this section, the sentence vector representation h output from the sentence coding layer undergoes an average pooling operation before being fed into the sigmoid activation function for relation extraction:

$$h_{avg} = Avgpool(h) \tag{3}$$
$$P_r = \sigma(W_r h_{avg} + b_r) \tag{4}$$
$$R = \{r_1, r_2, ..., r_m\} \tag{5}$$

where, h_{avg} is the vector representation of the sentence obtained after the average pooling operation of h. P_r is the probability of the relation label. W_r is the trainable weight. σ is the sigmoid activation function, which obtains the probability of all relation labels corresponding to the sentence. The existence of this relation label is determined by comparing the size of the set threshold with the relation label probability. If the probability is greater than the threshold, the node is 1, which means that this relation exists in the sentence; if the probability is less than the threshold, the node is 0, which means that this relation does not exist in the sentence. R is the relational representation set and m is the relational number. b_r is the bias vector.

3.5 Feature Fusion with Message Passing Mechanism Module

After obtaining the relational representation r_j and the sentence representation h, we input them into this module for feature fusion. The purpose of fusion is to make each token representation in a sentence contain information relevant to the representation of the relation, enhancing the interaction between the relation and each token in the sentence and their respective representational capabilities. This will be more conducive for the later entity extraction module to focus on selecting entities with high relevance to the relation and reduce the extraction of semantically irrelevant entities. We consider the relation representation r_j as a relation node and the token representation h_i in the sentence representation $h = [h_1, h_2, ..., h_n]$ as a word node. In order to fully integrate between word nodes and relation nodes, the node representation of each token in the sentence is updated through a message passing mechanism. This process is similar to the graph attention neural network, where the relation node r_j is considered as the neighbor node of all word nodes, and the importance between each node and its neighbors is dynamically computed by means of the attention mechanism, resulting in a more accurate node representation. The specific calculations are as follows:

$$e_{ij} = W_e[W_q h_i; W_k r_j] \tag{6}$$
$$\alpha_{ij} = \frac{Exp(e_{ij})}{\sum_{l \in n_i} Exp(e_{ij})} \tag{7}$$
$$h_i^{'} = h_i + \sum_{j \in n_i} \alpha_{ij} W_v r_j \tag{8}$$

where e_{ij} denotes the splicing operation of two nodes. W_e, W_q, W_k and W_v are trainable weights. α_{ij} denotes the attention weight between h_i and r_j.

In order to maintain the size of each dimension as well as the ability to maintain non-linearity, the fused output is obtained using the gate attention mechanism instead of the activation function as follows:

$$g_i = \sigma(W_g[h_i; h'_i]) \tag{9}$$

$$\tilde{h}_i = g_i \odot h'_i + (1 - g_i) \odot h_i \tag{10}$$

where g_i is the scalar. W_g is the trainable weight. $\tilde{h}_i$ denotes the updated token representation after fusion. $\odot$ represents the Hadamard product.

3.6 Entity Extraction Module

In this module, we identify the labels of each token using the BIO (Begin, Insize, and Outside) labeling scheme to extract the subject and object. For each candidate relation r, the triple classification distinction BIO is performed in two passes, the first to identify the BIO labels of the head entity and the second to identify the BIO labels of the tail entity. In predicting the entities of the corresponding relations, the input vector is the sentence vector $\tilde{h}$ updated by the Feature Fusion With Message Passing Mechanism Module, where the vector representation $\tilde{h}_i$ of each token is given different weights by the corresponding relations, effectively reducing the extraction of semantically irrelevant entities. The identification of the label of each token is computed as follows:

$$P_i^s = Softmax(W_s * \tilde{h}_i + b_s) \tag{11}$$

$$P_i^o = Softmax(W_o * \tilde{h}_i + b_o) \tag{12}$$

where P_i^s represents the probability that the i-th token corresponds to the BIO label in the head entity extraction. P_i^o represents the probability that the i-th token corresponds to the BIO label in the tail entity extraction. $\tilde{h}_i$ denotes the vector representation of the i-th token that has been updated by the Feature Fusion With Message Passing Mechanism Module. W_s and W_o are two trainable weights. b_s and b_o are two bias vectors.

3.7 Triples Verification Module

We obtain all the head and tail entities related to the candidate relation r through the entity extraction module. However, the relationship between these head and tail entities does not necessarily correspond to the candidate relation r. Therefore, in order to determine the correct triples, we verify the correctness of the head and tail entities extracted by the entity extraction module by using the results of the corresponding modules of the main and guest entities. After validation, the incorrect candidate triples are removed. As shown in Fig. 1, after the entity extraction module, the model extracts two possible triples (paclitaxel, treat, breast cancer), (paclitaxel, treat, breast pain). In the module of subject-object entity correspondence, the vertical coordinate "paclitaxel" and the horizontal coordinates "breast" and "cancer" are positioned at 1 in the two-dimensional matrix. That is, the head entity "paclitaxel" and the tail entity "breast cancer" belong to the relationship "treat" triples with high confidence. After validation, the incorrect candidate triple (paclitaxel, treat, breast pain) is removed and the correct triple (paclitaxel, treat, breast cancer) is retained.

3.8 Loss Function

In this paper, a joint extraction model is used to optimally combine the objective functions and share the parameters of the model encoder during the training process. The total loss function consists of three parts: the relationship extraction module, the entity extraction module, and the triples validation module. The outputs of these three modules are used to calculate their cross-entropy losses ($Loss_R$, $Loss_E$ and $Loss_V$) against the labeled data. The sum (Loss) is the total loss.

$$Loss_R = -\frac{1}{g_r} \sum_{i=1}^{g_r} (y_i log P_r + (1 - y_i) log(1 - P_r)) \tag{13}$$

$$Loss_E = -\frac{1}{2 \times n \times g_r^s} \sum_{b \in \{s,o\}} \sum_{j=1}^{g_r^s} \sum_{i=1}^{n} y_{i,j}^b log P_{i,j}^b \tag{14}$$

$$Loss_V = -\frac{1}{n^2} \sum_{i=1}^{n} \sum_{j=1}^{n} (y_{i,j} log C_{i,j} + (1 - y_{i,j}) log(1 - C_{i,j})) \tag{15}$$

$$Loss = Loss_R + Loss_E + Loss_V \tag{16}$$

where g_r is the number of all relations in the dataset. g_r^s is the number of relations in the predicted sentence.

4 Experiments

4.1 Dataset

In order to evaluate the effectiveness of the model, we conducted experiments on the dataset NYT10-HRL and NYT11-HRL. For these datasets further description is given below:

NYT10-HRL: NYT10 is a data set commonly used in the distantly supervised relation extraction task [20]. It was obtained by distant supervision from the New York Times corpus with the Freebase knowledge base, where at least 33% of FN cases exist. NYT10-HRL is NYT10 preprocessed with HRL, which removes training relations and "NA" sentences that do not appear in the test.

NYT11- HRL: NYT11 is a smaller version of NYT10, provided by CoType [21]. Wei et al. [16] point out that NYT11's test set misses a lot of triples, especially when multiple relationships appear in the same sentence, despite it being labeled by a human. NYT11-HRL was also preprocessed with HRL. Most of the samples in NYT11-HRL are labeled with only one relationship and have a false negative problem (Table 1).

Table 1. Statistics of datasets.

Datasets	Type	Train	Dev	Test
NYT10-HRL	Sentence	70,339	351	4,006
NYT11-HRL	Sentence	62,648	313	369

4.2 Hyperparameter

The parameters used in training the model are shown in Table 2. Among them, in order to prevent the model from overfitting, we use the early stopping method in the training process. The model is validated once on the validation set at the end of each epoch.

Table 2. Hyperparameter setting.

Hyperparameter	Value	Hyperparameter	Value
Max_seq_length	100	Downs_en_lr	0.001
Train_batch_size	64	Clip_grad	2
Val_batch_size	24	Drop_prob	0.3
Test_batch_size	64	Weight_decay_rate	0.01
Min_epoch_num	20	Warmup_proportion	0.1
Max_epoch_num	100	Gradient_accumulation_steps	2
fin_tuning_lr	0.0001		

4.3 Evaluation Metrics

We use three evaluation metrics, Precision, Recall, and F1 score, to measure model performance. Precision is the ratio of correctly predicted triples to all predicted triples. Recall is the ratio of correctly predicted triples to all actual triples. F1 score is the weighted summed average of Precision and Recall, which can comprehensively reflect the overall performance of the model.

4.4 Results and Analysis

In order to evaluate the effectiveness of the model and prove the feasibility of the model presented in this paper in alleviating FN problems caused by distant supervision, experimental analysis is carried out from the following aspects: (1) Comparative experiment and analysis of the model with NYT10-HRL dataset. (2) Comparative experiment and analysis of the model with NYT11-HRL dataset. (3) Ablation experiment.

Comparative Experiment and Analysis of Model with NYT10-HRL Dataset. Table 3 shows the comparative experimental results of our model with the NYT10-HRL dataset. 1) First, we observe that the F1 score of our method is superior to other comparison models, and the SOTA result is obtained on the NYT10-HRL dataset, which proves that the overall performance of our model is good. 2) The Precision, Recall, and F1 score of our method are 8.3%, 12.1%, and 10.5% higher than Baseline, respectively, indicating that our framework is more competitive and robust, even when using the same joint extraction model and the same paradigm. 3) Compared to the Pipeline method RERE, which recently achieved SOTA results on this dataset, our method is 4.25% and 0.95% higher than RERE on Precision and F1 score, respectively, illustrating the importance of effectively utilizing the connection between the two tasks. 4) With the exception of SPTree, all other models, including ours, have lower Recall than Precision, and we think that the likely reason for this is the presence of many false negatives on the NYT10-HRL test set. Among them, the KB Match method has the lowest Recall, which also suggests that distant supervision can lead to a large number of false negatives.

Table 3. Results on the NYT10-HRL dataset.

Model	Precision (%)	Recall (%)	F1 score (%)
KB Match [4]	38.10	32.38	34.97
SPTree [22]	49.2	55.7	52.2
NovelTagging [23]	59.3	38.1	46.4
CopyR [24]	56.9	45.2	50.4
HRL [18]	71.4	58.6	64.4
TPLinker [17]	**81.19**	65.41	72.45
CasRel [16]	77.7	68.8	73.0
RERE [4]	75.45	**72.50**	73.95
Ours	79.7	70.7	**74.9**

Comparative Experiments and Analysis of the Model with the NYT11-HRL Dataset. The results of the comparison experiments of our model with the NYT11-HRL dataset are shown in Table 4. 1) The experimental results show that our presented method outperforms the other comparative models in F1 score. Our method is 4.4% and 1.1% higher than the Baseline model in Recall and F1 score, respectively. This again demonstrates the effectiveness of our method. 2) By observation, we find that the simple distantly supervised method KB Match, which aligns the database to the corpus, is surprisingly better than the more complex distantly supervised relation extraction method MultiR, and the Precision is also similar to CoType. However, their Recall is low, which is consistent with the results on NYT10-HRL that distant supervision leads to many false negatives. 3) It can be seen that our method has a great improvement in Precision, Recall, and F1 score compared to some distant supervision models such as KB Match, MultiR, and CoType, which also shows the usefulness of our method for mitigating the effectiveness of the data FN problem caused by distant supervision.

Table 4. Results on the NYT11-HRL dataset.

Model	Precision (%)	Recall (%)	F1 score (%)
KB Match [4]	47.92	31.08	37.7
MultiR [25]	32.8	30.6	31.7
SPTree [22]	52.2	54.1	53.1
NovelTagging [23]	46.9	48.9	47.9
CoType [21]	48.6	38.6	43.0
CopyR [24]	34.7	53.4	42.1
HRL [18]	**53.8**	53.8	53.8
CasRel [16]	50.1	**58.4**	53.9
Ours	52.2	58.2	**55.0**

In conclusion, our model performs well on two datasets with false negative problems. This suggests that our model is beneficial in mitigating the FN problem arising from distant supervision. Moreover, our model performs better compared to the Pipeline approach with the same paradigm. The intrinsic connections and dependencies of the two subtasks can be effectively utilized through joint relation extraction.

Ablation Experiment. In order to verify the effectiveness of the model presented in this paper in mitigating the data FN problem caused by distant supervision, we created two comparative models for ablation experiments:

– W/O Feature Fusion With Message Passing Mechanism Module. Remove the Feature Fusion With Message Passing Mechanism Module and use only simple splicing to link sentence vectors to relation vectors.

– W/O Subject-Object Entity Correspondence Module. Remove the Subject-Object Entity Correspondence Module, i.e., there is no verification at the end that the triples are correct.

Table 5 shows the experimental results of ablation experiments on NYT10-HRL and NYT11-HRL datasets between the model presented in this paper and the above two comparative models for ablation experiments.

Table 5. Ablation experiment.

	Model	Precision (%)	Recall (%)	F1 score (%)
NYT10-HRL	Ours	79.7	70.7	74.9
	-Feature Fusion With Message Passing Mechanism Module	75.9	70.0	72.8
	-Subject-Object Entity Correspondence Module	73.7	69.8	71.7
NYT11-HRL	Ours	52.2	58.2	55.0
	-Feature Fusion With Message Passing Mechanism Module	53.7	53.5	53.6
	-Subject-Object Entity Correspondence Module	49.6	55.4	52.4

As shown in Table 5, when we remove the Feature Fusion With Message Passing Mechanism Module, the P, R, and F1 score decreased by 3.8%, 0.7%, and 2.1%, respectively, on the NYT10-HRL dataset; and the R, and F1 score decreased by 4.7%, and 1.4%, respectively, on the NYT11-HRL dataset. This proves that utilizing the Feature Fusion With Message Passing Mechanism of the sentence vectors and the relation vectors is effective. Because when we remove the Feature Fusion With Message Passing Mechanism Module and simply concatenate the sentence vector and the relational vector, each token representation in the sentence vector does not fully integrate the relational information. This will make the subsequent Entity Extraction Module cannot focus on extracting entities that are highly correlated with the relationship and cannot reduce the extraction of semantically independent entities. None of this is conducive to alleviating the data FN problem. When we remove the Subject-Object Entity Correspondence Module, the P, R, and F1 score drop by 6%, 0.9%, and 3.2% on the NYT10-HRL dataset; and the P, R, and F1 score drop by 2.6%, 2.8%, and 2.6% on the NYT11-HRL dataset, respectively. This proves the necessity of validating the triple by utilizing the results of the Subject-Object Entity Correspondence Module. Although the Feature Fusion With Message Passing Mechanism Module will reduce part of the redundant entity extraction in Entity Extraction module, the performance of the model will be degraded if the Subject-Object Entity Correspondence Module is removed, i.e., without verifying that the final extracted triples are correct. So, it shows that the two modules are not conflicting with each other and complement each other. In summary, the modules presented in this paper contribute to the validity of the final model.

5 Conclusion and Future Work

In this paper, we present a joint relation extraction model that utilizes the message passing mechanism to feature fusion to alleviate the FN problem arising from distant supervision. The model first extracts the relationships that may exist in the sentences, and then performs entity extraction on these relationships. This can effectively mitigate the FN problem generated by distant supervision, and also effectively utilize the intrinsic connections and dependencies of the two subtasks through the joint relationship extraction. Among them, the input vectors of the entity extraction module are sentence vectors that have been fused with the features of the relationship information through the message passing mechanism, so that the model will focus on selecting the entities that have high relevance to this relationship in the entity extraction module, thus reducing the extraction of semantically irrelevant entities and mitigating the redundant extraction problem. The experimental results show that our model consistently outperforms other comparative models on two datasets with false-negative problems obtained by distant supervision, and achieves SOTA results on the NYT10-HRL dataset. In our future work, we will focus on solving the problem of false negatives arising from distantly supervised relation extraction at the document level, to reduce the generation of document-level false negatives and to improve the effective utilization of document-level information.

Acknowledgment. This work was supported in part by Natural Science Foundation of Shandong Province (No. ZR2022MF328 and No. ZR2019LZH014), and in part by National Natural Science Foundation of China (No. 61602284 and No. 61602285).

References

1. Sun, X., Guo, Q., Ge, S.: GFN: a novel joint entity and relation extraction model with redundancy and denoising strategies. Knowl.-Based Syst. 112137 (2024)
2. Mintz, M., Bills, S., Snow, R., Jurafsky, D.: Distant supervision for relation extraction without labeled data. In: Proceedings of the Joint Conference of the 47th Annual Meeting of the ACL and the 4th International Joint Conference on Natural Language Processing of the AFNLP, pp. 1003–1011 (2009)
3. Liu, W., Yin, M., Zhang, J., Cui, L.: A joint entity relation extraction model based on relation semantic template automatically constructed. Comput. Mater. Continua **78**(1) (2024)
4. Xie, C., Liang, J., Liu, J., Huang, C., Huang, W., Xiao, Y.: Revisiting the negative data of distantly supervised relation extraction. arXiv preprint arXiv:2105.10158 (2021)
5. Xu, M., Pi, D., Cao, J., Yuan, S.: A novel entity joint annotation relation extraction model. Appl. Intell. **52**(11), 12754–12770 (2022)

6. Lin, Y., Shen, S., Liu, Z., Luan, H., Sun, M.: Neural relation extraction with selective attention over instances. In: Proceedings of the 54th Annual Meeting of the Association for Computational Linguistics (Volume 1: Long Papers), pp. 2124–2133 (2016)
7. Feng, J., Huang, M., Zhao, L., Yang, Y., Zhu, X.: Reinforcement learning for relation classification from noisy data. In: Proceedings of the AAAI Conference on Artificial Intelligence, vol. 32 (2018)
8. Jia, W., Dai, D., Xiao, X., Wu, H.: ARNOR: attention regularization based noise reduction for distant supervision relation classification. In: Proceedings of the 57th Annual Meeting of the Association for Computational Linguistics, pp. 1399–1408 (2019)
9. Li, D., Zhang, T., Hu, N., Wang, C., He, X.: HiCLRE: a hierarchical contrastive learning framework for distantly supervised relation extraction. arXiv preprint arXiv:2202.13352 (2022)
10. Zhang, J., Jiang, X., Sun, Y., Luo, H.: RS-TTS: a novel joint entity and relation extraction model. In: 2023 26th International Conference on Computer Supported Cooperative Work in Design (CSCWD), pp. 71–76. IEEE (2023)
11. Min, B., Grishman, R., Wan, L., Wang, C., Gondek, D.: Distant supervision for relation extraction with an incomplete knowledge base. In: Proceedings of the 2013 Conference of the North American Chapter of the Association for Computational Linguistics: Human Language Technologies, pp. 777–782 (2013)
12. Xu, W., Hoffmann, R., Zhao, L., Grishman, R.: Filling knowledge base gaps for distant supervision of relation extraction. In: Proceedings of the 51st Annual Meeting of the Association for Computational Linguistics (Volume 2: Short Papers), pp. 665–670 (2013)
13. Ritter, A., Zettlemoyer, L., Mausam, Etzioni, O.: Modeling missing data in distant supervision for information extraction. Trans. Assoc. Comput. Linguist. **1**, 367–378 (2013)
14. Wang, H., Focke, C., Sylvester, R., Mishra, N., Wang, W.: Fine-tune BERT for DocRED with two-step process. arXiv preprint arXiv:1909.11898 (2019)
15. Li, R., et al.: Joint extraction model of entity relations based on decomposition strategy. Sci. Rep. **14**(1), 1786 (2024)
16. Wei, Z., Su, J., Wang, Y., Tian, Y., Chang, Y.: A novel cascade binary tagging framework for relational triple extraction. arXiv preprint arXiv:1909.03227 (2019)
17. Wang, Y., Yu, B., Zhang, Y., Liu, T., Zhu, H., Sun, L.: TPLinker: single-stage joint extraction of entities and relations through token pair linking. arXiv preprint arXiv:2010.13415 (2020)
18. Takanobu, R., Zhang, T., Liu, J., Huang, M.: A hierarchical framework for relation extraction with reinforcement learning. In: Proceedings of the AAAI conference on artificial intelligence, vol. 33, pp. 7072–7079 (2019)
19. Zheng, H., et al.: PRGC: potential relation and global correspondence based joint relational triple extraction. arXiv preprint arXiv:2106.09895 (2021)
20. Riedel, S., Yao, L., McCallum, A.: Modeling relations and their mentions without labeled text. In: Balcázar, J.L., Bonchi, F., Gionis, A., Sebag, M. (eds.) ECML PKDD 2010. LNCS (LNAI), vol. 6323, pp. 148–163. Springer, Heidelberg (2010). https://doi.org/10.1007/978-3-642-15939-8_10
21. Ren, X., et al.: CoType: joint extraction of typed entities and relations with knowledge bases. In: Proceedings of the 26th International Conference on World Wide Web, pp. 1015–1024 (2017)
22. Miwa, M., Bansal, M.: End-to-end relation extraction using LSTMs on sequences and tree structures. arXiv preprint arXiv:1601.00770 (2016)

23. Zheng, S., Wang, F., Bao, H., Hao, Y., Zhou, P., Xu, B.: Joint extraction of entities and relations based on a novel tagging scheme. arXiv preprint arXiv:1706.05075 (2017)
24. Zeng, X., Zeng, D., He, S., Liu, K., Zhao, J.: Extracting relational facts by an end-to-end neural model with copy mechanism. In: Proceedings of the 56th Annual Meeting of the Association for Computational Linguistics (Volume 1: Long Papers), pp. 506–514 (2018)
25. Hoffmann, R., Zhang, C., Ling, X., Zettlemoyer, L., Weld, D.S.: Knowledge-based weak supervision for information extraction of overlapping relations. In: Proceedings of the 49th Annual Meeting of the Association for Computational Linguistics: Human Language Technologies, pp. 541–550 (2011)

Urban Traffic Management: A Predictive Approach Using Mobile Phone Data

Xiaolong Niu[1], Huachao Gao[1], Wentao Hu[1], Long Yan[1], Yawei Ren[1], Yang Li[1], Dazhong Li[1], Hao Ge[1], Chunxiao Li[2], Yanyan Xu[3], and Yulun Song[1](✉)

[1] Data Intelligence Division, China Unicom Digital Technology Co., Ltd., Beijing, China
{niux127,songyl100}@chinaunicom.cn
[2] School of SciTech Business, University of Science and Technology, Hefei, China
[3] AI Institute, Shanghai Jiao Tong University, Shanghai, China

Abstract. In pursuit of enhancing urban sustainability by mitigating traffic congestion, traditional efforts typically focus on optimizing transportation infrastructure. This paper introduces an innovative predictive framework for dynamic traffic assignment aimed at individual-level mobility optimization. By leveraging large-scale mobile phone data and road networks, travel demands are modeled and inter-zone traffic is forecasted using advanced deep learning techniques. These forecasts are integrated into a dynamic traffic assignment model, resulting in an iterative optimization framework that recommends socially optimal routes for individual travelers. Comprehensive experiments conducted in four Chinese cities demonstrate a significant reduction in road density and congestion during peak hours. To address the limitations of estimating road demand using mobile phone data, the framework incorporates base station deviation correction and sample expansion. This research makes a substantial contribution to urban sustainability by effectively alleviating congestion across the entire urban network.

Keywords: urban sustainability · Dynamic traffic assignment · Demand Prediction · User behavior prediction · Individual user re-routing · traffic congestion alleviation · real-time route recommendation

1 Introduction

With rapid economic growth, urban transportation has undergone significant development, leading to a surge in traffic demand and increased congestion in megacities [1]. Addressing this challenge is crucial for improving urban quality of life and reducing associated costs. Research in urban transportation, particularly in congestion alleviation and route recommendation, has gained prominence across disciplines [2].

Efforts to mitigate urban traffic congestion have included strategies such as road congestion fees, electronic toll collection, and high-occupancy vehicle (HOV) lanes [3,4]. The integration of big data and demand forecasting during large-scale events has enabled precise transportation planning [5]. Despite these advancements, limitations in existing research persist, particularly in accurately predicting traffic demand and individual travel behavior in real-time.

This paper introduces a predictive framework for dynamic traffic assignment, utilizing mobile phone data to forecast network demand and user behavior for real-time traffic management. The framework addresses gaps in current research by providing a more accurate and dynamic solution to traffic congestion. The mobile signaling data, provided by China Unicom, a telecommunications provider serving over 300 million users in China, ensures a comprehensive and representative dataset for analysis. This extensive dataset captures detailed and diverse traffic patterns, leading to more robust and reliable traffic predictions.

The primary scientific challenge addressed by this research is the development of a robust prediction model that can dynamically adapt to real-time data and optimize individual travel routes for congestion reduction. By incorporating deep learning techniques and iterative optimization, the framework offers a novel approach to urban traffic management.

The structure of this paper is as follows: Sect. 2 details the preprocessing of user mobile data and open street map data, while Sect. 3 explains the methodology of the prediction framework and its role in dynamic traffic assignment. Section 4 presents computational experiments conducted in four Chinese cities, demonstrating significant congestion reduction. Finally, Sect. 5 discusses the limitations of this study and outlines future research directions.

2 Data and Materials

2.1 Data Sources

The traffic network data utilized in this study were sourced from OpenStreetMap (OSM). Specifically, maps of the central urban areas of four Chinese cities—Beijing, Changchun, Xi'an, and Tangshan—were selected and downloaded. These cities were chosen to offer a comprehensive perspective on diverse urban road networks across different scales and regions. The maps encompass various types of roads, forming a detailed representation of each city's urban road network.

The mobile signaling data used in the study were provided by a telecommunications provider serving over 300 million users in China. This dataset comprises encrypted user IDs, timestamps, and geographical coordinates, enabling accurate recording of users' space-time trajectories. To ensure a robust dataset for in-depth analysis, we meticulously examined and assessed the signaling data. The high data density is illustrated in Fig. 1(a), which shows the distribution of base stations in Beijing's Zhongguancun area, indicating thorough network coverage. Figure 1(b) presents a histogram of signaling data record counts, demonstrating

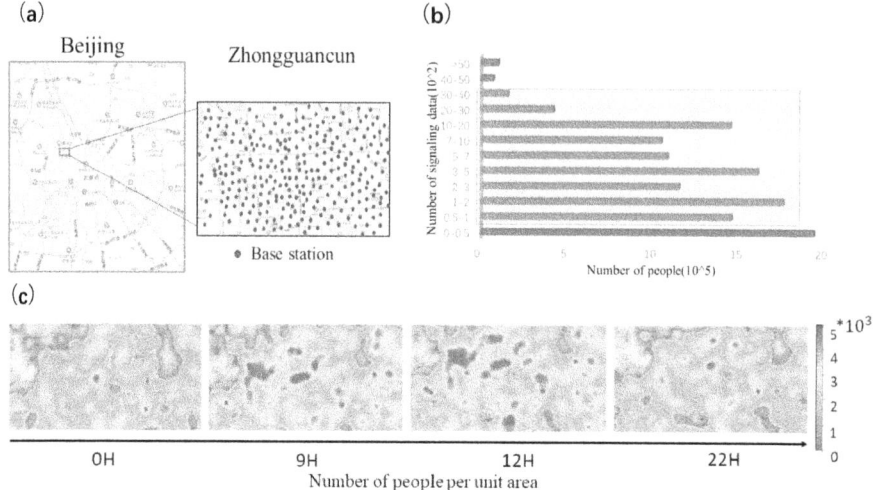

Fig. 1. Signaling data and user selection. (a) the base station distribution of zhongguancun area in Beijing (b) the histogram of signaling data record counts in Beijing's zhongguancun (c) Four heat maps of user density, respectively describing the population density distribution of Beijing Zhongguancun at 0:00, 9:00, 12:00 and 22:00.

the extensive data volume. Figure 1(c) features heat maps reflecting user density at different times, closely mirroring real population distribution trends.

3 Methodology

Traditional traffic prediction models, including historical average models, time series models like ARIMA, and classical machine learning models such as decision trees and support vector machines, each possess distinct strengths and limitations. Historical average models rely on past data for future predictions, offering simplicity but lacking adaptability to real-time changes [6]. Time series models capture temporal patterns but may struggle with non-linearities and abrupt shifts [7]. Classical machine learning models enhance accuracy but often necessitate extensive feature engineering and may not fully exploit complex temporal-spatial correlations [8].

The study employs a CNN-LSTM model chosen for its capability to handle both spatial and temporal intricacies in data. CNNs excel at extracting spatial features from road networks [9], while LSTMs are adept at capturing temporal dependencies in traffic flow [10]. This integration addresses the dynamic and non-linear nature of traffic patterns, providing a more robust prediction framework. Leveraging mobile phone data further enhances the model's capacity to capture real-time traffic dynamics and individual mobility patterns, offering a significant advantage over traditional methods [11].

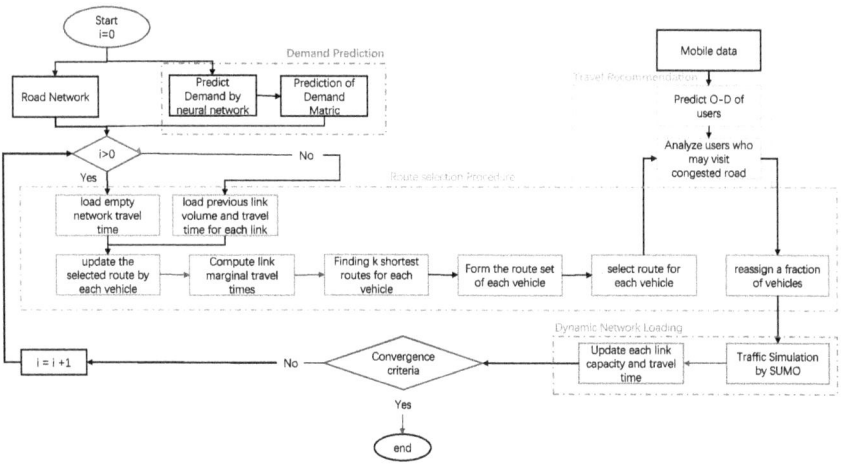

Fig. 2. Integrated Framework for Dynamic Traffic Assignment: Prediction and Equilibrium Algorithm Architecture

Figure 2 illustrates the comprehensive architectural design of our dynamic traffic assignment framework. This model synergizes advanced prediction techniques and dynamic equilibrium algorithms to achieve real-time traffic management. Key components include demand prediction, user behavior modeling, and the dynamic assignment algorithm. This architectural synergy aims to alleviate urban traffic congestion and enhance travel efficiency by recommending optimal routes for individual travelers based on predicted traffic conditions [12,13].

The predictive framework employs deep learning models and real-time mobile data to address the complex challenges of urban traffic management, offering an innovative solution for dynamic traffic assignment and congestion mitigation.

3.1 Demand Prediction

The innovative design of our CNN-LSTM architecture integrates spatial-temporal trajectory topology, combining the strengths of Convolutional Neural Networks (CNNs) and Long Short-Term Memory (LSTM) networks. CNNs are adept at extracting spatial features from road networks, employing dual-layered configurations optimized for both broad spatial patterns and localized features. The first CNN layer uses 16 filters with a 36×1 kernel to capture overarching spatial dynamics, while the subsequent layer employs 36 filters with a 1×90 kernel to refine finer-grained local features. This hierarchical approach ensures comprehensive coverage of traffic patterns across varying scales.

Complementing the CNN layers, the LSTM network consists of two stacked layers with 32 neurons each. These layers are pivotal for capturing temporal dependencies in traffic flow over time, enabling the model to learn from historical data and predict future traffic demands with precision. Dropout layers are strategically applied to enhance model robustness and prevent overfitting, ensuring reliable performance across diverse urban environments.

A significant innovation in our methodology is the integration of real-time mobile phone data, enriching the model's predictive capabilities with current traffic dynamics and individual mobility patterns. This adaptive approach allows the model to dynamically adjust to sudden traffic shifts and evolving urban conditions, surpassing traditional models that rely solely on static historical data. By continuously updating its understanding of traffic behaviors and patterns, our model provides actionable insights for optimizing traffic flow, resource allocation, and congestion mitigation strategies in real-time scenarios.

3.2 User Behavior Modeling

The methodology incorporates advanced User Behavior Modeling to enhance the accuracy and applicability of urban traffic management strategies. This component is pivotal in understanding how individual commuters interact with the transportation network, offering insights into their travel preferences, route choices, and flexibility in departure times.

At the heart of our approach is the integration of historical mobility patterns derived from real-time mobile phone data. This data enables us to capture nuanced behaviors such as commuting distances, preferred travel times, and route preferences based on daily routines, workdays, holidays, and peak travel periods. By integrating these insights into our CNN-LSTM framework, we enrich the model's understanding of user-centric traffic dynamics, surpassing traditional approaches that rely solely on aggregate traffic data.

The innovation lies in our model's ability to personalize traffic management strategies tailored to individual user needs. By analyzing and predicting user behaviors, the methodology optimizes route recommendations and traffic flow management, thereby improving overall travel efficiency and user satisfaction. This personalized approach not only enhances the accuracy of traffic predictions but also contributes to sustainable urban mobility by reducing congestion and travel times.

Furthermore, The User Behavior Modeling framework facilitates proactive decision-making in traffic management. By anticipating user preferences and behaviors, stakeholders can implement targeted interventions such as dynamic route guidance systems and adaptive traffic control measures. This proactive approach not only mitigates congestion but also enhances the overall reliability and efficiency of urban transportation systems.

3.3 Dynamic Assignment Algorithm

The methodology integrates a state-of-the-art Dynamic Assignment Algorithm (DAA) to achieve real-time traffic management and congestion mitigation. At its core, the algorithm is designed to dynamically allocate traffic demand across road networks while maintaining equilibrium conditions that minimize individual travel costs.

The innovation of our approach lies in the iterative refinement of traffic assignments based on real-time traffic conditions and user behaviors. Central to this algorithm is the concept of Dynamic User Equilibrium (DUE), where travelers adjust their routes to minimize travel times within specified departure time windows. This iterative process ensures that the traffic assignment remains responsive to changing network conditions, surpassing traditional static assignment methods.

Key components of our DAA include advanced routing algorithms such as Dijkstra's algorithm for shortest path calculations and path choice models like Logit to simulate user decision-making. These components work in tandem to optimize traffic flow by dynamically adjusting route assignments based on real-time traffic simulations and updates.

Additionally, our methodology employs dynamic network loading (DNL) algorithms that simulate the propagation of traffic flow over time. By modeling how traffic dynamically spreads across the network, DNL enables our DAA to predict and adapt to traffic patterns effectively, enhancing the accuracy of traffic management strategies.

The practical application of our Dynamic Assignment Algorithm extends beyond congestion mitigation to include proactive traffic management strategies. By continuously optimizing route assignments and traffic flows, stakeholders can implement adaptive traffic control measures and dynamic route guidance systems that improve overall travel efficiency and reduce environmental impact.

3.4 Indices Evaluating Traffic Congestion

In this section, two critical indices for assessing traffic conditions are introduced: Road Density (RD) and Traffic Congestion Index (TCI).

Road Density characterizes traffic congestion by quantifying the number of vehicles per kilometer of road at a given time t, measured in vehicles per kilometer (veh/km). RD is determined considering the distance $(dist)$ between vehicles, which adheres to safety standards defined by the following formula:

$$dist = \left(\frac{v}{10}\right)^2$$

During extreme congestion, when vehicles are nearly stationary with a car length of 5 m, the minimum distance between cars is approximately 2.5 m. Consequently, RD in such conditions can reach up to 130 veh/km. Conversely, on smooth roads where vehicles can travel at speeds of 100 km/h, the safe distance between vehicles would result in an RD not exceeding 10 veh/km.

The Traffic Congestion Index measures trip smoothness and is defined as the natural logarithm of the ratio between actual travel speed (v_{travel}) and road speed limit (v_{limit}), expressed as:

$$TCI = \ln\left(\frac{v_{limit}}{v_{travel}}\right)$$

Here, v_{travel} represents the actual travel speed, while v_{limit} denotes the road speed limit.

These indices are integral to our approach for evaluating congestion management effectiveness. Through the quantification of RD and TCI, accurate assessment of traffic conditions is facilitated, enabling informed feedback on congestion control strategies. This framework enhances the precision of traffic assessment and supports data-driven decision-making in urban traffic management, thereby enhancing efficiency and sustainability of transportation systems.

4 Results

In this study, research was conducted across four Chinese cities. To establish the analytical foundation, we initially partitioned the main urban areas of these cities into numerous grid cells, each covering one square kilometer. Mobile data was utilized to estimate traffic demand between these grid cells, resulting in the development of a demand matrix. The estimation is based on the formula:

$$d^{nm} = \frac{a^{nm} \cdot e}{s} \tag{1}$$

Here, $a^{nm} \in A^{nm}$ and $s \in [0,1]$ represent the market share of telecom operators providing mobile data. The variable e serves as an adjustment factor, essential for refining mobile phone data, taking into account potential errors in distinguishing user travel modes and positioning accuracy.

4.1 Demand Prediction

To enhance the precision and effectiveness of traffic demand prediction, we used historical network demand data spanning two years, from September 2020 to 2022. This dataset was divided into training samples, consisting of the first twenty months, and testing samples, comprising the remaining four months. Each sample corresponds to the OD traffic demand matrix during five-minute intervals within the peak hours (8–11 a.m.) per day.

The CNN-LSTM network, after being trained and validated using these samples, was employed to simulate the demand matrix of the urban network for future dates. This demand matrix represents user journeys from origin (O) to destination (D) at various time intervals. Figure 3 illustrates the four demand matrices at 8:00 a.m., 9:00 a.m., 10:00 a.m., and 11:00 a.m. for Beijing as an example.

The prediction model demonstrated high accuracy and stability, effectively capturing the temporal and spatial dynamics of traffic demand. However, potential biases could arise from inaccuracies in mobile data and unforeseen traffic incidents, which the model may not fully account for.

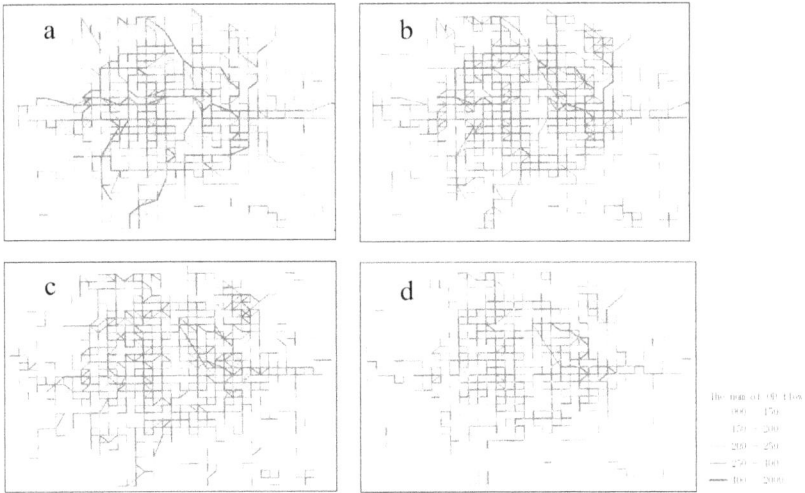

Fig. 3. Classification of traffic flow between origins and destinations (OD) during peak hours on September 30, 2022, in Beijing. (a) at 8:00 a.m.; (b) at 9:00 a.m.; (c) at 10:00 a.m.; (d) at 11:00 a.m.

4.2 Validation of Simulation

To ensure the fidelity of our simulation, we employed the Mann–Whitney U test to assess the distribution of travel times in comparison to real-world observations. Five hundred trips were randomly selected from our simulation, and real-world travel time estimates were obtained from the Gaode map API. The Mann–Whitney U results for Beijing, Changchun, Xian, and Tangshan are 43183 ($p > 0.05$), 17946 ($p > 0.05$), 18206 ($p > 0.05$), and 11290 ($p > 0.05$), indicating statistically consistent travel time distributions between simulation and reality. Pearson correlation coefficient results further confirm strong similarity with values of 0.95, 0.92, 0.97, and 0.84 for these cities. The 2D Histogram Contours in Fig. 4 further support the alignment between simulation and real-world travel times.

4.3 Alleviation Result

Following the validation of city simulations, we proceeded to iteratively calculate the assignment model's solutions. As shown in Fig. 5, setting the convergence criterion ϵ to 0.001, the iterative algorithm converges when the condition $RSD < \epsilon$ is met.

Comparing the city road density maps formed initially through urban flow simulation and after dynamic assignment algorithm convergence, as illustrated in Fig. 6, we observe a substantial reduction in congested roads within the optimized traffic network. This indicates significant alleviation of congestion in urban road networks after dynamic assignment.

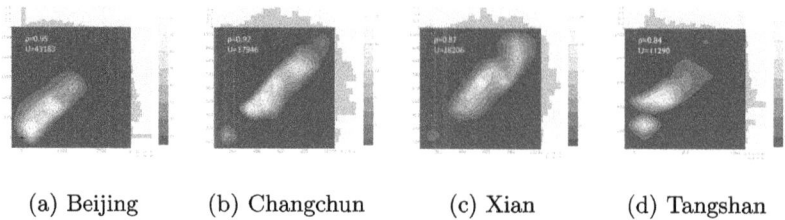

(a) Beijing (b) Changchun (c) Xian (d) Tangshan

Fig. 4. Validation of Simulation in Four Cities. (a)–(d) present validation results for Beijing, Changchun, Xian, and Tangshan, including 2D histogram contours comparing the travel times of selected trips in the simulation and the real world. The upper left corner of the figure provides statistical results for the Mann–Whitney U test and the Pearson correlation coefficient for travel times of selected trips.

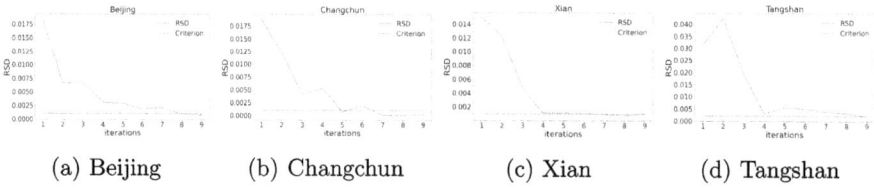

(a) Beijing (b) Changchun (c) Xian (d) Tangshan

Fig. 5. Relative Standard Deviation (RSD) of travel time within ten algorithmic iterations. (a) Demonstrates Beijing's DTA algorithm results with travel time convergence at the 8th iteration; (b) Represents Changchun's DTA algorithm results with travel time convergence at the 5th iteration; (c) Depicts Xian's DTA algorithm results with travel time convergence at the 4th iteration; (d) Shows Tangshan's DTA algorithm results with travel time convergence at the 9th iteration.

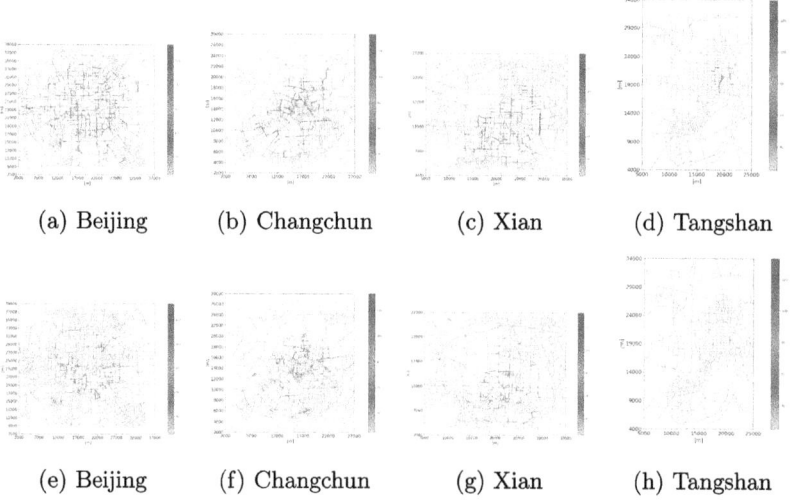

(a) Beijing (b) Changchun (c) Xian (d) Tangshan

(e) Beijing (f) Changchun (g) Xian (h) Tangshan

Fig. 6. Comparison of original (top row) and optimized (bottom row) road networks for Beijing, Changchun, Xian, and Tangshan.

Further analysis of the congestion road density quantiles, as shown in Fig. 7, reveals a significant reduction in both the average density and variance of congested roads post-optimization. This suggests that the optimized traffic network not only smoothens traffic flow but also makes it more stable and predictable for travelers.

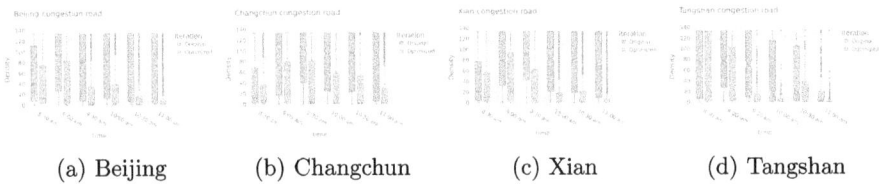

(a) Beijing　　(b) Changchun　　(c) Xian　　(d) Tangshan

Fig. 7. Comparison of congestion roads density quantiles between the original and optimized traffic networks during peak hours from 8:00 a.m. to 11:00 a.m. (a) Displays the density quantiles of congested roads in Beijing's original traffic network and the changed quantiles after traffic network optimization; (b) Illustrates the density quantiles of congested roads in Changchun's original traffic network and the changed quantiles after traffic network optimization; (c) Depicts the density quantiles of congested roads in Xian's original traffic network and the changed quantiles after traffic network optimization; (d) Shows the density quantiles of congested roads in Tangshan's original traffic network and the changed quantiles after traffic network optimization.

To evaluate the improvement in travel efficiency brought about by the road assignment model, we analyzed the TCI distribution for vehicles departing during peak hours. As shown in Fig. 8, there is a significant reduction in both the average value and variance of the TCI distribution across all four cities. This indicates that the model-assigned road network exhibits faster and more reliable traffic flow.

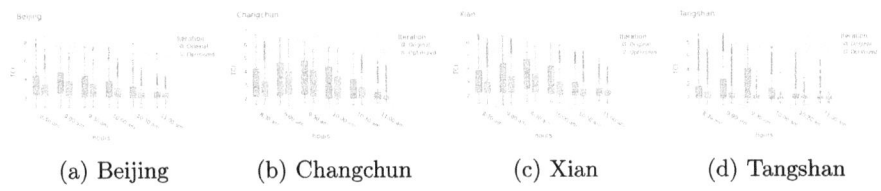

(a) Beijing　　(b) Changchun　　(c) Xian　　(d) Tangshan

Fig. 8. Comparison of Traffic Congestion Index (TCI) between the original and optimized traffic networks during peak hours from 8:00 a.m. to 11:00 a.m. (a)–(d) present the quantiles of trip TCIs in Beijing, Changchun, Xian, and Tangshan, comparing TCIs on the original and optimized traffic networks.

5 Discussion

In conclusion, this paper has introduced a predictive framework for dynamic traffic assignment, leveraging dynamic user equilibrium principles and deep learning algorithms. The practical implementation in four cities has demonstrated promising results, effectively mitigating urban traffic congestion and enhancing travel efficiency. However, in the event of abrupt occurrences like hurricanes, earthquakes and so on, which lead to the collapse of base stations, the model falls short in addressing them. Likewise, other methodologies available in the market also possess their inherent limitations. It is crucial to acknowledge limitations arising from discrepancies between mobile data and actual vehicle demand.

To address these challenges, market share coefficients and proportion coefficients for different urban travel modes were introduced, enabling a more nuanced estimation of road vehicle demand within simulations. Despite these efforts, reliance on data from a single mobile phone operator remains a limitation. Moving forward, expanding data sources with a direct focus on urban sustainability is imperative.

One promising approach involves integrating data from all mobile phone operators to achieve a comprehensive representation of users. Furthermore, the evolving landscape of the Internet of Things (IoT) presents significant potential. Enhanced capabilities of network service providers to differentiate and identify driving users can substantially improve the accuracy of data collection efforts. This aligns with our overarching goal of contributing to urban sustainability.

Successful execution of future projects will not only address current limitations but also pave the way for more robust and comprehensive solutions to urban transportation challenges, advancing our commitment to sustainable and efficient urban mobility.

Acknowledgments. The authors thank China United Network Communications Group Co., Ltd. provide data for research in this paper. We specially thank AirSage and Prof. Pu Wang for the data provided. This topic was supported by the key technology research of China Unicom 2022 Core Technology Research Project in trusted data resource space (Y915221FHF2000). Yanyan Xu was supported by the Shanghai Municipal Science and Technology Major Project (2021SHZDZX0102), the National Natural Science Foundation of China (62102258), and Shanghai Pujiang Program (21PJ1407300).

Funding. This topic was supported by the key technology research of China Unicom 2022 Core Technology Research Project in trusted data resource space (Y915221FHF2000). Yanyan Xu was jointly supported by the National Science Foundation of China (62102258), the Shanghai Pujiang Program (21PJ1407300), and the Shanghai Municipal Science and Technology Major Project (2021SHZDZX0102).

Availability of Data and Materials. All data needed to evaluate the conclusions in the paper are described in the paper and/or the Supplementary Materials. For contractual and privacy reasons, we cannot make the raw mobile phone data available.

Sample of the data may be requested from the corresponding author upon reasonable request.

Disclosure of Interests. The authors declare that they have no competing interests to declare that are relevant to the content of this article.

References

1. AutoNavi: Traffic analysis report of China's major cities in 2021 (2021)
2. Li, K., Shang, S.K.P.: Traffic congestion alleviation over dynamic road networks: continuous optimal route combination for trip query streams. In: International Joint Conferences on Artificial Intelligence Organization (IJCAI), pp. 1–13. International Joint Conferences on Artificial Intelligence Organization (2021)
3. De Palma, A., Lindsey, R.: Traffic congestion pricing methodologies and technologies. Transp. Res. Part C: Emerg. Technol. **19**(6), 1377–1399 (2011)
4. Yang, H., Hwang, H.-J.: Carpooling and congestion pricing in a multilane highway with high-occupancy-vehicle lanes. Transp. Res. Part A: Policy Pract. **33**, 139–155 (1999)
5. Xu, Y., González, M.C.: Collective benefits in traffic during mega events via the use of information technologies. J. R. Soc. Interface **14**(129), 20161041 (2017)
6. Ahmed, S.A., Cook, A.R.: An application of time series analysis to the problem of traffic forecasting. Transp. Res. Rec. **772**, 1–9 (1979)
7. Kumar, P., Vanajakshi, L.: Traffic flow forecasting using support vector regression. Intell. Transp. Syst. **9**(4), 289–296 (2015)
8. Ma, X., Tao, Z., Wang, Y., Yu, H., Wang, Y.: Large-scale transportation network congestion evolution prediction using deep learning theory. PLoS ONE **10**(3), 0119044 (2015)
9. Polson, N.G., Sokolov, V.O.: Deep learning for short-term traffic flow prediction. Transp. Res. Part C: Emerg. Technol. **79**, 1–17 (2017)
10. Lv, Y., Duan, Y., Kang, W., Li, Z., Wang, F.-Y.: Traffic flow prediction with big data: a deep learning approach. IEEE Trans. Intell. Transp. Syst. **16**(2), 865–873 (2015)
11. Ester, M., Kriegel, H.-P., Sander, J., Xu, X.: A density-based algorithm for discovering clusters in large spatial databases with noise. In: Proceedings of the Second International Conference on Knowledge Discovery and Data Mining (KDD), pp. 226–231 (1996)
12. Zhang, C., Zheng, Y., Qi, Y., Li, R., Li, X.: Graph convolutional networks for traffic forecasting: a survey. In: Proceedings of the 28th International Joint Conference on Artificial Intelligence (IJCAI), pp. 4423–4432 (2019)
13. Vaswani, A., et al.: Attention is all you need. In: Advances in Neural Information Processing Systems, vol. 30 (2017)

FedPSED: Federated Public Safety Emergency Detection Based on Adaptive Aggregation Strategy

Jiping Fan, Junping Du(✉), Zhe Xue, Ang Li, Zeli Guan, and Meiyu Liang

Beijing University of Posts and Telecommunications, Beijing, China
junpingdu@126.com, {xuezhe,david.lee,guanzeli,meiyu1210}@bupt.edu.cn

Abstract. Federated learning plays a significant role in utilizing multi-party data for public safety emergency detection while preserving privacy. However, due to the dispersion of public safety emergency data across different departments and the need for privacy protection, data islands have formed where data cannot be shared. Therefore, it is necessary to utilize federated learning to leverage multi-party data while protecting privacy. However, the heterogeneity of client data introduces noise during federated learning, which affects the performance of the global model. Thus, this paper proposes a federated public safety emergency detection method based on adaptive aggregation strategy (FedPSED). This method uses federated learning to train a public safety emergency detection model based on GAT and contrastive learning (PSED). When aggregating the global model, an adaptive aggregation strategy is employed to consider factors such as the topology of client data, enabling the FedPSED to utilize more high-quality data information and consider the performance of client models. This approach helps the global model parameters converge towards optimal parameters, enhancing the performance of the global model. Our proposed FedPSED has been shown to be effective through extensive experiments on multiple datasets.

Keywords: Public safety · emergency detection · federated learning · graph neural networks · contrastive learning

1 Introduction

Public safety emergencies pose serious threats to social order, the national economy, and other aspects. Detecting a public safety emergency is crucially significant. However, data regarding public safety emergencies is typically scattered across different institutions. Due to privacy protection, data sharing is restricted, resulting in insufficient data from individual institutions. Existing methods for event detection struggle to effectively utilize multi-source data for model training. Therefore, there is a need to employ federated learning [13] to train a global model using data from multiple sources while preserving the privacy of each client's data. This approach aims to enhance the performance metrics of the global model. In the real world, public safety emergencies often quickly spark discussions and posts on social platforms like Weibo and Zhihu. Collecting and organizing this data, using federated learning, allows these platforms to collaboratively train a model. This model can then be used for real-time detection of public safety emergencies, thereby improving the efficiency of handling such events. However, traditional federated learning struggles to address issues in federated public safety emergency detection scenarios due to the heterogeneity of local client data. Factors such as uneven data volumes and significant differences in label distributions can impact the performance of the global model. Therefore, current federated learning frameworks cannot be directly applied to federated public safety emergency detection.

Given these challenges, this paper proposes a novel approach for detecting public safety emergencies in a federated setting: federated public safety emergency detection method based on adaptive aggregation strategy (FedPSED). This method employs a local public safety emergency detection model based on graph neural networks [8,11,12,20] and contrastive learning [19,22–24] (PSED). Compared to traditional event detection models, Public Safety Emergency Detection model can utilize more data information and leverage contrastive loss to make features of similar events closer and features of different events farther apart, significantly enhancing event detection performance. During global model aggregation, an adaptive aggregation strategy is introduced that considers the topological structure of client data, such as relationships and associations between data points. This enables the FedPSED to utilize more high-quality data information, considering the performance of client models. Consequently, the global model parameters move closer to optimal parameters, thereby enhancing the performance of the global model.

We conducted extensive experiments on multiple datasets, and the results demonstrate that our proposed FedPSED outperforms baseline methods across these datasets. This confirms the effectiveness of FedPSED in public safety emergency detection.

The contributions of this paper are as follows:

- A method for detecting public safety events in a federated setting is proposed: FedPSED utilizes federated learning to train a public safety emergency detection model using data from multiple parties while protecting the privacy of individual client data. This approach addresses the impact of heterogeneous client data on the performance of the global model, achieving accurate detection of public safety emergencies.
- A federated learning global model aggregation strategy based on adaptive aggregation strategy is proposed, which considers information such as the topology of graph data. This enables the FedPSED to utilize more high-quality data information, taking into account the performance of the models. Consequently, the global model parameters move closer to optimal parameters, enhancing the performance of the global model.

- We conducted extensive experiments on multiple datasets, which showed that our proposed FedPSED outperforms baseline methods across these datasets. This validates the effectiveness of FedPSED as demonstrated in our experiments.

2 Related Work

2.1 Event Detection

Event detection methods for public safety emergencies can be classified into clustering-based, statistical-based, probability model-based, and deep learning-based methods. The first three categories typically require manual feature extraction and employ human-designed strategies to detect emergencies, resulting in significant workload and often suboptimal effectiveness. Deep learning-based anomaly detection methods integrate feature extraction with event detection, training an event detector end-to-end using existing data to detect specific types of events. Sabokrou et al. [18] proposed a method for video anomaly detection and localization based on fully convolutional neural networks (FCN). Burel et al. [2] introduced the Sem-CNN model for identifying crisis-related social media content. Popescu et al. [17] employed machine learning and text analysis techniques to achieve this. Gaglio et al. [7] presented a variant of the frequent pattern mining algorithm, which enables real-time detection of sudden events in dynamic environments. Cao et al. [3] proposed the KPGNN model based on GAT for dynamic event detection in streaming Twitter data. Peng et al. [16] introduced a fine-grained social event classification model based on Pairwise Popularity Graph Convolutional Networks (PP-GCN). However, deep learning methods for event detection heavily rely on high-quality data. In the context of public safety emergency detection, the data distribution of public safety emergency is scattered across various agencies, with each agency possessing only a small amount of data. Due to privacy concerns, this situation has led to data silos. Existing event detection methods cannot be directly applied to the detection of public safety emergency.

2.2 Federated Learning

In federated learning, clients first train their local models on their local datasets, then upload the model parameters to the server for model aggregation. This process iterates until the global model converges. Model aggregation is one of the most important steps in federated learning. FedAvg [15] is the most commonly used method, which computes the arithmetic average of model weights across all nodes. Li et al. proposed FedProx [14], which introduces a proximal term to constrain the deviation of client's locally trained model parameters from the server's global model parameters in heterogeneous data scenarios. This accelerates model convergence, enhances algorithm stability, and can be viewed as a generalization and reparameterization of FedAvg. Karimireddy et al. proposed SCAFFOLD

[10], which utilizes control variates (variance reduction) to correct for "client drift" in its local updates, reducing the number of communication rounds and remaining unaffected by data heterogeneity or client sampling. Acar et al. proposed FedDyn [1], where in each training round, a penalty term sent from the server is added to the client's learning objective. This ensures that each client's model converges towards the global optimum, thus accelerating model convergence. FedBE [4] utilizes Bayesian inference for model aggregation, effectively integrating Bayesian models into federated learning. Ye et al. [21] introduced a selective model aggregation method, where "good" local DNN models are chosen based on the evaluation of local image quality and computational capabilities. These selected models are then sent to the central server to enhance the accuracy and efficiency of model aggregation. Fu et al. [6] proposed a node selection strategy for efficient federated learning, considering node reputation, local model quality, and time overhead before uploading local models. This strategy aims to select as many high-quality users as possible while considering communication resources and time constraints. However, due to the significant heterogeneity in public safety emergency data across different clients, it severely impacts the performance of the global model. Therefore, current federated learning frameworks cannot be directly applied to public safety emergency detection.

3 Federated Public Safety Emergency Detection Method Based on Adaptive Aggregation Strategy

In this paper, we propose a method for detecting public safety emergency in a federated scenario: FedPSED. This method consists of two parts: the local Public Safety Event Detection model (PSED) and the global model aggregation framework based on adaptive aggregation strategies. The FedPSED process consists of three steps. Firstly, each client trains the local Public Safety Event Detection model (PSED) on its own local dataset. Then, the server calculates the contribution of each client's model to the global model based on the quality of their local data and the performance of their local models, and updates the global model accordingly. Finally, the local models are updated using the global model, and this process iterates until the global model converges.

3.1 Public Safety Emergency Detection Model Based on GAT and Contrastive Learning

Public safety emergencies are graph-structured events. To better utilize the topological and semantic information of the data for more accurate detection of public safety emergencies, Therefore, inspired by [3], we adopt a PSED as our local model, to effectively utilize the local data of each client. The network architecture is illustrated in Fig. 1. The PSED consists of two components: a feature extractor and an event detector. The feature extractor is constructed using GAT combined with contrastive learning to thoroughly extract information from the

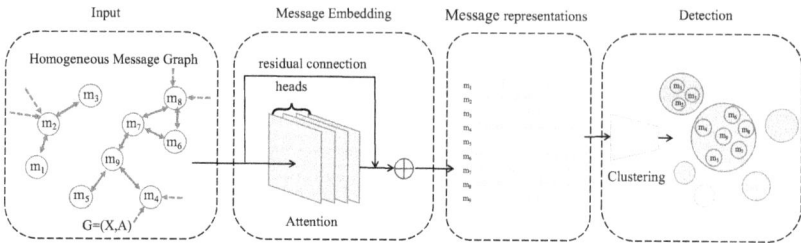

Fig. 1. PSED network architecture

event graph. The event detector utilizes clustering techniques to detect public safety emergency.

We trained an encoder $E: H^{N \times d} \times \{0,1\}^{N \times N} \to H^{N \times d'}$ based on Graph Attention Networks (GAT), with the propagation formula for each layer as shown in Eq. (1):

$$h_i^l = \|_{k=1}^{K} (e_{i,i}^{l-1} W^{l-1} h_i^{l-1} + \sum_{j \in N(i)} e_{i,j}^{l-1} W^{l-1} h_i^{l-1}) \quad (1)$$

where, h_i^l represents the representation of node i at layer l-th, $\|$ represents the concatenation operation, K is the number of heads, $e_{i,j}^l$ represents the attention scores between node i and node j at layer l-th, W^l represents the trainable parameters at layer l-th.

To better distinguish events, we employ contrastive learning to construct the loss. For each event m_i, we sample a positive event m_i^+ (from the same class of events) and a negative event m_i^- (from a different class of events), forming a triplet (m_i, m_i^+, m_i^-). The contrastive loss function pushes positive samples closer to the anchor point and pushes negative samples away from the anchor point. The contrastive loss is formulated as shown in Eq. (2):

$$L_t = \sum_{(m_i, m_i^+, m_i^-) \in T} \text{ReLU}\left(d(h_{m_i}, h_{m_i^+}) - d(h_{m_i}, h_{m_i^-})\right) \quad (2)$$

where, $d(\cdot)$ denotes the Euclidean distance between two vectors, ReLU is the ReLU activation function, T represents a set of online sampled triplets satisfying the conditions of $d(h_{m_i}, h_{m_i^-}) \leq d(h_{m_i}, h_{m_i^+})$, which can lead to faster convergence.

During the detection phase, messages are clustered based on the learned message representations. Distance-based clustering algorithms such as K-Means can be easily applied to cluster the representations.

3.2 Global Model Aggregation Based on Adaptive Aggregation Strategy

Aggregation Framework. The general process of federated learning involves the server distributing a unified model to each client. Each client then trains the model locally on its own data. Selected clients upload their local model parameters to the server for global model parameter aggregation. After aggregation, the next round of training begins, and this process iterates until the model converges. The aggregation formula is as shown in Eq. (3):

$$W_t = \sum_{k=1}^{K} q_t^k W_t^k \qquad (3)$$

where, K is the number of selected clients during the federated learning communication process, W_t^k represents the parameters of the local public safety emergency incident detection model from each client, W_t represents the parameters of the global model. q_t^k represents the aggregation weight of the k-th client.

Due to the heterogeneity of client data in federated learning, which can degrade the performance of the global model, we propose an adaptive aggregation strategy for federated learning. This strategy considers information such as the topological structure of each client's local data. By doing so, FedPSED can leverage more high-quality data and take into account the performance of each client's model. This approach helps the global model parameters converge towards optimal values, thereby enhancing the overall performance of the global model.

Adaptive Aggregation Strategy. In graph federated learning, each client locally maintains a message graph. Assuming that in the message graph, the richness of the topology increases with more nodes, more label categories, and more edges between nodes, then the richer the message graph's topology, the higher the data quality. Therefore, we consider that the contribution of a client's model to the global model is proportional to the number of nodes, edges, and label categories in its local data. Thus, we obtain the client's data quality contribution d^k as shown in Eq. (4):

$$d^k = \frac{c^k}{\sum_{k=1}^{K} c^k} + \frac{e^k}{\sum_{k=1}^{K} e^k} + \frac{l^k}{\sum_{k=1}^{K} l^k} \qquad (4)$$

where, c^k is the number of nodes in the data of the k-th client, e^k is the number of edges in the data of the k-th client, l^k is the number of label categories in the data of the k-th client.

Furthermore, considering that in each communication round, under the same loss calculation mechanism, when the model's loss value is lower, the better the performance of the local model on the local validation set. This indicates a stronger ability of local public safety emergency incident detection. Therefore, we consider that the contribution of the local model is inversely proportional to the loss value and directly proportional to the performance metric. Thus, we obtain the model performance contribution m_t^k as shown in Eq. (5):

$$m_t^k = \frac{\frac{1}{f_t^k}}{\sum_{k=1}^{K} \frac{1}{f_t^k}} + \frac{n_t^k}{\sum_{k=1}^{K} n_t^k} \qquad (5)$$

where, f_t^k is the loss of client k in the t-th round of federated training, n_t^k is the NMI (Normalized Mutual Information) metric of client k in the t-th round of federated training.

The aggregation weight q_t^k of each client during federated learning model aggregation is calculated based on the contribution as shown in Eq. (6):

$$q_t^k = \frac{e^{d^k + m_t^k}}{\sum_{k=1}^{K} e^{d^k + m_t^k}} \tag{6}$$

3.3 The Algorithm

Algorithm 1 outlines the implementation process of FedPSED. Input client set K, communication rounds T, local epochs E. For each round of federated training. First, each client updates its local model parameters W_t^k using the global model parameters W_t issued by the server. Secondly, local model training is conducted on the local training dataset to update the local model parameters W_t^k. Thirdly, the local model is validated on the local validation dataset to update the data quality contribution d^k and the model performance contribution m_t^k. Fourthly, after local training is completed, the local model parameters W_t^k, data quality contribution d^k, and model performance contribution m_t^k are sent to the server. Finally, the server calculates the weights q_t^k for each client based on their data quality contribution d^k and model performance contribution m_t^k, and updates the global model parameters W_t.

Algorithm 1. The proposed FedPSED

Input: FedPSED parameters including communication round T, client set K, local epochs E;
Output: Global model performance W_t;
1: **for** each round $t = 1, 2, 3..., T$ **do**
2:     Server sends global model parameters W_t to K clients;
3:     **for** client $k \in [K]$ **do**
4:         Update local model parameters W_t^k using global model W_t;
5:         **for** $i = 1, 2, 3..., E$ **do**
6:             Use local training sets for training;
7:             Update local model parameters W_t^k;
8:             Use local validation set for validation;
9:             Update the model performance contribution m_t^k based on Equation (5);
10:        **end for**
11:        Calculate the data quality contribution d^k based on Equation (4);
12:        Send local model performance W_t^k, data quality contribution d^k, model performance contribution m_t^k to the server;
13:     **end for**
14:     Server calculates the weight q_t^k of all clients based on Equation (6);
15:     Server calculates global model performance W_t based on Equation (3);
16: **end for**

4 Experiment

4.1 Experimental Setup

Dataset. Since there are currently no available public safety emergency event graph datasets. To validate the effectiveness of FedPSED that we propose, we utilize a heterogeneous information network (HIN) [9] to construct the public safety emergency incident graph dataset WeiBo. First, we crawled a large number of Weibo posts related to public safety emergencies from the Weibo platform. Next, for each Weibo post m_i related to a public safety emergency, we extracted its users, named entities, keywords, and topic elements. Then, we incorporated m_i and its elements as different types of nodes into a HIN. Subsequently, we calculated similarities between these elements, and if the similarity exceeded a certain threshold, we added relationships between the corresponding nodes. Finally, we constructed the WeiBo graph data of public safety emergencies by forming event-element-event meta-paths [5] to establish relationships between events. The WeiBo dataset consists of a total of 5,124 nodes, where each node represents a Weibo post related to a public safety emergency. There are 9 types in total, and each post is represented by a 300-dimensional vector. The edges in the dataset represent the relationships between the Weibo posts.

In addition, we conducted extensive experiments on the WeiBo dataset, Core dataset, and CiteSeer dataset. The Core dataset consists of 2,708 samples, where each sample is a scientific paper categorized into one of 8 classes. Each paper is represented by a 1,433-dimensional word vector, and the edges in the dataset denote citation relationships between the papers. The CiteSeer dataset consists of 3,327 nodes, where each node represents a scientific paper categorized into one of 6 classes. Each paper is represented by a 3,703-dimensional vector, and the edges in the dataset capture citation or co-citation relationships between the papers. The statistics of the datasets are summarized in Table 1.

To simulate real-world federated public safety emergency detection, we divided each dataset into 6 non-independent and identically distributed parts to mimic the scenario of 6 clients participating in the FedPSED process.

Table 1. Statistical Information Table for the Datasets

Dataset	Nodes	Edges	Features	Classes
WeiBo	5,124	3,928,476	300	9
Core	2,708	13,264	1,433	7
CiteSeer	3,327	12,431	3,703	6

Baseline and Evaluation Metrics. To validate the effectiveness of FedPSED, we conducted extensive experiments comparing it with PSEED, FedAvg [15], FedProx [14], SCAFFOLD [10], and FedDyn [1] on the WeiBo dataset, Core dataset, and CiteSeer dataset. The evaluation metrics used include NMI (Normalized Mutual Information), AMI (Adjusted Mutual Information), ARI (Adjusted Rand Index), and PURITY. PSED uses the PSEED local model without federated training. The final result is the weighted average of performance metrics on the local test set after each client's local model has converged. FedAvg, FedProx, SCAFFOLD, and FedDyn use the PSEED local model with federated training, where each federated framework adopts FedAvg, FedProx, SCAFFOLD, and FedDyn, respectively. The final result is the weighted average of performance metrics on each client's local test set after the global model has converged.

Table 2. The results of FedPSED and the baseline on the WeiBo, Core and CiteSeer dataset

Model		PSEED	Fedavg	Fedprox	Scaffold	Feddyn	**FedPSED**
WeiBo	NMI	0.7730	0.8655	0.8593	0.8687	0.8759	**0.8794**
	AMI	0.7669	0.8601	0.8545	0.8654	0.8726	**0.8755**
	ARI	0.7691	0.8651	0.8768	0.8529	0.8907	**0.8909**
	PURITY	0.7457	0.8561	0.8653	0.8609	0.8665	**0.8690**
Core	NMI	0.5104	0.5817	0.5927	0.5858	0.5607	**0.6049**
	AMI	0.4568	0.5262	0.5419	0.5329	0.5100	**0.5577**
	ARI	0.4050	0.4655	0.4845	0.4866	0.4915	**0.4948**
	PURITY	0.5438	0.6224	0.6416	0.6200	0.5863	**0.6488**
CiteSeer	NMI	0.3901	0.4633	0.4810	0.4794	0.4089	**0.4853**
	AMI	0.3465	0.4179	0.4397	0.4347	0.3663	**0.4445**
	ARI	0.3585	0.4190	0.4504	0.4466	0.3729	**0.4513**
	PURITY	0.3999	0.4770	0.4856	0.4965	0.4171	**0.4968**

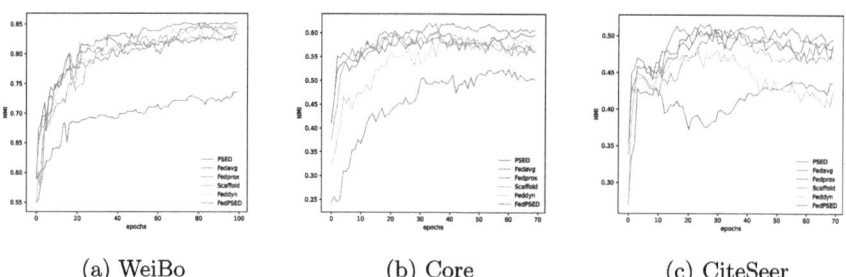

(a) WeiBo (b) Core (c) CiteSeer

Fig. 2. The training curves of FedPSED and the baseline on the WeiBo, Core and CiteSeer dataset

4.2 Results

Table 2 and Fig. 2 illustrate the comparison between FedPSED and baseline methods based on performance metrics. From the figures, we observe:

Our proposed FedPSED outperforms all baseline methods on three datasets, demonstrating its effectiveness as the global model's performance consistently surpasses that of the baseline methods.

Compared to FedProx, Scaffold, and Feddyn, the FedPSED demonstrates that adopting an adaptive federated learning global model aggregation strategy is more conducive to improving the performance of federated public safety emergency detection than methods that do not adopt an adaptive global model aggregation strategy.

The FedPSED, along with FedProx, Scaffold, and Feddyn, outperforms Fedavg in terms of global model performance. This indicates that by introducing additional parameters to constrain the deviation between client models and the global model, it is possible to enhance the performance of the global model.

FedPSED, FedProx, Scaffold, FedDyn, and FedAvg outperform the non-federated method PSED on the test set in terms of performance metrics. This indicates that federated learning, which aggregates data from multiple parties to train a global model, can enhance performance by leveraging a larger amount of data.

FedPSED, compared to FedAvg, is an ablation experiment. It demonstrates that employing an adaptive aggregation strategy for global models in federated learning is more beneficial for improving the performance of detecting public safety emergencies in federated settings than directly averaging parameters.

FedPSED outperforms baseline methods on both the public safety emergency dataset WeiBo and the academic paper graph datasets Core and CiteSeer. This indicates that our proposed FedPSED exhibits strong generalization capabilities across different types of datasets, demonstrating superior performance in detecting public safety emergencies as well as in academic contexts.

5 Conclusion

In this paper, we proposed FedPSED, which enables public safety emergency detection in a federated learning setting. In this method, an adaptive federated learning global model aggregation strategy is proposed, which considers information such as the topological structure of client data, allowing the FedPSED to utilize more high-quality data. By considering the performance of client-side local models, the global model parameters are guided towards optimal values, enhancing the performance of the global model. We conducted extensive experiments on multiple datasets, demonstrating that our proposed FedPSED outperforms baseline methods in terms of model performance and exhibits strong generalization capabilities. For future work, we aim to further explore federated public safety emergency event detection on social media data.

Acknowledgments. This work was supported by the National Natural Science Foundation of China (62192784, U22B2038).

Disclosure of Interests. The authors have no competing interests to declare that are relevant to the content of this article.

References

1. Acar, D.A.E., Zhao, Y., Navarro, R.M., Mattina, M., Whatmough, P.N., Saligrama, V.: Federated learning based on dynamic regularization. arXiv preprint arXiv:2111.04263 (2021)
2. Burel, G., Saif, H., Alani, H.: Semantic wide and deep learning for detecting crisis-information categories on social media. In: d'Amato, C., et al. (eds.) ISWC 2017. LNCS, vol. 10587, pp. 138–155. Springer, Cham (2017). https://doi.org/10.1007/978-3-319-68288-4_9
3. Cao, Y., Peng, H., Wu, J., Dou, Y., Li, J., Yu, P.S.: Knowledge-preserving incremental social event detection via heterogeneous GNNs. In: Proceedings of the Web Conference 2021, pp. 3383–3395 (2021)
4. Chen, H.Y., Chao, W.L.: FedBE: making Bayesian model ensemble applicable to federated learning. arXiv preprint arXiv:2009.01974 (2020)
5. Fu, X., Zhang, J., Meng, Z., King, I.: MAGNN: metapath aggregated graph neural network for heterogeneous graph embedding. In: Proceedings of the Web Conference 2020, pp. 2331–2341 (2020)
6. Fu, Y., Li, C., Yu, F.R., Luan, T.H., Zhang, Y.: A selective federated reinforcement learning strategy for autonomous driving. IEEE Trans. Intell. Transp. Syst. **24**(2), 1655–1668 (2022)
7. Gaglio, S., Re, G.L., Morana, M.: Real-time detection of twitter social events from the user's perspective. In: 2015 IEEE International Conference on Communications (ICC), pp. 1207–1212. IEEE (2015)
8. Hamilton, W., Ying, Z., Leskovec, J.: Inductive representation learning on large graphs. In: Advances in Neural Information Processing Systems, vol. 30 (2017)
9. Han, J.: Mining heterogeneous information networks by exploring the power of links. In: International Conference on Discovery Science, pp. 13–30. Springer (2009)
10. Karimireddy, S.P., Kale, S., Mohri, M., Reddi, S., Stich, S., Suresh, A.T.: Scaffold: stochastic controlled averaging for federated learning. In: International Conference on Machine Learning, pp. 5132–5143. PMLR (2020)
11. Kipf, T.N., Welling, M.: Semi-supervised classification with graph convolutional networks. arXiv preprint arXiv:1609.02907 (2016)
12. Kipf, T.N., Welling, M.: Variational graph auto-encoders. arXiv preprint arXiv:1611.07308 (2016)
13. Konečný, J., McMahan, H.B., Ramage, D., Richtárik, P.: Federated optimization: distributed machine learning for on-device intelligence. arXiv preprint arXiv:1610.02527 (2016)
14. Li, T., Sahu, A.K., Zaheer, M., Sanjabi, M., Talwalkar, A., Smith, V.: Federated optimization in heterogeneous networks. Proc. Mach. Learn. Syst. **2**, 429–450 (2020)
15. McMahan, B., Moore, E., Ramage, D., Hampson, S., Arcas, B.A.: Communication-efficient learning of deep networks from decentralized data. In: Artificial Intelligence and Statistics, pp. 1273–1282. PMLR (2017)

16. Peng, H., et al.: Fine-grained event categorization with heterogeneous graph convolutional networks. arXiv preprint arXiv:1906.04580 (2019)
17. Popescu, A.M., Pennacchiotti, M., Paranjpe, D.: Extracting events and event descriptions from twitter. In: Proceedings of the 20th International Conference Companion on World Wide Web, pp. 105–106 (2011)
18. Sabokrou, M., Fayyaz, M., Fathy, M., Moayed, Z., Klette, R.: Deep-anomaly: fully convolutional neural network for fast anomaly detection in crowded scenes. Comput. Vis. Image Underst. **172**, 88–97 (2018)
19. Suresh, S., Li, P., Hao, C., Neville, J.: Adversarial graph augmentation to improve graph contrastive learning. Adv. Neural. Inf. Process. Syst. **34**, 15920–15933 (2021)
20. Velickovic, P., et al.: Graph attention networks. Stat **1050**(20), 10–48550 (2017)
21. Ye, D., Yu, R., Pan, M., Han, Z.: Federated learning in vehicular edge computing: a selective model aggregation approach. IEEE Access **8**, 23920–23935 (2020)
22. You, Y., Chen, T., Sui, Y., Chen, T., Wang, Z., Shen, Y.: Graph contrastive learning with augmentations. Adv. Neural. Inf. Process. Syst. **33**, 5812–5823 (2020)
23. You, Y., Chen, T., Wang, Z., Shen, Y.: Bringing your own view: graph contrastive learning without prefabricated data augmentations. In: Proceedings of the Fifteenth ACM International Conference on Web Search and Data Mining, pp. 1300–1309 (2022)
24. Zhu, Y., Xu, Y., Yu, F., Liu, Q., Wu, S., Wang, L.: Graph contrastive learning with adaptive augmentation. In: Proceedings of the Web Conference 2021, pp. 2069–2080 (2021)

Evaluation of Retrieval-Augmented Generation: A Survey

Hao Yu[1,2], Aoran Gan[3], Kai Zhang[3], Shiwei Tong[1(✉)], Qi Liu[3], and Zhaofeng Liu[1]

[1] Tencent Company, Shenzhen, China
{shiweitong,zhaofengliu}@tencent.com
[2] McGill University, Montreal, Canada
hao.yu2@mail.mcgill.ca
[3] State Key Laboratory of Cognitive Intelligence, University of Science and Technology of China, Hefei, China
gar@mail.ustc.edu.cn, {kkzhang08,qiliuql}@ustc.edu.cn
https://github.com/YHPeter/Awesome-RAG-Evaluation

Abstract. Retrieval-Augmented Generation (RAG) has recently gained traction in natural language processing. Numerous studies and real-world applications are leveraging its ability to enhance generative models through external information retrieval. Evaluating these RAG systems, however, poses unique challenges due to their hybrid structure and reliance on dynamic knowledge sources. To better understand these challenges, we conduct *A Unified Evaluation Process of RAG* (*Auepora*) and aim to provide a comprehensive overview of the evaluation and benchmarks of RAG systems. Specifically, we examine and compare several quantifiable metrics of the Retrieval and Generation components, such as relevance, accuracy, and faithfulness, within the current RAG benchmarks, encompassing the possible output and ground truth pairs. We then analyze the various datasets and metrics, discuss the limitations of current benchmarks, and suggest potential directions to advance the field of RAG benchmarks.

1 Introduction

Retrieval-Augmented Generation (RAG) [34] efficiently enhances the performance of generative language models through integrating information retrieval techniques. It addresses a critical challenge faced by standalone generative language models: the tendency to produce responses that, while plausible, may not be grounded in facts. By retrieving relevant information from external sources, RAG significantly reduces the incidence of hallucinations [23] or factually incorrect outputs, thereby improving the content's reliability and richness. [73] This fusion of retrieval and generation capabilities enable the creation of responses that are not only contextually appropriate but also informed by the most current and accurate information available, making RAG a development in the pursuit of more intelligent and versatile language models [64,73].

Numerous studies of RAG systems have emerged from various perspectives since the advent of Large Language Models (LLMs) [16,41,42,45,55,59,69]. The RAG system comprises two primary components: ***Retrieval*** and ***Generation***. The retrieval component aims to extract relevant information from various external knowledge sources.

It involves two main phases, *indexing* and *searching*. Indexing organizes documents to facilitate efficient retrieval, using either inverted indexes for sparse retrieval or dense vector encoding for dense retrieval [12,16,28]. The searching component utilizes these indexes to fetch relevant documents on the user's query, often incorporating the optional rerankers [4,6,39,52] to refine the ranking of the retrieved documents. The generation component utilizes the retrieved content and question query to formulate coherent and contextually relevant responses with the prompting and inferencing phases. As the "Emerging" ability [59] of LLMs and the breakthrough in aligning human commands [42], LLMs are the best performance choices model for the generation stage. Prompting methods like Chain of Thought (CoT) [60], Tree of Thought [65], Rephrase and Respond (RaR) [8] guide better generation results. In the inferencing step, LLMs interpret the prompted input to generate accurate and in-depth responses that align with the query's intent and integrate the extracted information [9,35] without further finetuning, such as fully finetuning [1,16,67,68] or LoRA [21]. Appendix A details the complete RAG structure. Figure 1 illustrates the structure of the RAG systems as mentioned.

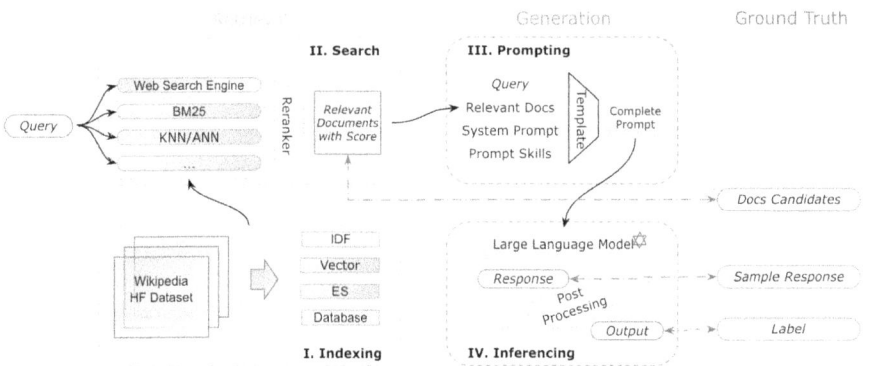

Fig. 1. The structure of the RAG system with retrieval and generation components and corresponding four phrases: indexing, search, prompting and inferencing. The pairs of "Evaluable Outputs" (EOs) and "Ground Truths" (GTs) are highlighted in read frame and green frame, with brown dashed arrows. (Color figure online)

The importance of evaluating RAG is increasing in parallel with the advancement of RAG-specific methodologies. On the one hand, RAG is a complex system intricately tied to specific requirements and language models, resulting in various evaluation methods, indicators, and tools, particularly given the black-box LLM generation. Evaluating RAG systems involves specific components and the complexity of the overall system assessment. On the other hand, the complexity of RAG systems is further compounded by the external dynamic database and the various downstream tasks, such as content creation or open domain question answering [16,70]. These challenges necessitate the development of comprehensive evaluation metrics that can effectively capture the interplay between retrieval accuracy and generative quality [2,7]. To clarify the elements further, we try to address the current gaps in the area, which differs from the prior RAG surveys [16,24,74] that predominantly collected specific RAG methods or data. We have compiled 12 distinct evaluation frameworks, encompassing a range of aspects of the RAG system. Following the procedure of making benchmarks, we analyze through targets, datasets and metrics mentioned in these benchmarks and summarize them into **A Unified Evaluation Process of RAG** (*Auepora*) as three corresponding phases.

For this paper, we contribute in the following aspects:

1. **Challenge of Evaluation**: This is the first work that summarizes and classifies the challenges in evaluating RAG systems through the structure of RAG systems, including three parts retrieval, generation, and the whole system.
2. **Analysis Framework**: In light of the challenges posed by RAG systems, we introduce an analytical framework, referred to as *A Unified Evaluation Process of RAG* (*Auepora*), which aims to elucidate the unique complexities inherent to RAG systems and guide for readers to comprehend the effectiveness of RAG benchmarks across various dimensions
3. **RAG Benchmark Analysis**: With the help of *Auepora*, we comprehensively analyze existing RAG benchmarks, highlighting their strengths and limitations and proposing recommendations for future developments in RAG system evaluation.

2 Challenges in Evaluating RAG Systems

Evaluating hybrid RAG systems entails evaluating retrieval, generation and the RAG system as a whole. These evaluations are multifaceted, requiring careful consideration and analysis. Each of them encompasses specific difficulties that complicate the development of a comprehensive evaluation framework and benchmarks for RAG systems.

Retrieval. The retrieval component is critical for fetching relevant information that informs the generation process. One primary challenge is the dynamic and vast nature of potential knowledge bases, ranging from structured databases to the entire web. This vastness requires evaluation metrics that can effectively measure the precision, recall, and relevance of retrieved documents in the context of a given query [32,52]. Moreover, the temporal aspect of information, where the relevance and accuracy of data can change over time, adds another layer of complexity to the evaluation process [6]. Additionally, the diversity of information sources and the possibility of retrieving misleading or low-quality information poses significant challenges in assessing the effectiveness of filtering and selecting the most pertinent information [39]. The traditional evaluation indicators for retrieval, such as Recall and Precision, cannot fully capture the nuances of RAG retrieval systems, necessitating the development of more nuanced and task-specific evaluation metrics [49].

Generation. The generation component, powered by LLMs, produces coherent and contextually appropriate responses based on the retrieved content. The challenge here lies in evaluating the faithfulness and accuracy of the generated content to the input data. This involves not only assessing the factual correctness of responses but also their relevance to the original query and the coherence of the generated text [49,75]. The subjective nature of certain tasks, such as creative content generation or open-ended question answering, further complicates the evaluation, as it introduces variability in what constitutes a "correct" or "high-quality" response [48].

RAG System as a Whole. Evaluating the whole RAG system introduces additional complexities. The interplay between the retrieval and generation components means that the entire system's performance cannot be fully understood by evaluating each component in isolation [14,49]. The system needs to be assessed on its ability to leverage retrieved information effectively to improve response quality, which involves measuring the added value of the retrieval component to the generative process. Furthermore, practical considerations such as response latency and the ability to handle ambiguous or complex queries are also crucial for evaluating the system's overall effectiveness and usability [6,39].

Conclusion. Evaluating the target shift from traditional absolute numeric metrics to multi-source and multi-target generation evaluation, along with the intricate interplay between retrieval and generation components, poses significant challenges. [5,50] Searches in a dynamic database may lead to misleading results or contradict the facts. Diverse and comprehensive datasets that accurately reflect real-world scenarios are crucial. Challenges also arise in the realm of metrics, encompassing generative evaluation criteria for distinct downstream tasks, human preferences, and practical considerations within the RAG system. Most prior benchmarks predominantly tackle one or several aspects of the RAG assessment but lack a comprehensive, holistic analysis.

3   A Unified Evaluation Process of RAG (*Auepora*)

To facilitate a deeper understanding of RAG benchmarks, we introduce *A Unified Evaluation Process of RAG* (*Auepora*), which focuses on three key questions of benchmarks: *What to Evaluate? How to Evaluate? How to Measure?* which correlated to *Target*, *Dataset*, and *Metric* respectively. We aim to provide a clear and accessible way for readers to comprehend the complexities and nuances of RAG benchmarking.

The *Target* module is intended to determine the evaluation direction. The *Dataset* module facilitates the comparison of various data constructions in RAG benchmarks. The final module, *Metrics*, introduces the metrics that correspond to specific targets and datasets used during evaluation. Overall, it is designed to provide a systematic methodology for assessing the effectiveness of RAG systems across various aspects by covering all possible pairs at the beginning between the "Evaluable Outputs" (EOs) and "Ground Truths" (GTs). In the following section, we will explain thoroughly *Auepora* and utilize it for introducing and comparing the RAG benchmarks.

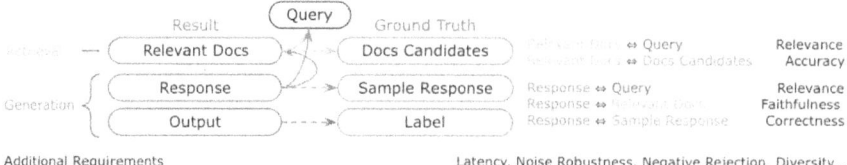

Fig. 2. The *Target* modular of the *Auepora*.

3.1 Evaluation Target (*What to Evaluate?*)

The combination of EOs and GTs in the RAG system can generate all possible targets, which is the fundamental concept of the *Auepora* (as shown in Fig. 1). Once identified, these targets can be defined based on a specific pair of EOs or EO with GT, as illustrated in Fig. 2, and used to analyze all aspects of current RAG benchmarks.

Retrieval. The EOs are the relevant documents for evaluating the retrieval component depending on the query. Then we can construct two pairwise relationships for the retrieval component, which are *Relevant Documents ↔ Query*, *Relevant Documents ↔ Documents Candidates*.

- **Relevance** (*Relevant Documents ↔ Query*) evaluates how well the retrieved documents match the information needed expressed in the query. It measures the precision and specificity of the retrieval process.
- **Accuracy** (*Relevant Documents ↔ Documents Candidates*) assesses how accurate the retrieved documents are in comparison to a set of candidate documents. It is a measure of the system's ability to identify and score relevant documents higher than less relevant or irrelevant ones.

Generation. The similar pairwise relations for the generation components are listed below. The EOs are the generated text and phrased structured content. Then we need to compare these EOs with the provided GTs and labels.

- **Relevance** (*Response ↔ Query*) measures how well the generated response aligns with the intent and content of the initial query. It ensures that the response is related to the query topic and meets the query's specific requirements.
- **Faithfulness** (*Response ↔ Relevant Documents*) evaluates if the generated response accurately reflects the information contained within the relevant documents and measures the consistency between generated content and the source documents.
- **Correctness** (*Response ↔ Sample Response*) Similar to the accuracy in the retrieval component, this measures the accuracy of the generated response against a sample response, which serves as a ground truth. It checks if the response is correct in terms of factual information and appropriate in the context of the query.

The targets of Retrieval and Generation components are introduced. Table 1 lists the relative work on improving and evaluating RAG and its benchmarks cut off in June

Table 1. The evaluating targets and corresponding metrics across various frameworks for evaluating RAG systems. The presentation distinguishes between the core areas of Retrieval and Generation considered in the evaluation. The different aspects of the evaluation are set as different colours in the table: Relevance, *Accuracy* of Retrieval and *Faithfulness*, Correctness and Relevance of Generation. The consideration of the *Additional Requirements* beyond the retrieval and generation component is also collected. Noted that quite a few of the works employed multiple methods or evaluated multiple aspects simultaneously.

Category	Framework	Time	Raw Targets	Retrieval	Generation
Tool	TruEra RAG Triad [54]	2023.10	Context Relevance Answer Relevance *Groundedness*	LLM as a Judge	LLM as a Judge
Tool	LangChain Bench. [32]	2023.11	*Accuracy* *Faithfulness* *Execution Time* *Embed. CosDistance*	*Accuracy*	LLM as a Judge
Tool	Databricks Eval [33]	2023.12	Correctness *Readability* *Comprehensiveness*	-	LLM as a Judge
Benchmark	RAGAs [14]	2023.09	Context Relevance Answer Relevance *Faithfulness*	LLM as a Judge	LLM Gen + CosSim LLM as a Judge
Benchmark	RECALL [38]	2023.11	Response Quality *Robustness*	-	BLEU, ROUGE-L
Benchmark	ARES [49]	2023.11	Context Relevance *Answer Faithfulness* Answer Relevance	LLM + Classifier	LLM + Classifier LLM + Classifier
Benchmark	RGB [6]	2023.12	*Information Integration* *Noise Robustness* *Negative Rejection* *Counterfactual Robustness*	-	*Accuracy*
Benchmark	MultiHop-RAG [52]	2024.01	*Retrieval Quality* Response Correctness	MAP, MRR, Hit@K	LLM as a Judge
Benchmark	CRUD-RAG [39]	2024.02	CREATE, READ UPDATE, DELETE	-	ROUGE, BLEU RAGQuestEval
Benchmark	MedRAG [61]	2024.02	Accuracy	-	Accuracy
Benchmark	FeB4RAG [57]	2024.02	*Consistency* Correctness Clarity Coverage	-	Human Evaluation Human Evaluation
Benchmark	CDQA [62]	2024.03	Accuracy	-	F1
Benchmark	DomainRAG [58]	2024.06	Correctness *Faithfulness* *Noise Robustness* *Structural Output*	-	F1, Exact-Match Rouge-L LLM as a Judge
Benchmark	ReEval [66]	2024.06	*Hallucination*	-	F1, Exact-Match LLM as a Judge Human Evaluation
Research	FiD-Light [20]	2023.07	*Latency*	-	-
Research	Diversity Reranker [4]	2023.08	*Diversity*	Cosine Distance	-

2024. Table 1 portrays this information, where each evaluation criterion is represented by a different colour. For example, FeB4RAG [57], the fourth from the last, has posited four standards based on [17] that comprise Consistency, Correctness, Clarity, and Coverage. Correctness is equivalent to accuracy in retrieval, and Consistency is tantamount to faithfulness in the generation component. While accuracy in retrieval gauges the correctness of the retrieved information, we posit that Coverage pertains to the coverage rate and is more associated with diversity. Therefore, we consider *Coverage* to be linked with diversity and an additional requirement in our proposed evaluation framework, which will be introduced subsequently. The remaining standard, *Clarity*, is also classified as an additional requirement in our proposed framework. The other tools and benchmarks are processed similarly.

Tools and benchmarks offer varying degrees of flexibility in evaluating datasets for RAG systems. Tools, which specify only evaluation targets, provide a versatile framework capable of constructing complete RAG applications and evaluation pipelines, as seen in works like [32,33,54]. Benchmarks, on the other hand, focus on different aspects of RAG evaluation with specific emphasis on either retrieval outputs or generation targets. For instance, RAGAs [14] and ARES [49] assess the relevance of retrieval documents, while RGB and MultiHop-RAG [6,52] prioritize accuracy, necessitating comparison with GTs. The [66] focuses on the Hallucination, which is a combination of faithfulness and correctness. All benchmarks consider generation targets due to their critical role in RAG systems, though their focus areas vary.

Additional Requirement. In addition to evaluating the two primary components outlined, a portion of the works also addressed some additional requirements of RAG (Black and *Italics* targets in Table 2). The requirements are as follows:

- **Latency** [20,32] measures how quickly the system can find information and respond, crucial for user experience.
- **Diversity** [4,32] checks if the system retrieves a variety of relevant documents and generates diverse responses.
- **Noise Robustness** [6] assesses how well the system handles irrelevant information without affecting response quality.
- **Negative Rejection** [6] gauges the system's ability to refrain from providing a response when the available information is insufficient.
- **Counterfactual Robustness** [6] evaluates the system's capacity to identify and disregard incorrect information, even when alerted about potential misinformation.
- **More**: For more human preferences considerations, there can be more additional requirements, such as readability [33,57], toxicity, perplexity [33], etc.

For the exception, CRUD-RAG [39] introduces a comprehensive benchmark addressing the broader spectrum of RAG applications beyond question-answering, categorized into Create, Read, Update, and Delete scenarios. This benchmark evaluates RAG systems across diverse tasks, including text continuation, question answering, hallucination modification, and multi-document summarization. It offers insights for optimizing RAG technology across different scenarios. DomainRAG [58] identifies six complex abilities for RAG systems: conversational, structural information, faithfulness, denoising, time-sensitive problem solving, and multi-doc understanding. ReEval [66] specifically targets hallucination evaluation by employing a cost-effective LLM-based framework that utilizes prompt chaining to create dynamic test cases.

Table 2. The evaluation datasets used for each benchmark. The dataset without citation was constructed by the benchmark itself.

Benchmark	Dataset
RAGAs [14]	WikiEval
RECALL [38]	EventKG [19], UJ [22]
ARES [49]	NQ [29], Hotpot [63], FEVER [53], WoW [11], MultiRC [10], ReCoRD [71]
RGB [6]	Generated (Source: News)
MultiHop-RAG [52]	Generated (Source: News)
CRUD-RAG [39]	Generated (Source: News) UHGEval [36]
MedRAG [61]	MIRAGE
FeB4RAG [57]	FeB4RAG, BEIR [26]
CDQA [62]	Generation (Source: News), Labeller
DomainRAG [58]	Generation (Source: College Admission Information)
ReEval [66]	RealTimeQA [27], NQ [15,29])

3.2 Evaluation Dataset (*How to Evaluate?*)

In Table 2, distinct benchmarks employ varying strategies for dataset construction, ranging from leveraging existing resources to generating entirely new data tailored for specific evaluation aspects. Several benchmarks draw upon the part of KILT (Knowledge Intensive Language Tasks) benchmark [44] (Natural Questions [29], HotpotQA [63], and FEVER [53]) and other established datasets such as SuperGLUE [56] (MultiRC [10], and ReCoRD [71]) [49]. However, the drawback of using such datasets can't solve the challenges in dynamic real-world scenarios. A similar situation can be observed in WikiEval, from Wikipedia pages post 2022, constructed by RAGAs [14].

The advent of powerful LLMs has revolutionized the process of dataset construction. With the ability to design queries and ground truths for specific evaluation targets using these frameworks, authors can now create datasets in the desired format with ease. Benchmarks like RGB, MultiHop-RAG, CRUD-RAG, and CDQA [6,39,52,62] have taken this approach further by building their own datasets using online news articles to test RAG systems' ability to handle real-world information beyond the training data of LM frameworks. Most recently, DomainRAG [58] combines various types of QA datasets with single-doc, multi-doc, single-round, and multi-round. These datasets are generated from the yearly changed information from the college website for admission and enrollment, which forces the LLMs to use the provided and updated information.

In summary, the creation and selection of datasets are crucial for evaluating RAG systems. Datasets tailored for specific metrics or tasks improve evaluation accuracy and guide the development of adaptable RAG systems for real-world information needs.

3.3 Evaluation Metric (*How to Quantify?*)

Navigating the intricate terrain of evaluating RAG systems necessitates a nuanced understanding of the metrics that can precisely quantify the evaluation targets. However, creating evaluative criteria that align with human preferences and address practical considerations is challenging. Each component within the RAG systems requires a tailored evaluative approach that reflects its distinct functionalities and objectives.

Retrieval Metrics. Various targets can be evaluated with various metrics that correspond to the given datasets. This section will introduce several commonly used metrics for retrieval and generation targets. The metrics for additional requirements can also be found in these commonly used metrics. The more specifically designed metrics can be explored in the original paper via Table 1 as a reference.

For the retrieval evaluation, the focus is on metrics that can accurately capture the relevance, accuracy, diversity, and robustness of the information retrieved in response to queries. These metrics must not only reflect the system's precision in fetching pertinent information but also its resilience in navigating the dynamic, vast, and sometimes misleading landscape of available data. The deployment of metrics like *Misleading Rate*, *Mistake Reappearance Rate*, and *Error Detection Rate* within the [38] benchmark underscores a heightened awareness of RAG systems' inherent intricacies. The integration of *MAP@K*, *MRR@K*, and *Tokenization with F1* into benchmarks like [52,62] mirrors a deepening comprehension of traditional retrieval's multifaceted evaluation. While the [17] also emphasizes that this ranking-based evaluation methodology is not unsuitable for the RAG system, and should have more RAG-specific retrieval evaluation metrics. These metrics not only capture the precision and recall of retrieval systems but also account for the diversity and relevance of retrieved documents, aligning with the complex and dynamic nature of information needs in RAG systems. The introduction of LLMs as evaluative judges, as seen in [14], further underscores the adaptability and versatility of retrieval evaluation, offering a comprehensive and context-aware approach to assessing retrieval quality.

Non-rank Based Metrics. often assess binary outcomes—whether an item is relevant or not—without considering the position of the item in a ranked list. Notice, that the following formula is just one format of these metrics, the definition of each metric may vary by the different evaluating tasks.

- **Accuracy** is the proportion of true results (both true positives and true negatives) among the total number of cases examined.
- **Precision** is the fraction of relevant instances among the retrieved instances,

$$\text{Precision} = \frac{TP}{TP + FP}$$

  where TP represents true positives and FP represents false positives.
- **Recall** at k ($Recall@k$) is the fraction of relevant instances that have been retrieved over the total amount of relevant cases, considering only the top k results.

$$Recall@k = \frac{|RD \cap Top_{kd}|}{|RD|}$$

where RD is the relevant documents, and Top_{kd} is the top-k retrieved documents.

Rank-Based Metrics evaluate the order in which relevant items are presented, with higher importance placed on the positioning of relevant items at the ranking list.

- **Mean Reciprocal Rank (MRR)** is the average of the reciprocal ranks of the first correct answer for a set of queries.

$$MRR = \frac{1}{|Q|} \sum_{i=1}^{|Q|} \frac{1}{rank_i}$$

where $|Q|$ is the number of queries and $rank_i$ is the rank position of the first relevant document for the i-th query.
- **Mean Average Precision (MAP)** is the mean of the average precision scores for each query.

$$MAP = \frac{1}{|Q|} \sum_{q=1}^{|Q|} \frac{\sum_{k=1}^{n}(P(k) \times rel(k))}{|\text{relevant documents}_q|}$$

where $P(k)$ is the precision at cutoff k in the list, $rel(k)$ is an indicator function equaling 1 if the item at rank k is a relevant document, 0 otherwise, and n is the number of retrieved documents.

Generation Metrics. In the realm of generation, evaluation transcends the mere accuracy of generated responses, venturing into the quality of text in terms of coherence, relevance, fluency, and alignment with human judgment. This necessitates metrics that can assess the nuanced aspects of language production, including factual correctness, readability, and user satisfaction with the generated content. The traditional metrics like *BLEU, ROUGE*, and *F1 Score* continue to play a crucial role, emphasizing the significance of precision and recall in determining response quality. Yet, the advent of metrics such as *Misleading Rate, Mistake Reappearance Rate*, and *Error Detection Rate* highlights an evolving understanding of RAG systems' distinct challenges [38].

The evaluation done by humans is still a very significant standard to compare the performance of generation models with one another or with the ground truth. The approach of employing LLMs as evaluative judges [75] is a versatile and automatic method for quality assessment, catering to instances where traditional ground truths may be elusive [14]. This methodology benefits from employing prediction-powered inference (PPI) and context relevance scoring, offering a nuanced lens through which LLM output can be assessed. [49] The strategic use of detailed prompt templates ensures a guided assessment aligned with human preferences, effectively standardizing evaluations across various content dimensions [1]. This shift towards leveraging LLMs as arbiters marks a significant progression towards automated and context-responsive evaluation frameworks, enriching the evaluation landscape with minimal reliance on reference comparisons.

- **ROUGE** Recall-Oriented Understudy for Gisting Evaluation (ROUGE) [37] is a set of metrics designed to evaluate the quality of summaries by comparing them to human-generated reference summaries. ROUGE can be indicative of the content overlap between the generated text and the reference text. The variants of ROUGEs measure the overlap of n-grams (ROUGE-N, ROUGGE-W), word subsequences (ROUGE-L, ROUGGE-S), and word pairs between the system-generated summary and the reference summaries.
- **BLEU** Bilingual Evaluation Understudy (BLEU) [43] is a metric for evaluating the quality of machine-translated text against one or more reference translations. BLEU calculates the precision of n-grams in the generated text compared to the reference text and then applies a brevity penalty to discourage overly short translations. BLEU has limitations, such as not accounting for the fluency or grammaticality of the generated text.
- **BertScore** BertScore [72] leverages the contextual embedding from pre-trained transformers like BERT to evaluate the semantic similarity between generated text and reference text. BertScore computes token-level similarity using contextual embedding and produces precision, recall, and F1 scores. Unlike n-gram-based metrics, BertScore captures the meaning of words in context, making it more robust to paraphrasing and more sensitive to semantic equivalence.
- **LLM as a Judge** Using "LLM as a Judge" for evaluating generated text is a more recent approach. [75] In this method, LLMs are used to score the generated text based on criteria such as coherence, relevance, and fluency. The LLM can be optionally finetuned on human judgments to predict the quality of unseen text or used to generate evaluations in a zero-shot or few-shot setting. This approach leverages the LLM's understanding of language and context to provide a more nuanced text quality assessment. For instance, [1] illustrates how providing LLM judges with detailed scoring guidelines, such as a scale from 1 to 5, can standardize the evaluation process. This methodology encompasses critical aspects of content assessment, including coherence, relevance, fluency, coverage, diversity, and detail - both in the context of answer evaluation and query formulation.

Additional Requirements. These additional requirements, such as latency, diversity, noise robustness, negative rejection, and counterfactual robustness, are used to ensure the practical applicability of RAG systems in real-world scenarios aligned with human preference. This section delves into the metrics used for evaluating these additional requirements, highlighting their significance in the comprehensive assessment of RAG systems.

Latency measures the time taken by the RAG system to finish the response of one query. It is a critical factor for user experience, especially in interactive applications such as chatbots or search engines [20]. *Single Query Latency*: The mean time taken to process a single query, including both retrieval and generating phases.

Diversity evaluates the variety and breadth of information retrieved and generated by the RAG system. It ensures that the system can provide a wide range of perspectives and avoid redundancy in responses [4]. *Cosine Similarity/Cosine Distance*: The cosine similarity/distance calculates embeddings of retrieved documents or generated responses. [30] Lower cosine similarity scores indicate higher diversity, suggesting that the system can retrieve or generate a broader spectrum of information.

Noise Robustness measures the RAG system's ability to handle irrelevant or misleading information without compromising the quality of the response [38]. The metrics *Misleading Rate* and *Mistake Reappearance Rate* are described in [38], providing detailed descriptions tailored to the specific dataset and experimental setup [58].

Negative Rejection evaluates the system's capability to withhold responses when the available information is insufficient or too ambiguous to provide an accurate answer [6]. *Rejection Rate*: The rate at which the system refrains from generating a response.

Counterfactual Robustness Counterfactual robustness assesses the system's ability to identify and disregard incorrect or counterfactual information within the retrieved documents [39]. *Error Detection Rate*: The ratio of counterfactual statements detected in retrieved information.

4 Discussion

For RAG systems, traditional Question Answering (QA) datasets and metrics remain a common format for interaction. [6,14,38,49,58,61,62,66] While these provide a basic verification of RAG's capabilities, it becomes challenging to distinguish the impact of retrieval components when faced with strong Language Models (LLMs) capable of excelling in QA benchmarks. To comprehensively evaluate the performance of entire RAG systems, there is a need for diverse and RAG-specific benchmarks. Several papers offer guidance on improving QA format benchmarks, including variations in question types: from simple Wikipedia filling questions to multi-hop [52], multi-document questions [66] and single-round to multi-round dialogue [39,58]. For answers, aspects such as structural output [58], content moderation [6,54], and hallucination [66] can be considered when evaluating relevance, faithfulness, and correctness. In addition to these, RAG systems require additional requirements such as robustness to noisy documents, language expression, latency, and result diversity. [4,6,20,32,33,38,39,57,58] Furthermore, research is needed on performance changes involving intermediate outputs and retrieved documents, as well as the relationship and analysis between retrieval metrics and final generation outputs.

Regarding *datasets*, creating a universal dataset was challenging due to the target-specific nature of different RAG benchmarks. Tailored datasets [14,38,39,49,57] are necessary for a thorough evaluation, but this approach increases the effort and resources required. Moreover, the diversity of datasets, from news articles to structured databases [66], reflects the adaptability required of RAG systems but also poses a barrier to streamlined evaluation. Recently, with the cutting-edge performance of LLMs, complex data processing and automatic QA pair generation can be automated to achieve daily or finer-grained time resolution, preventing LLMs from cheating and evaluating the robustness of RAG systems in rapidly changing data [6,39,52,58,62,66].

When it comes to *metrics*, the use of LLMs as automatic evaluative judges signifies a burgeoning trend, promising versatility and depth in generative outputs with reasoning on a large scale compared to human evaluation. However, using "LLMs as a Judge" [75] for responses presents challenges in aligning with human judgment, establishing effective grading scales, and applying consistent evaluation across varied use cases. Determining correctness, clarity, and richness can differ between automated and human assessments. Moreover, the effectiveness of example-based scoring can vary, and there's no universally applicable grading scale and prompting text, complicating the standardization of "LLM as a Judge" [33].

In addition to the challenges mentioned above, it is important to consider the resource-intensive nature [76] of using Large Language Models (LLMs) for data generation and validation. RAG benchmarks must balance the need for thorough evaluation with the practical constraints of limited computational resources. As such, it is desirable to develop evaluation methodologies that can effectively assess RAG systems using smaller amounts of data while maintaining the validity and reliability of the results.

5 Conclusion

This survey systematically explores the complexities of evaluating RAG systems, highlighting the challenges in assessing their performance. Through the proposed *A Unified Evaluation Process of RAG*, we outline a structured approach to analyzing RAG evaluations, focusing on targets, datasets and measures. Our analysis emphasizes the need for targeted benchmarks that reflect the dynamic interplay between retrieval accuracy and generative quality and practical considerations for real-world applications. By identifying gaps in current methodologies and suggesting future research directions, we aim to contribute to more effective, and user-aligned benchmarks of RAG systems.

A Structure of RAG System

A.1 Retrieval Component

The retrieval component of RAG systems in Fig. 1 can be categorized into three types: sparse retrieval, dense retrieval [77], and web search engine. The standard for evaluation is the output of *relevant documents* with numerical scores or rankings.

Before the introduction of neural networks, *sparse retrievals* are widely used for retrieving relative text content. Methods like TF-IDF [46] and BM25 [47] rely on keyword matching and word frequency but may miss semantically relevant documents without keyword overlap.

By leveraging deep learning models such as BERT [9], *dense retrieval* can capture the semantic meaning of texts, which allows them to find relevant documents even when keyword overlap is minimal. This is crucial for complex queries that require a contextual understanding to retrieve accurate information. With advanced fusion structure for queries and documents [28] and the more efficient implementation of K-Nearest Neighbors (KNN) [51], Approximate Nearest Neighbor (ANN) [12,25] search techniques, dense retrieval methods have become practical for large-scale use.

Web search engine employs the complex online search engine to provide relevant documents, such as Google Search [18], Bing Search [40], DuckDuckGo [13]. RAG systems can traverse the web's extensive information, potentially returning a more diverse and semantically relevant set of documents via the API of the search provider.

The black box of the search engine and the expense of large-scale search are not affordable sometimes.

It is observed that dense retrieval techniques, particularly those leveraging embeddings, stand out as the preferred choice within the RAG ecosystem. These methods are frequently employed in tandem with sparse retrieval strategies, creating a hybrid approach that balances precision and breadth in information retrieval. Moreover, the adoption of sophisticated web search engines for benchmark assessment underscores their growing significance in enhancing the robustness and comprehensiveness of evaluations.

Indexing. The indexing component processes and indexes document collections, such as HuggingFace datasets or Wikipedia pages. Chunking before indexing can improve retrieval by limiting similarity scores to individual chunks, as semantic embedding is less accurate for long articles, and desired content is often brief [32]. Index creation is designed for fast and efficient search. For example, the inverted index for sparse retrieval and the ANN index for dense retrieval.

Sparse Retrieval involves calculating IDF for each term and storing values in a database for quick look-up and scoring when queried.

Dense Retrieval encodes documents into dense vectors using a pre-trained language model like BERT. These vectors are then indexed using an Approximate Nearest Neighbor (ANN) search technique, like graph-based Hierarchical Navigable Small World (HNSW) or Inverted File Index (IVF) [12]. This process allows for the efficient retrieval of "closed" items by given predefined distance metrics.

Search. This step is responsible for retrieving relevant documents based on a given query. Queries are submitted using the respective API to retrieve relevant documents for web search engine retrieval. For local resources, the query component is responsible for formatting the query in the format required by different sparse or dense retrieval methods. Then, the query is submitted to the retrieval system, which returns a set of relevant documents along with their scores.

In both local and web-based scenarios, an optional reranker can be employed to refine the ranking of retrieved documents further. The reranker usually comprises a more complex and larger model that considers additional features of the documents and the given query. These additional features often include the semantic relationship between the query and the document content, document importance or popularity, and other custom measures specific to the information need at hand.

A.2 Generation Component

The evaluable output for the generation component is the *response* of LLMs and the *structured or formatted output* from the phrased response.

Prompting. The generation process critically hinges on prompting, where a query, retrieval outcomes, and instructions converge into a single input for the language model.

Research showcases various strategic prompting tactics such as the Chain of Thought (CoT) [60], Tree of Thought (ToT) [3], and Self-Note [31], each significantly shaping the model's output. These methods, especially the step-by-step approach, are pivotal in augmenting LLMs for intricate tasks.

Prompting innovations have introduced methods like Rephrase and Respond (RaR) [8], enhancing LLMs by refining queries within prompts for better comprehension and response. This technique has proven to boost performance across diverse tasks. The latest RAG benchmarks [61,62] in the specific domains start to evaluate the robustness of various prompting engineering skills, including CoT, RaR, etc.

Inference. The final input string prepared in the prompting step is then passed on to the LLMs as input, which generates the output. The inference stage is where the LLM operates on the input derived from the retrieval and the prompting stages in the pipeline to generate the final output. This is usually the answer to the initial query and is used for downstream tasks.

Depending on the specifics of the task or expected output structure, a post-processing step may be implemented here to format the generated output suitably or extract specific information from the response. For example, the classification problems (multichoice questions) or if the task requires the extraction of specific information from the generated text, this step could involve additional named entity recognition or parsing operations.

References

1. Balaguer, A., et al.: RAG vs fine-tuning: pipelines, tradeoffs, and a case study on agriculture. Technical report (2024). arXiv:2401.08406 [cs] type: article
2. Barnett, S., Kurniawan, S., Thudumu, S., Brannelly, Z., Abdelrazek, M.: Seven failure points when engineering a retrieval augmented generation system (2024). https://doi.org/10.48550/ARXIV.2401.05856
3. Besta, M., et al.: Graph of thoughts: solving elaborate problems with large language models. In: Proceedings of the AAAI Conference on Artificial Intelligence 2024 (AAAI 2024) (2023). https://doi.org/10.48550/ARXIV.2308.09687
4. Blagojevic, V.: Enhancing RAG Pipelines in Haystack: Introducing DiversityRanker and LostInTheMiddleRanker (2023). https://towardsdatascience.com/enhancing-rag-pipelines-in-haystack-45f14e2bc9f5
5. Chang, Y., et al.: A survey on evaluation of large language models. ACM Trans. Intell. Syst. Technol. **15**(3), 1–45 (2024)
6. Chen, J., Lin, H., Han, X., Sun, L.: Benchmarking large language models in retrieval-augmented generation (2023). https://doi.org/10.48550/ARXIV.2309.01431
7. Cuconasu, F., et al.: The power of noise: redefining retrieval for rag systems (2024). https://doi.org/10.48550/ARXIV.2401.14887
8. Deng, Y., Zhang, W., Chen, Z., Gu, Q.: Rephrase and respond: let large language models ask better questions for themselves (2023). https://doi.org/10.48550/ARXIV.2311.04205
9. Devlin, J., Chang, M.W., Lee, K., Toutanova, K.: BERT: pre-training of deep bidirectional transformers for language understanding. In: Burstein, J., Doran, C., Solorio, T. (eds.) Proceedings of the 2019 Conference of the North American Chapter of the Association for Computational Linguistics: Human Language Technologies (Volume 1: Long and Short Papers),

pp. 4171–4186. Association for Computational Linguistics, Minneapolis, Minnesota (2019). https://doi.org/10.18653/v1/N19-1423. https://aclanthology.org/N19-1423
10. DeYoung, J., et al.: Eraser: a benchmark to evaluate rationalized NLP models
11. Dinan, E., Roller, S., Shuster, K., Fan, A., Auli, M., Weston, J.: Wizard of Wikipedia: knowledge-powered conversational agents. In: Proceedings of the International Conference on Learning Representations (ICLR) (2019)
12. Douze, M., et al.: The faiss library (2024)
13. DuckDuckGo: DuckDuckGo—Privacy, simplified (2024). https://duckduckgo.com//home
14. Es, S., James, J., Espinosa-Anke, L., Schockaert, S.: Ragas: automated evaluation of retrieval augmented generation (2023). https://doi.org/10.48550/ARXIV.2309.15217
15. Fisch, A., Talmor, A., Jia, R., Seo, M., Choi, E., Chen, D.: MRQA 2019 shared task: evaluating generalization in reading comprehension. In: Fisch, A., Talmor, A., Jia, R., Seo, M., Choi, E., Chen, D. (eds.) Proceedings of the 2nd Workshop on Machine Reading for Question Answering, pp. 1–13. Association for Computational Linguistics, Hong Kong, China (2019). https://doi.org/10.18653/v1/D19-5801. https://aclanthology.org/D19-5801
16. Gao, Y., et al.: Retrieval-augmented generation for large language models: a survey. Technical report (2024). arXiv:2312.10997 [cs] type: article
17. Gienapp, L., et al.: Evaluating generative ad hoc information retrieval. Technical report (2023). arXiv:2311.04694 [cs] type: article
18. Google: Programmable Search Engine | Google for Developers (2024). https://developers.google.com/custom-search
19. Gottschalk, S., Demidova, E.: Eventkg: a multilingual event-centric temporal knowledge graph (2018). https://doi.org/10.48550/ARXIV.1804.04526
20. Hofstätter, S., Chen, J., Raman, K., Zamani, H.: FiD-light: efficient and effective retrieval-augmented text generation. In: Proceedings of the 46th International ACM SIGIR Conference on Research and Development in Information Retrieval, SIGIR 2023, pp. 1437–1447. Association for Computing Machinery, New York (2023). https://doi.org/10.1145/3539618.3591687
21. Hu, E.J., et al.: LoRA: low-rank adaptation of large language models. Technical report (2021). arXiv:2106.09685 [cs] type: article
22. Huang, J., Shao, H., Chang, K.C.C., Xiong, J., Hwu, W.: Understanding jargon: combining extraction and generation for definition modeling. In: Proceedings of EMNLP (2022)
23. Huang, L., et al.: A survey on hallucination in large language models: principles, taxonomy, challenges, and open questions (2023). https://doi.org/10.48550/ARXIV.2311.05232
24. Huang, Y., Huang, J.: A survey on retrieval-augmented text generation for large language models (2024). https://doi.org/10.48550/ARXIV.2404.10981
25. Johnson, J., Douze, M., Jégou, H.: Billion-scale similarity search with GPUs. IEEE Trans. Big Data **7**(3), 535–547 (2019)
26. Kamalloo, E., Thakur, N., Lassance, C., Ma, X., Yang, J.H., Lin, J.: Resources for brewing beir: reproducible reference models and an official leaderboard (2023)
27. Kasai, J., et al.: Realtime QA: what's the answer right now? (2022). https://doi.org/10.48550/ARXIV.2207.13332
28. Khattab, O., Zaharia, M.: Colbert: efficient and effective passage search via contextualized late interaction over BERT (2020). https://doi.org/10.48550/ARXIV.2004.12832
29. Kwiatkowski, T., et al.: Natural questions: a benchmark for question answering research. Trans. Assoc. Comput. Linguist. **7**, 453–466 (2019). https://doi.org/10.1162/tacl_a_00276
30. Lahitani, A.R., Permanasari, A.E., Setiawan, N.A.: Cosine similarity to determine similarity measure: study case in online essay assessment. In: 2016 4th International Conference on Cyber and IT Service Management, pp. 1–6 (2016). https://doi.org/10.1109/CITSM.2016.7577578

31. Lanchantin, J., Toshniwal, S., Weston, J., Szlam, A., Sukhbaatar, S.: Learning to reason and memorize with self-notes (2023). https://doi.org/10.48550/ARXIV.2305.00833
32. LangChain: Evaluating rag architectures on benchmark tasks (2023). https://langchain-ai.github.io/langchain-benchmarks/notebooks/retrieval/langchain_docs_qa.html
33. Leng, Q., Uhlenhuth, K., Polyzotis, A.: Best Practices for LLM Evaluation of RAG Applications (2023). https://www.databricks.com/blog/LLM-auto-eval-best-practices-RAG
34. Lewis, P., et al.: Retrieval-augmented generation for knowledge-intensive NLP tasks. In: Proceedings of the 34th International Conference on Neural Information Processing Systems, NIPS 2020, pp. 9459–9474. Curran Associates Inc., Red Hook (2020)
35. Lewis, P., et al.: Retrieval-augmented generation for knowledge-intensive NLP tasks. Technical report (2021). arXiv:2005.11401 [cs] type: article
36. Liang, X., et al.: Uhgeval: benchmarking the hallucination of Chinese large language models via unconstrained generation. arXiv preprint arXiv:2311.15296 (2023)
37. Lin, C.Y.: ROUGE: a package for automatic evaluation of summaries. In: Text Summarization Branches Out, pp. 74–81. Association for Computational Linguistics, Barcelona, Spain (2004). https://aclanthology.org/W04-1013
38. Liu, Y., et al.: Recall: a benchmark for LLMs robustness against external counterfactual knowledge (2023). https://doi.org/10.48550/ARXIV.2311.08147
39. Lyu, Y., et al.: Crud-rag: a comprehensive Chinese benchmark for retrieval-augmented generation of large language models (2024). https://doi.org/10.48550/ARXIV.2401.17043
40. Microsoft: Web Search API | Microsoft Bing. https://www.microsoft.com/en-us/bing/apis/bing-web-search-api
41. OpenAI, Achiam, J., et al.: GPT-4 Technical Report (2023). https://doi.org/10.48550/ARXIV.2303.08774
42. Ouyang, L., et al.: Training language models to follow instructions with human feedback. Technical report (2022). arXiv:2203.02155 [cs] type: article
43. Papineni, K., Roukos, S., Ward, T., Zhu, W.J.: Bleu: a method for automatic evaluation of machine translation. In: Isabelle, P., Charniak, E., Lin, D. (eds.) Proceedings of the 40th Annual Meeting of the Association for Computational Linguistics, pp. 311–318. Association for Computational Linguistics, Philadelphia, Pennsylvania, USA (2002). https://doi.org/10.3115/1073083.1073135. https://aclanthology.org/P02-1040
44. Petroni, F., et al.: KILT: a benchmark for knowledge intensive language tasks. In: Proceedings of the 2021 Conference of the North American Chapter of the Association for Computational Linguistics: Human Language Technologies, pp. 2523–2544. Association for Computational Linguistics, Online (2021). https://doi.org/10.18653/v1/2021.naacl-main.200. https://aclanthology.org/2021.naacl-main.200
45. Radford, A., Wu, J., Child, R., Luan, D., Amodei, D., Sutskever, I., et al.: Language models are unsupervised multitask learners. OpenAI Blog **1**(8), 9 (2019)
46. Ramos, J., et al.: Using TF-IDF to determine word relevance in document queries. In: Proceedings of the First Instructional Conference on Machine Learning, vol. 242, pp. 29–48. Citeseer (2003)
47. Robertson, S., Zaragoza, H., et al.: The probabilistic relevance framework: BM25 and beyond. Found. Trends® Inf. Retrieval **3**(4), 333–389 (2009)
48. Rosset, C., et al.: Researchy questions: a dataset of multi-perspective, decompositional questions for LLM web agents (2024). https://doi.org/10.48550/ARXIV.2402.17896
49. Saad-Falcon, J., Khattab, O., Potts, C., Zaharia, M.: Ares: an automated evaluation framework for retrieval-augmented generation systems (2023). https://doi.org/10.48550/ARXIV.2311.09476
50. Sai, A.B., Mohankumar, A.K., Khapra, M.M.: A survey of evaluation metrics used for NLG systems. ACM Comput. Surv. (CSUR) **55**(2), 1–39 (2022)

51. Shahabi, C., Kolahdouzan, M.R., Sharifzadeh, M.: A road network embedding technique for k-nearest neighbor search in moving object databases. In: Proceedings of the 10th ACM International Symposium on Advances in Geographic Information Systems, pp. 94–100 (2002)
52. Tang, Y., Yang, Y.: Multihop-rag: benchmarking retrieval-augmented generation for multi-hop queries (2024). https://doi.org/10.48550/ARXIV.2401.15391
53. Thorne, J., Vlachos, A., Christodoulopoulos, C., Mittal, A.: FEVER: a large-scale dataset for fact extraction and VERification. In: NAACL-HLT (2018)
54. TruLens: TruLens (2023). https://www.trulens.org/trulens_eval/getting_started/quickstarts/quickstart/
55. Vaswani, A., et al.: Attention is all you need (2017). https://doi.org/10.48550/ARXIV.1706.03762
56. Wang, A., et al.: SuperGLUE: a stickier benchmark for general-purpose language understanding systems. arXiv preprint 1905.00537 (2019)
57. Wang, S., Khramtsova, E., Zhuang, S., Zuccon, G.: Feb4rag: evaluating federated search in the context of retrieval augmented generation (2024). https://doi.org/10.48550/ARXIV.2402.11891
58. Wang, S., et al.: Domainrag: a Chinese benchmark for evaluating domain-specific retrieval-augmented generation (2024). https://doi.org/10.48550/ARXIV.2406.05654
59. Wei, J., et al.: Emergent abilities of large language models (2022). https://doi.org/10.48550/ARXIV.2206.07682
60. Wei, J., et al.: Chain-of-thought prompting elicits reasoning in large language models (2022). https://doi.org/10.48550/ARXIV.2201.11903
61. Xiong, G., Jin, Q., Lu, Z., Zhang, A.: Benchmarking retrieval-augmented generation for medicine (2024). https://doi.org/10.48550/ARXIV.2402.13178
62. Xu, Z., et al.: Let LLMs take on the latest challenges! a Chinese dynamic question answering benchmark (2024). https://doi.org/10.48550/ARXIV.2402.19248
63. Yang, Z., et al.: HotpotQA: a dataset for diverse, explainable multi-hop question answering. In: Conference on Empirical Methods in Natural Language Processing (EMNLP) (2018)
64. Yao, J.Y., Ning, K.P., Liu, Z.H., Ning, M.N., Yuan, L.: LLM lies: hallucinations are not bugs, but features as adversarial examples. arXiv preprint arXiv:2310.01469 (2023)
65. Yao, S., et al.: Tree of thoughts: deliberate problem solving with large language models (2023)
66. Yu, X., Cheng, H., Liu, X., Roth, D., Gao, J.: ReEval: automatic hallucination evaluation for retrieval-augmented large language models via transferable adversarial attacks. In: Duh, K., Gomez, H., Bethard, S. (eds.) Findings of the Association for Computational Linguistics: NAACL 2024, pp. 1333–1351. Association for Computational Linguistics, Mexico City, Mexico (2024). https://aclanthology.org/2024.findings-naacl.85
67. Zhang, K., et al.: EATN: an efficient adaptive transfer network for aspect-level sentiment analysis. IEEE Trans. Knowl. Data Eng. **35**(1), 377–389 (2021)
68. Zhang, K., Zhang, H., Liu, Q., Zhao, H., Zhu, H., Chen, E.: Interactive attention transfer network for cross-domain sentiment classification. In: Proceedings of the AAAI Conference on Artificial Intelligence, vol. 33, pp. 5773–5780 (2019)
69. Zhang, K., et al.: Incorporating dynamic semantics into pre-trained language model for aspect-based sentiment analysis. arXiv preprint arXiv:2203.16369 (2022)
70. Zhang, Q., et al.: A survey for efficient open domain question answering. In: Rogers, A., Boyd-Graber, J., Okazaki, N. (eds.) Proceedings of the 61st Annual Meeting of the Association for Computational Linguistics (Volume 1: Long Papers), pp. 14447–14465. Association for Computational Linguistics, Toronto, Canada (2023). https://doi.org/10.18653/v1/2023.acl-long.808. https://aclanthology.org/2023.acl-long.808

71. Zhang, S., Liu, X., Liu, J., Gao, J., Duh, K., Van Durme, B.: Record: bridging the gap between human and machine commonsense reading comprehension (2018). https://doi.org/10.48550/ARXIV.1810.12885
72. Zhang, T., Kishore, V., Wu, F., Weinberger, K.Q., Artzi, Y.: BERTScore: evaluating text generation with BERT. In: 8th International Conference on Learning Representations, ICLR 2020, Addis Ababa, Ethiopia, 26–30 April 2020. OpenReview.net (2020). https://openreview.net/forum?id=SkeHuCVFDr
73. Zhang, Y., Khalifa, M., Logeswaran, L., Lee, M., Lee, H., Wang, L.: Merging generated and retrieved knowledge for open-domain QA. In: Bouamor, H., Pino, J., Bali, K. (eds.) Proceedings of the 2023 Conference on Empirical Methods in Natural Language Processing, pp. 4710–4728. Association for Computational Linguistics, Singapore (2023). https://doi.org/10.18653/v1/2023.emnlp-main.286. https://aclanthology.org/2023.emnlp-main.286
74. Zhao, P., et al.: Retrieval-augmented generation for AI-generated content: a survey (2024). https://doi.org/10.48550/ARXIV.2402.19473
75. Zheng, L., et al.: Judging LLM-as-a-judge with MT-bench and chatbot arena (2023). https://doi.org/10.48550/ARXIV.2306.05685
76. Zhou, Y., et al.: On the opportunities of green computing: a survey (2023)
77. Zhu, F., Lei, W., Wang, C., Zheng, J., Poria, S., Chua, T.S.: Retrieving and reading: a comprehensive survey on open-domain question answering. Technical report (2021). arXiv:2101.00774 [cs] type: article

CMNet: Fast Time Series Forecasting Based on Hybrid Convolution-MLP Architecture

Yikun Yang[1], Kailiang Chen[1], Shufen Chen[1], Jiaen Chen[1], Renzhong Niu[1], Wenbin Chen[2], and Zhigang Li[1(✉)]

[1] Shihezi University, Shihezi 832003, China
lizhigang1998@163.com
[2] Xinjiang Tianfu Information Co., Ltd., Shihezi 832003, China

Abstract. Real-world long-term time series are challenging to model due to their complex temporal patterns. While popular RNN-based, Transformer-based, or linear modeling approaches achieve excellent results in long-term time series forecasting, they often fail to capture both global and local views of the time series and ignore its intrinsic properties. This paper proposes CMNet, which combines the stochastic downsampling strategy (SDS) and the hybrid fusion of convolution and MLP (CM). Specifically, the SDS module addresses the intrinsic limitation of representation ability in one-dimensional time series by effectively downsampling them into a more compact format. Immediately, we use a hybrid method of convolution and MLP to extract information from the local and global perspectives of the sequence and then focus on potential relationships between the multiple variables. Experimentally, CMNet outperforms the state-of-the-art methods on five out of seven widely used benchmark datasets, what's more, CMNet is less than them in terms of the number of parameters.

Keywords: Time Series Forecasting · Long-Term Forecasting

1 Introduction

Long-term multivariate time series forecasting is a key application task in the fields of economy, transportation, energy, and meteorology, such as real Gross Domestic Product (GDP) growth forecasting [7], photovoltaic solar power generation and electric load forecasting [8], and meteorological data forecasting [14]. Multivariate time series is a fundamental and ubiquitous data type encountered in real-world applications. It involves the simultaneous observation of multiple variables, each possessing unique characteristics, including periodicity, trend, and stochasticity. Furthermore, these variables are often interconnected, exhibiting potential interdependencies and mutual influence. Traditional methods, such as ARIMA [1], are frequently inadequate in capturing the inherent complexity presented in multivariate time series data.

The advent of deep learning significantly enhances our capacity to process intricate information, leading to notable advancements in modeling long sequences. RNN-based methods [6,9,18], which leverage Markov assumptions to model consecutive data points, demonstrate promising results in this task. However, these methods are confronted with challenges such as the issue of vanishment or explosion of gradients, which hampers their training process. Convolutional neural networks (CNNs) demonstrate their effectiveness in extracting temporal information from time series data by leveraging the time dimension [2]. While CNNs excel at simulating changes at adjacent points in time, they are not good at effectively capturing broader contextual relationships. The Transformer model [21], originally renowned for its breakthroughs in Natural Language Processing (NLP), is adapted for time series analysis, demonstrating remarkable performance in capturing long-term dependencies through its distinctive attention mechanism. Subsequent advancements in the field of time series analysis lead to the development of models like Informer [30] and Autoformer [24], which aim to not only relatively reduce the computational burden but also leverage the inherent capacity of these models to effectively capture intricate temporal patterns that exhibit long-term dependencies. MTGNN [25], StemGNN [4], and IGMTF [27] use graphical neural networks to explicitly model dependencies between variables. Although deep learning-based models offer impressive modeling capabilities, their training and inference processes involve significant computational costs. Given the aforementioned circumstances, we pose the following question:

How can a network structure be designed to capture the dependencies in complex time series in a comprehensive and efficient way?

This paper introduces the CMNet, a lightweight deep learning network architecture designed for multivariate long-term time series forecasting task. CMNet combines both convolution and MLP techniques to enhance its modeling capabilities. Notably, CMNet uses Stochastic Downsampling strategy (SDS), which allows the same sequence to be downsampled multiple times, and then the generated multiple subsequence segments are compressed through folding and reconstruction before used in downstream tasks. The hybrid convolution and MLP (CM) method is employed to extract features from 2D tensors acquired through SDS. These tensors undergo convolution to capture local information, while the MLP method is utilized to capture global information across the temporal and channel dimensions of the sequence. The primary contributions of this work can be summarized as follows:

- In order to make full use of the information in the limited time series, stochastic downsampling is used. This process involves reconstructing the time series into 2D tensors, which helps capture complex temporal patterns.
- To handle the reorganized 2D tensors, we introduce the CM, which plays a crucial role in extracting and capturing features. The CM utilizes convolution operations to extract local features from the input sequence, creating feature mappings that highlight relevant patterns.

– The experimental results on seven benchmark datasets demonstrate our proposed model outperforms other state-of-the-art models with less parameters and faster inference speed.

2 Related Work

Time series forecasting entails the essential tasks of identifying inherent patterns within the data and forecasting future trends, carrying substantial academic and practical implications [12].

In recent years, many deep learning models have been proposed and made remarkable progress in time series forecasting tasks. The MLP-based model [26,28,29] takes the historical observation of time series as input and uses the hidden layer for nonlinear transformation and feature extraction. The CNN-based model [3,10,19,20,22] uses convolution layers to capture local and global relations in time series and uses residual connection and hole convolution techniques to enhance the representation ability of the model. RNN-based models [6,13,17,18] preserve and propagate historical information through connections, enabling them to capture long-term dependencies. Transformer-based networks [5,21] first shone on NLP tasks and later proved to be excellent at capturing complex relationships in multivariable time series [11]. Informer [30] applies ProbSparse Self-Attention and the generative decoder to address the shortcomings of primitive attention design that results in reduced error accumulation and decreased memory usage for long-term forecasting. Autoformer [24] utilizes a distinctive autocorrelation mechanism and alternate decomposition scheme to optimize the preprocessing design and attention structure. Nonstationary Transformer [16] introduces Series Stationarization and De-Stationary Attention to tackle non-stationary time series.

However, Ailing Zeng et al. propose a simple MLP-based architecture Dlinear for multistep forecasting, which significantly outperforms the previous Transformer-based models. Tianping Zhang et al. propose LightTS, a simple MLP network [29], which uses two delicate downsampling methods to demonstrate comparable performance on benchmark datasets. TimesNet [23] expands time-varying analyses to 2D space through the proposal of the TimesNet model, which uses TimesBlock as a task-generic backbone for time series analysis. The proposed method can effectively identify multiple periodic patterns and extract complex time-varying changes from the 2D tensors. Moreover, the model consistently achieves state-of-the-art results across multiple datasets.

The above models take different approaches to address the challenges of high hardware requirements and long-term prediction complexity in time series modeling. However, due to the complexity of these models, their upper bounds of efficiency are limited, and the modeling accuracy is not high. Therefore, we try to develop a new type of time series prediction model to improve the accuracy and efficiency on the time series prediction task. Compared with most of the above models, our model adopts a more lightweight network structure to reduce the computational cost and the number of parameters. Our simple strategy of transforming time series data extraction into 2D tensors using the SDS module allows us to overcome the inherent constraints of expressing information in a single time series. In addition, we use the CM module to enhance the attention ability to capture local information and global information in time series from 2D tensors, thereby improving the model accuracy.

3 Model Structure

Long-term time series forecasting is a challenging task that involves a sequence-to-sequence problem. The time dimension of the input sequence is represented as T, and the channel dimension is denoted as C, which can be expressed as $\mathbf{X} = \{X_1, ..., X_T\} \in \mathbb{R}^{T \times C}$. Similarly, the time dimension of the output sequence is denoted as F, and the channel dimension remains as C, which can be represented as $\mathbf{Y} = \{X_{T+1}, ..., X_{T+F}\} \in \mathbb{R}^{F \times C}$. In this paper, we present CMNet, an architecture that significantly simplifies model computation requirements.

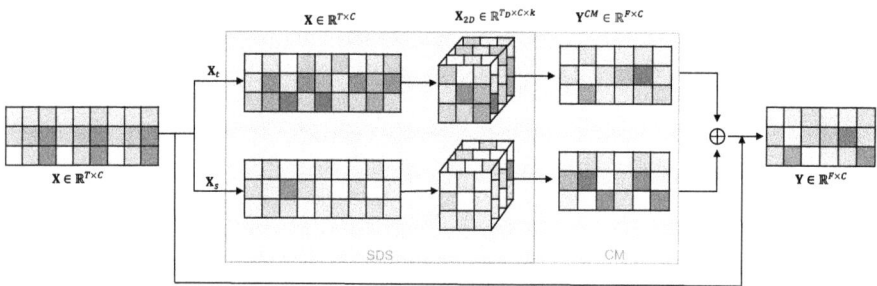

Fig. 1. The overall architecture of CMNet. By utilizing the SDS, we transform the 1D time series into 2D tensors. We then employ the CM module to extract various features within the subsequence.

3.1 Architecture Overview

The overall architecture of CMNet is presented in Fig. 1. The SDS effectively captures several temporal patterns for the time series and transforms them into a 2D tensor, which is elaborated in Sect. 3.2. Moreover, we develop the CM that combines convolution operations and MLP to efficiently model the time series, as explained in Sect. 3.3.

3.2 Stochastic Downsampling Strategy

This section provides a detailed explanation of the SDS utilized in CMNet. The architecture of the SDS is shown in Fig. 2. As empirically proven in LightTS [29], downsampling works well for time series, especially for multivariate real-world time series. The sampling method selects time points with different time span relationships for downstream time series forecasting tasks. We stochastic downsample the same time series segments multiple times with a sampling length T_D smaller than the time series segment length T. Above-mentioned strategies to analyze aggregated sampled segments can help capture more global and local information.

The decomposition of the periodic trend plays a crucial role in time series forecasting tasks [24]. To address this, we employ the moving average method to separate the trend and seasonal components. For the input series $\mathbf{X} \in \mathbb{R}^{T \times C}$, the process is:

$$\begin{aligned} \mathbf{X}_t &= \text{AvgPooling}(\text{Padding}(\mathbf{X}))_{kernel} \\ \mathbf{X}_s &= \mathbf{X} - \mathbf{X}_t, \end{aligned} \quad (1)$$

where: $\mathbf{X}_t, \mathbf{X}_s \in \mathbb{R}^{T \times C}$ denotes the trend and seasonal components, respectively, and the Padding(·) function to maintain a constant sequence length. Subsequently, the trend and seasonal components are processed separately. For convenience, we denote $\mathbf{X}_{:n}$ as the entire time series after component decomposition, where n is the trend or seasonal component.

For a time series $\mathbf{X}_{:n} \in \mathbb{R}^{T \times C}$ of length T and variable C, we use the following stochastic downsampling approach:

$$\mathbf{X}_{1D}^i = \text{SDS}(\mathbf{X}_{:n}) \in \mathbb{R}^{T_D \times C}, i \in \{1, ..., k\} \quad (2)$$

where k represents the number of samples in the time series and T_D is the sampling length. Here, SDS(·) denotes the process of stochastic downsampling the time series. To capture a wider range of information within the time series, we utilize a straightforward stochastic sampling approach that avoids repetition. This sampling method considers the presence of multiple periodic patterns and the inherent randomness observed in real-world time series data.

Then, We integrate $\{\mathbf{X}_{1D}^1, ..., \mathbf{X}_{1D}^k\}$ as a whole to obtain distinguish semantic information within different sampled time series segments. This process can be formulated as:

$$\mathbf{X}_{2D} = \text{Concatenation}(\mathbf{X}_{1D}^i) \in \mathbb{R}^{T_D \times C \times k}, i \in \{1, ..., k\} \quad (3)$$

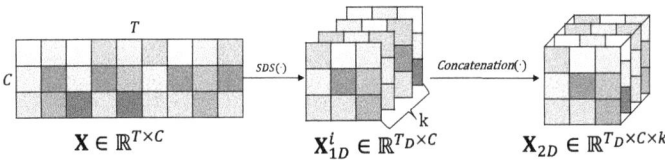

Fig. 2. The architecture of the SDS module.

Time subseries exhibit variations in their distributions based on their starting positions. To address this issue, we employ the SDS to generate multiple subseries with distinct distributions. Our proposed sampling method improves the capability of information extraction by effectively integrating supplementary semantic relationships of different segments. Our ablation experiment results demonstrate that the SDS improves the generalization capability of our proposed model and yields superior forecasts for data with unknown distributions. In the following section, we present the CM module as a novel framework specifically designed to extract informative features from the 2D tensor obtained through the SDS.

3.3 CM-Blocks

We obtain more comprehensive representations after SDS method. In order to furthermore fully leverage distinctive characteristics both in time and channel dimensions, we propose a hybrid module combining convolution and MLP, which effectively captures both the interaction and feature information embedded within the sub-sequences. The detailed processing of CM is illustrated in Fig. 3.

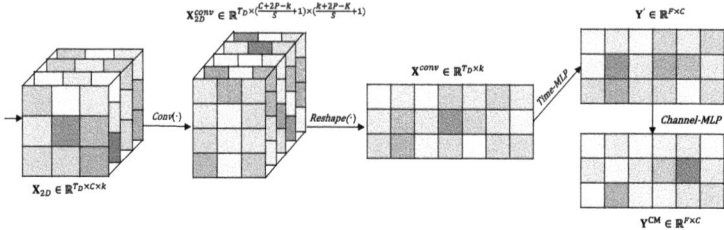

Fig. 3. The architecture of the CM module.

CM-Convolution. To effectively extract local information and features from the time series, we employ convolution to process the 2D tensor. The convolution operation, referred to as $\text{Conv}(\cdot)$, utilizes kernel size K, padding size P, and stride size S as follows:

$$\mathbf{X}_{2D}^{conv} = \text{Conv}(\mathbf{X}_{2D}) \in \mathbb{R}^{T_D \times (\frac{C+2P-K}{S}+1) \times (\frac{k+2P-K}{S}+1)} \qquad (4)$$

Subsequently, we reshape $\mathbf{X}_{2D}^{conv}$ into $\mathbf{X}^{conv} \in \mathbb{R}^{T_D \times k}$ to facilitate further processing by the MLP. In the $\text{Reshape}(\cdot)$, we stack batches in channels dimension to change the data dimension. The reshaping procedure is carried out as follows:

$$\mathbf{X}^{conv} = \text{Reshape}(\mathbf{X}_{2D}^{conv}) \in \mathbb{R}^{T_D \times k} \qquad (5)$$

CM-MLP. Lastly, our MLP structure allows us to capture the global features of the time series. We initiate the feature extraction process along the time dimension, referred to as Time-MLP, which maps the length of the time series to the forecasted length F. Subsequently, to extract feature information between channels, we extract the features along the channel dimension, known as Channel-MLP. This extraction process provides us with the desired final output.

$$\mathbf{Y}^{'} = \text{Time-MLP}(\mathbf{X}^{conv}) \in \mathbb{R}^{F \times C} \qquad (6)$$

$$\mathbf{Y}^{CM} = \text{Channel-MLP}(\mathbf{Y}^{'}) \in \mathbb{R}^{F \times C} \tag{7}$$

According to Fig. 1, $\mathbf{X}_t$ and $\mathbf{X}_s$ generate the corresponding prediction results $\mathbf{Y}_t^{CM}$ and $\mathbf{Y}_s^{CM}$. Then we conduct pixel-wise addition on $\mathbf{Y}_t^{CM}$ and $\mathbf{Y}_s^{CM}$ to fuse the trending and seasonal prediction results. Finally, the residual technique is applied at the end of CM, which can be formulated as:

$$\mathbf{Y} = (\mathbf{Y}_s^{CM} + \mathbf{Y}_t^{CM}) + \text{Linear}(\mathbf{X}) \tag{8}$$

where the $\mathbf{X}$ is the original input series, Linear denotes the linear mapping process.

4 Experiments

We evaluate CMNet on seven popular multivariate long-term time series forecasting benchmark datasets. The statistics for these datasets are shown in Table 1. Given the underwhelming results of traditional methods like ARIMA and RNN-based approaches, we compare our model against state-of-the-art Transformer-based methods such as FEDformer [31], Nonstationary Transformer [16], Pyraformer [15], Autoformer [24], and Informer [30], as well as various variants of MLP-based existing methods such as Dlinear [28], TimesNet [23], and LightTS [29]. We also integrate the sequence stationarity module in the Nonstationary Transformer, which contains two stages: normalization and denormalization, and finally constitutes the normalization-CMNet-denormalization structure. All experiments are conducted three times using PyTorch and a single Nvidia GeForce RTX 3090 GPU.

Table 1. Statistical data from 7 benchmark datasets.

	ETTh1 ETTh2	ETTm1 ETTm2	Weather	Traffic	Exchange Rate
Variates(C)	7	7	21	862	8
Timesteps	17,420	69,680	52,696	17,544	7,588
Granularity	1 h	15 min	10 min	1 h	1 day

4.1 Main Results

For improving comparability, we adopt previous experimental configurations (FEDformer, Autoformer, etc.) with an input length T of 96 and a forecastion length $F \in \{96, 192, 336, 720\}$ for both training and evaluation. Table 2 indicates that our CMNet excels in long-term forecasting tasks. CMNet achieves state-of-the-art performance in 21 out of all cases on the complete long-term forecasting benchmark datasets, while also achieving the second-best performance in 11 cases.

Various Tracking Window Sizes. To gain a deeper understanding of CMNet's ability to extract sequence information, we conduct experiments on the Weather dataset. We assess $T \in \{96, 336, 480, 720\}$, using a sampling length that corresponds to 15% of the tracking window. The results are illustrated in Fig. 4. Notably, CMNet demonstrates optimal results when the tracking window size is set to 336.

Selection of Different SDS Parameters. This section delves into the performance evaluation of various SDS parameter choices. In the final version, we set the sampling number parameter k to 18. Furthermore, for comparative analysis, we choose a smaller k, as specified in Table 3. In general, augmenting the number of samples enables a more comprehensive comprehension of the sequence's distribution, ultimately leading to enhanced performance.

Table 2. Evaluation results for 7 long-term forecastion datasets with input length T = 96 and forecastion length $F \in \{96, 192, 336, 720\}$. Results are taken from [23,29] (for results from LightTS to Autoformer). The best number in each row is highlighted in red, and the second best number is in bold.

		CMNet		LightTS		Dlinear		TimesNet		FEDformer		Nonstationary Transformer		Autoformer	
		MSE	MAE	MSE	MAE	MSE	MAE	MSE	MAE	MSE	MAE	MSE	MAE	MSE	MAE
ETTh1	96	0.409	0.414	0.424	0.432	0.386	0.400	**0.384**	**0.402**	0.376	0.419	0.513	0.491	0.449	0.459
	192	0.459	0.444	0.475	0.462	0.437	**0.432**	**0.436**	0.429	0.420	0.448	0.534	0.504	0.500	0.482
	336	0.504	0.469	0.518	0.488	**0.481**	0.459	0.491	0.469	0.459	**0.465**	0.588	0.535	0.521	0.496
	720	**0.508**	0.490	0.547	0.533	0.519	0.516	0.521	**0.500**	0.506	0.507	0.643	0.616	0.514	0.512
ETTh2	96	0.312	0.359	0.397	0.437	**0.333**	0.387	0.340	**0.374**	0.358	0.397	0.476	0.458	0.346	0.388
	192	0.394	0.409	0.520	0.504	0.477	0.476	**0.402**	**0.414**	0.429	0.439	0.512	0.493	0.456	0.452
	336	0.437	0.444	0.626	0.559	0.594	0.541	**0.452**	**0.452**	0.496	0.487	0.552	0.551	0.482	0.486
	720	0.441	0.454	0.863	0.672	0.831	0.657	**0.462**	**0.468**	0.463	0.474	0.562	0.560	0.515	0.511
ETTm1	96	**0.345**	0.379	0.374	0.400	**0.345**	0.372	0.338	**0.375**	0.379	0.419	0.386	0.398	0.505	0.475
	192	0.382	0.397	0.400	0.407	**0.380**	**0.389**	0.374	0.387	0.426	0.441	0.459	0.444	0.553	0.496
	336	0.416	0.417	0.438	0.438	**0.413**	**0.413**	0.410	0.411	0.445	0.459	0.495	0.464	0.621	0.537
	720	**0.478**	**0.451**	0.527	0.502	0.474	0.453	**0.478**	0.450	0.543	0.490	0.585	0.516	0.671	0.561
ETTm2	96	0.179	0.262	0.209	0.308	0.193	0.292	**0.187**	**0.267**	0.203	0.287	0.192	0.274	0.255	0.339
	192	0.245	0.304	0.311	0.382	0.284	0.362	**0.249**	**0.309**	0.269	0.328	0.280	0.339	0.281	0.340
	336	0.308	0.346	0.442	0.466	0.369	0.427	**0.321**	**0.351**	0.325	0.366	0.334	0.361	0.339	0.372
	720	0.479	0.451	0.675	0.587	0.554	0.522	0.408	0.403	0.421	0.415	**0.417**	**0.413**	0.433	0.432
Weather	96	0.175	0.224	0.182	0.242	0.196	0.255	0.172	0.220	0.217	0.296	**0.173**	**0.223**	0.266	0.336
	192	**0.221**	**0.264**	0.227	0.287	0.237	0.296	0.219	0.261	0.276	0.336	0.245	0.285	0.307	0.367
	336	0.279	0.304	0.282	0.334	0.283	0.335	**0.280**	**0.306**	0.339	0.380	0.321	0.338	0.359	0.395
	720	**0.349**	0.350	0.352	0.386	0.345	0.381	0.365	**0.359**	0.403	0.428	0.414	0.410	0.419	0.428
ExchangeRate	96	**0.094**	0.216	0.116	0.262	0.088	**0.218**	0.107	0.234	0.148	0.278	0.111	0.237	0.197	0.323
	192	**0.182**	0.306	0.215	0.359	0.176	**0.315**	0.226	0.344	0.271	0.380	0.219	0.335	0.300	0.369
	336	**0.334**	0.422	0.377	0.466	0.313	**0.427**	0.367	0.448	0.460	0.500	0.421	0.476	0.509	0.524
	720	0.855	**0.699**	0.831	**0.699**	0.839	0.695	0.964	0.746	1.195	0.841	1.092	0.769	1.447	0.941
Traffic	96	0.674	0.446	0.615	0.391	0.650	0.396	**0.593**	0.321	0.587	0.366	0.612	**0.338**	0.613	0.388
	192	0.633	0.421	**0.601**	0.382	0.598	0.370	0.617	0.336	0.604	0.373	0.613	**0.340**	0.616	0.382
	336	0.631	0.414	**0.613**	0.386	0.605	0.373	0.629	**0.336**	0.621	0.383	0.618	0.328	0.622	0.337
	720	0.662	0.432	0.658	0.407	0.645	0.394	**0.640**	0.350	0.626	0.382	0.653	**0.355**	0.660	0.408
Count		21		1		11		16		6		1		0	

4.2 Showcase

To demonstrate the performance comparison among different models, we present the results obtained using the ETT dataset with a forecasting length of 192, as depicted in Fig. 5. In contrast, our CMNet achieves accurate forecasting of data mutations and exhibits superior capability in detecting variability in trend characteristics and abrupt data changes, regardless of their magnitude.

4.3 Comparative Efficiency

Table 4 presents a comparison between the prevailing linear model-based approach and the Transformer-based approach. It is evident that although FEDformer boasts a superior structural design, it introduces additional design elements that lead to the slowest training times. Our CMNet has the same training time as LightTS, but in terms of performance, CMNet emerges as the clear victor. When compared to the latest popular TimesNet, CMNet exhibits superior performance in memory usage, time efficiency, and parameter count, while also maintaining a slight lead in overall performance.

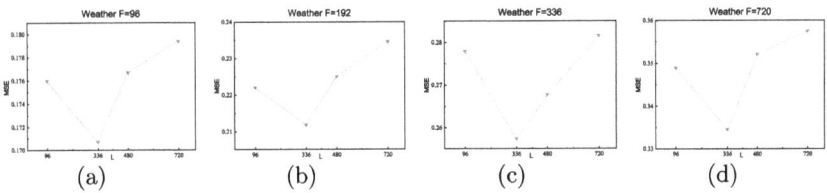

Fig. 4. Performance comparison of Weather datasets over various tracking window $T \in \{96, 336, 480, 720\}$.

Table 3. Comparative analysis of performance across different values of the sampling times parameter k. CMNet-SDS-9 has reduced the parameters k to only 9, which is half of the original CMNet. The optimal results in each row are indicated in bold.

		CMNet		CMNet-SDS-9	
		MSE	MAE	MSE	MAE
ETTh1	96	**0.409**	**0.414**	0.411	0.416
	192	**0.459**	**0.444**	0.460	0.445
	336	0.504	0.469	**0.503**	**0.467**
	720	**0.508**	**0.490**	0.513	0.933
Exchange-Rate	96	**0.094**	**0.216**	0.097	0.222
	192	**0.182**	**0.306**	0.183	0.307
	336	**0.334**	**0.422**	0.357	0.437
	720	**0.855**	**0.699**	0.882	0.711

Table 4. Comparison of the efficiency of different methods on the ETTh1 dataset with an input length of 96 and a forecasting range of 720 steps. The results are averaged over 3 runs.

	CMNet	LightTS	Dlinear	TimesNet	FEDformer
GPU Memory Usage(MiB)	1474	1122	1118	3668	4494
Time(s)	188.4	182.4	139.0	434.2	1777.0
Parameter(M)	0.18	0.13	0.02	0.58	16.05

4.4 Ablation Studies

In this section, we perform ablation experiments to validate the efficacy of the SDS and the CM. We evaluate two ablation variants of CMNet: CMNet V1, which retains only the MLP structure for residual connections and excludes the SDS and the CM; and CMNet V2, which utilizes the SDS and the CM while removing the convolution component in the CM. According to the results in Table 5, CMNet V1 exhibits the poorest performance. While CMNet V2 shows similar performance to CMNet in the ETTm1 dataset, CMNet outperforms the other models in the ETTm2 dataset. Thus, these results bolster the validity of the CM and SDS designs.

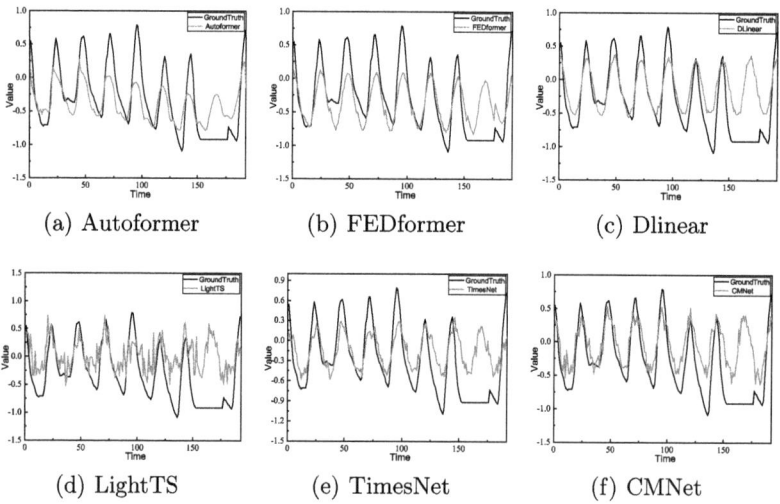

(a) Autoformer (b) FEDformer (c) Dlinear

(d) LightTS (e) TimesNet (f) CMNet

Fig. 5. The case of forecastion length 192 for the multivariate ETTh2 dataset under Autoformer, FEDformer, Dlinear, LightTS, TimesNet, and CMNet.

Table 5. Ablation study. The best results are shown in bold.

		CMNet v1		CMNet v2		CMNet	
		MSE	MAE	MSE	MAE	MSE	MAE
ETTm1	96	0.382	0.391	0.347	**0.375**	**0.345**	0.379
	192	0.411	0.404	0.388	**0.392**	**0.382**	0.397
	336	0.440	0.423	**0.413**	**0.410**	0.416	**0.417**
	720	0.494	0.454	0.479	**0.443**	**0.478**	0.451
ETTm2	96	0.187	0.270	0.183	0.267	**0.179**	**0.262**
	192	0.250	0.309	0.246	0.307	**0.245**	**0.304**
	336	0.311	0.347	0.311	**0.346**	**0.308**	**0.346**
	720	0.494	0.454	**0.479**	**0.443**	**0.479**	0.451

5 Conclusions

We introduce CMNet, a novel architectural network designed for time series forecasting tasks, consisting of a Stochastic Downsampling strategy(SDS) and a hybrid Convolution-MLP(CM). SDS samplings subsequences in different time spans to enhance the time series representation. CM utilizes the convolution operations to capture local information and MLP to global features along time and channel dimensions based on enhanced representations. CMNet outperforms state-of-the-art methods on five out of seven widely used benchmark datasets with less parameters and faster inference speed, demonstrating its value in time forecasting tasks and showcasing significant potential, which paves the way for the development of more competitive and innovative network architectures in the future.

Acknowledgments. This work was supported by the Innovative Development Project of Shihezi University under Grant (CXFZ202101).

References

1. Anderson, O.D., Box, G.E.P., Jenkins, G.M.: Time series analysis: forecasting and control. The Statistician, p. 265 (1978)
2. Bai, S., Kolter, J.Z., Koltun, V.: An empirical evaluation of generic convolutional and recurrent networks for sequence modeling. arXiv preprint arXiv:1803.01271 (2018)
3. Borovykh, A., Bohte, S., Oosterlee, C.W.: Conditional time series forecasting with convolutional neural networks. arXiv preprint arXiv:1703.04691 (2017)
4. Cao, D., et al.: Spectral temporal graph neural network for multivariate time-series forecasting. Adv. Neural. Inf. Process. Syst. **33**, 17766–17778 (2020)
5. Devlin, J., Chang, M.W., Lee, K., Toutanova, K.: BERT: pre-training of deep bidirectional transformers for language understanding. arXiv preprint arXiv:1810.04805 (2018)

6. Du, Y., et al.: Adarnn: adaptive learning and forecasting of time series. In: Proceedings of the 30th ACM International Conference on Information & Knowledge Management, pp. 402–411 (2021)
7. Giannone, D., Reichlin, L., Small, D.: Nowcasting: the real-time informational content of macroeconomic data. J. Monet. Econ. **55**(4), 665–676 (2008)
8. Koprinska, I., Wu, D., Wang, Z.: Convolutional neural networks for energy time series forecasting. In: 2018 International Joint Conference on Neural Networks (IJCNN), pp. 1–8 (2018)
9. Lai, G., Chang, W.C., Yang, Y., Liu, H.: Modeling long-and short-term temporal patterns with deep neural networks. In: The 41st International ACM SIGIR Conference on Research & Development in Information Retrieval, pp. 95–104 (2018)
10. Lara-Benítez, P., Carranza-García, M., Luna-Romera, J.M., Riquelme, J.C.: Temporal convolutional networks applied to energy-related time series forecasting. Appl. Sci. **10**(7), 2322 (2020)
11. Li, S., et al.: Enhancing the locality and breaking the memory bottleneck of transformer on time series forecasting. In: Advances in Neural Information Processing Systems, vol. 32 (2019)
12. Lim, B., Zohren, S.: Time-series forecasting with deep learning: a survey. Phil. Trans. R. Soc. A **379**(2194), 20200209 (2021)
13. Lim, B., Zohren, S., Roberts, S.: Recurrent neural filters: learning independent Bayesian filtering steps for time series prediction. In: 2020 International Joint Conference on Neural Networks (IJCNN), pp. 1–8. IEEE (2020)
14. Lin, H., Gao, Z., Xu, Y., Wu, L., Li, L., Li, S.Z.: Conditional local convolution for spatio-temporal meteorological forecasting. In: Proceedings of the AAAI Conference on Artificial Intelligence, vol. 36, pp. 7470–7478 (2022)
15. Liu, S., et al.: Pyraformer: low-complexity pyramidal attention for long-range time series modeling and forecasting. In: International Conference on Learning Representations (2021)
16. Liu, Y., Wu, H., Wang, J., Long, M.: Non-stationary transformers: exploring the stationarity in time series forecasting. Adv. Neural. Inf. Process. Syst. **35**, 9881–9893 (2022)
17. Salinas, D., Bohlke-Schneider, M., Callot, L., Medico, R., Gasthaus, J.: High-dimensional multivariate forecasting with low-rank Gaussian copula processes. In: Advances in Neural Information Processing Systems, vol. 32 (2019)
18. Salinas, D., Flunkert, V., Gasthaus, J., Januschowski, T.: Deepar: probabilistic forecasting with autoregressive recurrent networks. Int. J. Forecast. **36**(3), 1181–1191 (2020)
19. Sen, R., Yu, H.F., Dhillon, I.S.: Think globally, act locally: a deep neural network approach to high-dimensional time series forecasting. In: Advances in Neural Information Processing Systems, vol. 32 (2019)
20. Shen, L., Wang, Y.: TCCT: tightly-coupled convolutional transformer on time series forecasting. Neurocomputing **480**, 131–145 (2022)
21. Vaswani, A., et al.: Attention is all you need. In: Advances in Neural Information Processing Systems, vol. 30 (2017)
22. Wibawa, A.P., Utama, A.B.P., Elmunsyah, H., Pujianto, U., Dwiyanto, F.A., Hernandez, L.: Time-series analysis with smoothed convolutional neural network. J. Big Data **9**(1), 44 (2022)
23. Wu, H., Hu, T., Liu, Y., Zhou, H., Wang, J., Long, M.: Timesnet: temporal 2D-variation modeling for general time series analysis. arXiv preprint arXiv:2210.02186 (2022)

24. Wu, H., Xu, J., Wang, J., Long, M.: Autoformer: decomposition transformers with auto-correlation for long-term series forecasting. Adv. Neural. Inf. Process. Syst. **34**, 22419–22430 (2021)
25. Wu, Z., Pan, S., Long, G., Jiang, J., Chang, X., Zhang, C.: Connecting the dots: multivariate time series forecasting with graph neural networks. In: Proceedings of the 26th ACM SIGKDD International Conference on Knowledge Discovery & Data Mining, pp. 753–763 (2020)
26. Xiao, X., Liu, J., Liu, D., Tang, Y., Zhang, F.: Condition monitoring of wind turbine main bearing based on multivariate time series forecasting. Energies **15**(5), 1951 (2022)
27. Xu, W., Liu, W., Bian, J., Yin, J., Liu, T.Y.: Instance-wise graph-based framework for multivariate time series forecasting. arXiv preprint arXiv:2109.06489 (2021)
28. Zeng, A., Chen, M., Zhang, L., Xu, Q.: Are transformers effective for time series forecasting? In: Proceedings of the AAAI Conference on Artificial Intelligence, vol. 37, pp. 11121–11128 (2023)
29. Zhang, T., et al.: Less is more: fast multivariate time series forecasting with light sampling-oriented MLP structures. arXiv preprint arXiv:2207.01186 (2022)
30. Zhou, H., et al.: Informer: beyond efficient transformer for long sequence time-series forecasting. In: Proceedings of the AAAI Conference on Artificial Intelligence, vol. 35, pp. 11106–11115 (2021)
31. Zhou, T., Ma, Z., Wen, Q., Wang, X., Sun, L., Jin, R.: Fedformer: frequency enhanced decomposed transformer for long-term series forecasting. In: International Conference on Machine Learning, pp. 27268–27286. PMLR (2022)

A Domain Adaptive Based Reinforcement Learning Algorithm for Job Shop Scheduling Problems

Kuan Miao, Lilan Peng, Wuyang Zhang, Jibao Pan, Chongshou Li[✉], and Tianrui Li

School of Computing and Artificial Intelligence, Southwest Jiaotong University, Chengdu, Sichuan, China
{llpeng,zwyyyds}@my.swjtu.edu.cn, {lics,trli}@swjtu.edu.cn

Abstract. The Job Shop Scheduling Problem (JSSP) is a well-known combinatorial optimization problem in operations research, commonly found in fields like manufacturing and transportation. There are two main challenges: 1) a limited number of data instances, and 2) the inconsistent distributions in the scheduling process. This paper develops a domain adaptive-based reinforcement learning algorithm (DA-L2D) for job shop scheduling problems. In particular, it comprises two main parts: a domain adaptation module is designed to learn and comprehend the characteristics of data instance distribution, and an L2D model is employed to understand and manage the dynamic graph relationships. The combination of these two components enables the DA-L2D model to adeptly address intricate scheduling problems by mining the data's characteristics. We conduct experiments on the Taillard and DMU datasets. Experimental results demonstrate the superiority of the domain adaptation to the scheduling problems.

Keywords: Job Shop Scheduling Problem · Domain Adaptive · Reinforcement Learning

1 Introduction

Job Shop Scheduling Problem (JSSP) is a classic combinatorial optimization problem in operations research, widely present in industries such as manufacturing and transportation. The static JSSP problem consists of n jobs and m machines, where each job is composed of m operations that require processing in a specific order on different machines. By scheduling these operations effectively, the objective is to minimize the completion time. For instance, considering a job shop instance with 3 jobs and 3 machines, the scheduling results in the completion of all operations at time 14, as shown in Fig. 1. It can be observed that at time 14, all operations are completed, making the maximum completion time for this instance 14.

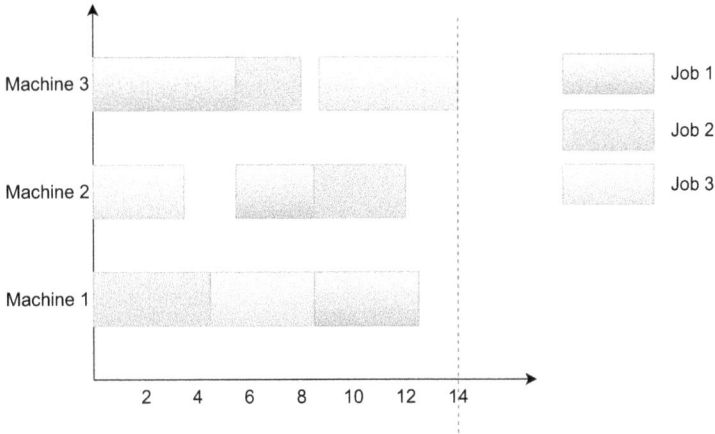

Fig. 1. JSSP Dispatch Example

In the past, to obtain higher-quality solutions, researchers have designed various priority dispatching rules (PDR) and attempted to solve them using heuristic algorithms. However, designing PDR rules and heuristic algorithms requires a significant amount of domain knowledge and repeated experiments. With the widespread use of reinforcement learning in routing problems, researchers have started utilizing reinforcement learning to solve JSSP. Zhang et al. [30] designed a reinforcement learning framework to obtain effective representations and learning mechanisms for JSSP. Following this idea, researchers have made further modifications and improvements such as model structure [15], graph-edge relationships [14], training data [6].

Although the scheduler has strong generalization ability, when facing a limited number of data instances, it often exhibits more suboptimal selection, leading to longer completion times. In the current research, most studies validate algorithm performance through publicly available datasets, which are limited in number and have a fixed data distribution. Therefore, even though a scheduler model with strong generalization ability can achieve lower completion times on these public datasets, a significant amount of suboptimal choices due to inconsistent distributions still exist in the scheduling process, limiting the performance of the scheduler.

In this paper, we develop a domain adaptation reinforcement learning algorithm L2D to automatically learn scheduling rules and distributions for JSSP. Specifically, we design a size-independent policy network with a graph neural network with learning scheduling rules and a domain adaptation module to capture distribution variances. The contributions of this work are listed as follows:

1. We develop a domain adaptation module to learn the distribution characteristics of the training and testing data for optimizing the model's training process. Its design is independent of graph edge relationship and thus can be combined with other algorithms for JSSP scheduling.

2. We integrate GRU networks and attention networks for the domain adaptation component. By including inter-domain variations in the loss function during training, we refine the model parameters successfully. This allows the algorithm to explore a wider variety of dynamic probability distributions and enhance the stability of the training process.
3. We conducted evaluations by combining the adaptive module with the popular neural architecture model L2D, and the experiments conducted on the Taillard and DMU datasets effectively validated the effectiveness of the domain adaptation module.

2 Related Work

2.1 Job Shop Scheduling Problems

In Job Shop Scheduling, schedulers arrange schedules based on given operation times and machines, often solving similar problems repeatedly when data changes but the structure remains the same. Khalil et al. [7] addressed this by employing graph neural networks and reinforcement learning for problems like minimum vertex cover and maximum cut. Zhang et al. [30] proposed Graph Isomorphism Networks [25] and the PPO algorithm [18] to treat operations as nodes in a disjunctive graph, achieving promising results. Park et al. [15] solved the sparse state matrix issue in job shop scheduling by constructing state representations based on statistical features. Additionally, they introduced delayed rewards based on the maximum completion time property in job shop scheduling instances, leveraging a graph attention network to build a strategy and achieve superior results [14].

While many studies utilize graph neural networks as embedding layers, some researchers employ other networks. Chen et al. [1] combined attention mechanisms and exclusive graph embeddings to learn long-distance dependencies, validating their method on large-scale scheduling problems. Monaci et al. [12] proposed a dual-layer LSTM network [5] to handle task variability. To deal with training difficulties with large models, researchers train on smaller instances before testing on larger ones. Puiseau et al. [17] exploited Curriculum Learning to sort training instances, improving model convergence and solution quality. Iklassov et al. [6] developed an adaptive step-by-step curriculum learning strategy, adjusting difficulty levels dynamically for better performance. Zeng et al. [29] used eight PDR rules as actions for scheduling operations based on strategies. Ni et al. [13] combined reinforcement learning with the Iterated Greedy algorithm, generating initial solutions with heuristic searches and iteratively refining them. Tassel et al. [20] constructed a JSSP environment and used online training for reinforcement learning scheduling, achieving results close to Google's OR-Tools [16].

Despite improvements to the training process, a significant gap remains between the maximum completion time and the optimal value due to distribution differences between training and testing data, leading to suboptimal selection behavior. Thus, how to reduce the second-best choice has emerged as a crucial issue among researchers.

2.2 Domain Adaptation

The domain adaptation strategy based on differences leverages fine-tuning to minimize the transfer between small datasets. Yosinski et al. [28] proved that due to fragile co-adaptation and representation specificity, the transferable features learned by deep neural networks are limited. Tzeng et al. [23] integrated adaptation layers and Maximum Mean Discrepancy (MMD) to measure domain confusion loss to learn domain-invariant representations. Long et al. [10] assumed distributional invariance by multi-kernel MMD instead of traditional MMD and selecting multiple fully connected layers to reduce differences. Long et al. [11] aligned changes in the joint distribution of input features and output labels in multiple specific domain layers based on the Joint Maximum Mean Discrepancy criterion. Hinton et al. [4] employed hard-coded pre-trained large models to compute soft labels and designed the soft label loss function to the small model, thereby obtaining most of the information for learning the function. Tzeng et al. [21] developed a shared network for feature extraction, separately using domain confusion loss and classification loss to measure intra-domain distance and inter-domain distance. Yan et al. [26] believed that domain adaptation based on MMD ignores changes in class prior distributions, thus proposing a weighted MMD model.

Based on adversarial deep domain adaptation methods, Tzeng et al. [22] proposed a unified framework based on adversarial methods. By utilizing the strategy of generative models with the help of source data, it is possible to generate an infinite amount of synthetic target data using a generator. The core idea of CoGAN [9] was to generate synthetic target data paired with synthetic source data. To enhance representation learning in the target domain, Liang et al. [8] assumed that the parameters of the distributor could be shared between the source and target domains, and presented a novel two-stage self-supervised pseudo-labeling strategy. Yao et al. [27] introduced a discriminative manifold distribution alignment method that enhances feature transferability by aligning global and local distributions. Cui et al. [2] designed a progressive domain adaptation method that reduces the transfer difficulty by equipping gradually disappearing bridges on the generator and discriminator as intermediate layers. Xie et al. [24] simultaneously considered cross-domain global distribution statistics and cross-sample local semantic features, proposing a collaborative alignment framework.

3 Problem Definition

The JSSP is a classical combinatorial optimization problem which is defined as follows: Given a set of workpieces $J = 1, 2, \cdots, n$ to be machined on a set of machines $M = 1, 2, 3, \cdots, m$. Each workpiece consists of a sequence of operations, each of which is processed on a different machine. In addition, these operations must be composed in a specific order and processed on a specific machine for a specified time. The objective of the job shop scheduling problem is to find a scheduling solution that minimizes the completion time of all operations. $\min C_{max}$ represents the overall objective, i.e., to minimize the maximum completion time. $C_{ijk} - p_{ijk} \leq C_{i'j'k}$ or $C_{ijk} \leq C_{i'j'k}$ $\forall i! = i'$ is the machine constraint, which means that only one workpiece can be processed by the same machine at the same time, and $C_{\max} \geq C_{ijk}$ indicates that the completion time of all workpieces is the completion time of the last workpiece. $C_{ij} + p_{ij} \leq C_{i(j+1)}$ indicates that on a single workpiece, the processes need to be processed sequentially. $i, i' \in \{1, \cdots, n\}, j, j' \in \{1, \cdots, m\}, k \in \{1, \cdots, m\}$ is the range of variables. where p_{ijk} denotes the processing time of the jth process of the workpiece J_i on the machine k, n denotes the number of workpieces, and m is the number of machines/processes. C_{ijk} denotes the completion time of the jth process of the workpiece J_i on the machine k, and C_{max} is the final completion time of the whole instance.

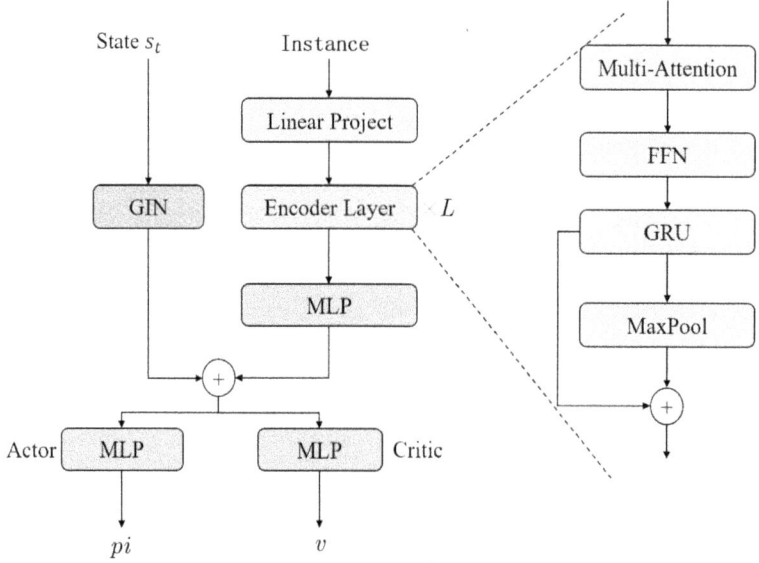

Fig. 2. DA-L2D Model

4 DA-L2D

The framework of proposed model DA-L2D is shown in Fig. 2. The left part of the model is the L2D model, which inputs the state of the time s_t into the graph isomorphism network for feature encoding, and in the right part, it is the DA model, which maps the input instances into vectors. After that, we combine the DA and L2D and deliver into the layers *Actor* and *Critic* of the multilayer perceptual, respectively, to obtain the distribution probabilities of the actions pi and the evaluation v of $<s_t, a_t>$.

4.1 L2D Model

The state-dependent disjunctive graph $G(t)$ is a hybrid graph with directed arcs that describe priority constraints on the machine and machining sequential constraints between workpieces. Graph isomorphism networks are utilized here to identify and encode graph structure. Graph isomorphism networks are initially undirected graphs, so two directed arcs with connections in exactly opposite directions are designed to replace the undirected arcs of $G(t)$, resulting in a fully directed graph denoted as $G_D(t)$. Furthermore, for iterative updating in a GIN, the neighbourhood of the nodes involved can be expressed as Eq. (1), where $E(\cdot)$ is the set of arcs of the graph, and $N(v)$ is the set of arcs that contains v.

$$N(v) = \{u|(u,v) \in E(G_D(t))\} \tag{1}$$

During the scheduling process, the model takes the approach of gradually adding two undirected arcs connected in opposite directions to represent the graph edge relationship $G_D(t)$ during the scheduling process. As the scheduling process proceeds, $G_D(t)$ becomes larger and eventually becomes a fully directed graph. Figure 3(a) is a complete disjunctive graph, when the instance scheduling has been completed to obtain a complete solution, a similar graph structure can be obtained. Figure 3(b) is a partial graph structure, the current scheduling has been completed for the two processes of O_{11} and O_{31}, and the next process to be scheduled for the O_{32}, at this time, O_{11} has been completed scheduling, and both of them are processed in the machine M_1, therefore, the establishment of two-way arcs for the two processes of O_{11} and O_{31}.

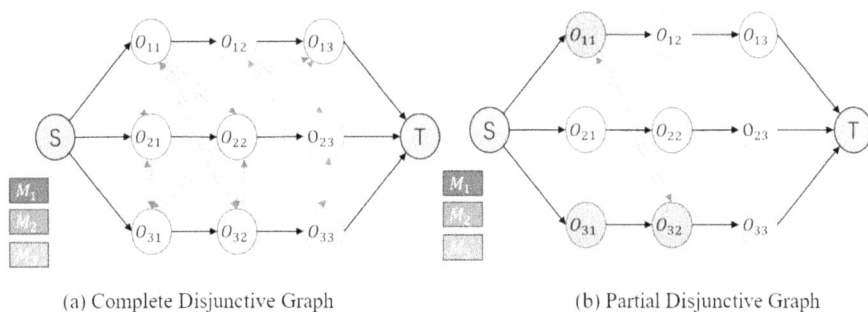

(a) Complete Disjunctive Graph   (b) Partial Disjunctive Graph

Fig. 3. The example of complete and partial disjunctive graph

4.2 DA Model

How to map the instances into vectors and fuse them into the training process of the L2D model is the key point of this model. The inputs to the DA module are the processing time of the instance and the corresponding processing machine, and connect them into a vector matrix of size $(b, L, 2)$. Where L is the number of processes. Inspired by the structure of the Transformer and the work [23], we integrate the multi-head attention network and GRU network to encode the instances, and then use the multi-layer perceptual to decode them and map them to the vector space for distance calculation. The vector x is dimensionally transformed through a linear layer to obtain the vector x^l, and x^l is the input of the multi-head self-attention layer, which captures the important and distributional information in the instances, and produces an output of $Q^l \in \mathbb{R}^{b \times L \times d}$, which is computed as follows. Q, K, and V are query vectors, key vectors, and value vectors, respectively, $head_i$ is the ith attention head, h denotes the number of attention heads, and W^o, W_i^Q, W_i^K and W_i^V are linear mapping matrices, which are responsible for the one-dimensional transformation of vectors.

$$Q^l = MultiHead(x^l, x^l, x^l) \tag{2}$$

$$head_i = Attention(QW_i^Q, KW_i^K, VW_i^V) \tag{3}$$

$$MultiHead(Q, K, V) = concat(head_1, head_2, \cdots, head_h) W^o \tag{4}$$

$$Attention(Q, K, V) = soft\max\left(\frac{QK^T}{\sqrt[2]{d_k}}\right) V \tag{5}$$

Considering the fact that only using the attention network does not fit the complex process well enough to represent the instance features effectively, the Feed Forward Network (FFN) layer is leveraged to further extract the data features. Q^l goes through the FFN layer to get vector Q^F.

$$Q^F = FFN(Q^l) \tag{6}$$

$$FFN(x) = W_2(relu(W_1 x + b_1)) + b_2 \tag{7}$$

$$relu(x) = \max(0, x) \tag{8}$$

After the FFN layer, the key information in the instance has been captured. To capture the distribution information, the vector Q^F is passed through the GRU layer and the vector Q^g is obtained. To further enhance the key features, we sum Q^g and the maximum pooling of Q^g. We can obtain the vector of the instance by multiple encoding layers and the multilayer perceptron network. We set the vector of the generated data as Q_1 and the vector of the test data as Q_2, thus the distance of them is the inter-domain loss L_{domain}. The PPO algorithm loss is L_{ppo}, the overall loss is L.

$$L_{domain} = MMD(Q_1, Q_2) \tag{9}$$

$$MMD(X_S, X_T) = \left\| \frac{1}{|X_S|} \sum_{x_s \in X_s} \phi(x_s) - \frac{1}{|X_T|} \sum_{x_t \in X_T} \phi(x_t) \right\| \tag{10}$$

$$L = L_{ppo} + \varepsilon L_{domain} \tag{11}$$

5 Experiment

5.1 DataSet and Baseline

We conducted the experiment through the 70 instances (15×15, 20×15, 20×20, 30×15, 30×20, 50×15, 50×20) from Taillard dataset [19] and 40 instances (20×15, 20×20, 30×15, 30×20) from the DMU dataset [3]. Here, $n \times m$ means the number of n jobs and m is the number of machines.

To validate the experimental performance of the proposed models, we compared with six baselines, including OPT[1](the best-known completion time optimal value), four traditional PDR methods reported in [30] (Shortest Processing Time (SPT), Most Work Remaining (MWKR), Minimum ratio of Flow Due Date to Most Work Remaining (FDD/MWKR), Most Operations Remaining (MOR)), and L2D model [30].

5.2 Metric and Implementation

The objective of the JSSP problem is to minimize the maximum completion time. Therefore, this paper uses the maximum completion time of models or strategies as the evaluation metric.

We keep the parameters in the L2D module the same as in the original paper, with the following specifications: GIN has 3 layers, 4 environments, both linear layers in the GIN and the DA modules have 2 layers, hidden dimension is 32, 10000 update steps, CLIP loss function coefficient is 1, square loss function coefficient is 2, and entropy loss function coefficient is 0.01. In the DA module, the attention function dimension is 128, the number of heads is 4, and the coefficient for the encoding layer is 3. Additionally, the model's learning rate is $2e^{-5}$, using the Adam optimizer, and the scheduler selects StepLR, decaying by a factor of 0.9 every 2000 steps. For the parameters of the PPO algorithm, the discount factor γ is 1, training collects data for 3 epochs each time, the clipping parameter is 0.2, and the loss coefficient for domain adaptation is 10 (Table 1).

Table 1. Experimental Results of the Taillard Dataset

Methods	15×15	20×15	20×20	30×15	30×20	50×15	50×20
OPT	1228.0	1364.6	1588.0	1788.0	1948.0	2733.8	2843.9
SPT	1902.6	2253.6	2655.8	2888.4	3234.7	4194.7	4532.2
MWKR	1927.5	2653.8	2518.6	2728.0	3193.3	3907.8	4375.1
FDD/MWKR	1808.6	2054	2387.2	2590.8	3045	3736.3	4022.1
MOR	1782.3	2015.8	2309.9	2601.3	2888.1	3608.0	3920.0
L2D	1547.4	1774.7	2128.1	2378.8	2603.9	3393.8	3593.9
DA-L2D	**1497.5**	**1744.6**	**2086.1**	**2325.3**	**2579.3**	**3346.6**	**3524.8**

---

[1] http://jobshop.jjvh.nl/.

Table 2. Experimental Results of the DMU Dataset

Methods	20 × 15	20 × 20	30 × 15	30 × 20
OPT	3025.7	3473.0	3884.8	4257.0
SPT	4951.5	5690.5	6306.2	7036.0
MWKR	4909.9	5489.0	6252.9	6925.0
FDD/MWKR	4666.3	5298.2	6016.5	6827.3
MOR	4513.2	5052.3	5742.8	6491.9
L2D	4215.3	4804.5	5557.9	5967.4
DA-L2D	**4079.1**	**4695.3**	**5433.9**	**5766.1**

5.3 Experimental Result

Experimental results of the DA-L2D model on the Taillard and DMU datasets are shown in Tables 2 and 3, respectively. The DA-L2D model outperforms traditional priority scheduling rules and the L2D model in all instances in two datasets. In particular, the comparison of time results with the L2D model validates the effectiveness of the DA-L2D model in reducing inter-domain losses and minimizing suboptimal selection. It is worth noting that the performance improvement of the DA-L2D model on the DMU dataset surpasses that on the Taillard dataset. This difference may be attributed to the greater disparity between the distribution of the DMU dataset and the training dataset, where the data distribution mapping capability of the domain adaptation module proves to be more effective in assisting the scheduler in making more appropriate scheduling decisions. These experimental results demonstrate that domain adaptation strategies can effectively enhance the performance of the model.

5.4 Parameter Analysis

This section elaborates on two hyperparameters: the number of layers in the encoder of the DA model and the ratio of inter-domain distances on Taillard dataset.

The Layers of Encoder. In the DA model, multiple encoding layers are designed to extract features from instances. The number of stacked layers is an important parameter in the DA-L2D model, which has a significant impact on the total number of model parameters and the model's ability to represent domain invariance. To observe the influence of encoding layers on the overall model's experimental results, five sets of experiments with encoding layer numbers 1, 2, 3, 4, and 5 were designed, as shown in Fig. 4. From the figure, it can be seen that the model performs best when the number of encoding layers is 3; therefore, 3 is the final choice.

Inter-domain Loss Ratio. In domain adaptation models, domain loss is leveraged to measure the distance between the generated data and the target data set,

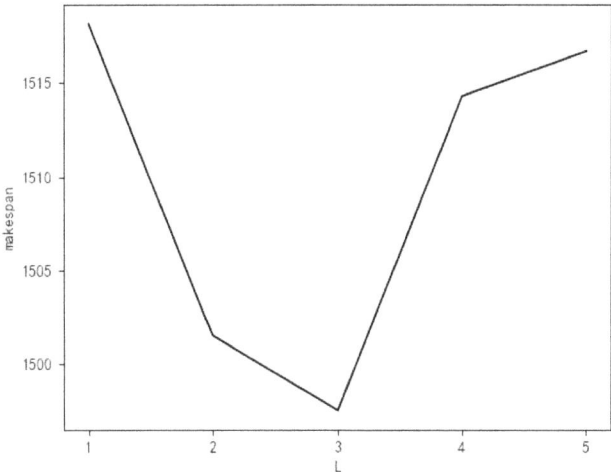

Fig. 4. Comparison Experiment on the Number of Stacked Layers in the DA Model

and the ratio is used to measure the relative importance of domain loss to PPO loss. If the ratio is too small, it affects the convergence of the domain adaptation model and its ability to learn domain-invariant knowledge. If the ratio is too large, it affects the search ability of the original model. Therefore, this section verifies the impact of 6 different ratios ε on the model's convergence ability. The experimental results are shown in Fig. 5.

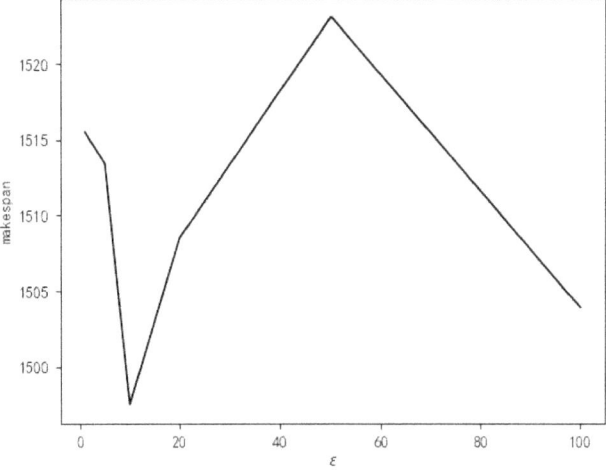

Fig. 5. Inter-domain loss ratio comparison experiment

6 Conclusion

Based on the concept of domain adaptation, this paper developed a new scheduling model called DA-L2D. This model effectively improved the performance and efficiency of scheduler decision-making by integrating instance distribution information. DA-L2D consisted of two parts: a domain adaptation module responsible for learning and understanding the characteristics of data instance distribution, and an L2D model was employed to learn and handle dynamic graph relationships. The combination of these two parts allowed the DA-L2D model to effectively handle complex scheduling problems while understanding the characteristics of the data. To further enhance the model's performance and generalization ability, this paper introduced the Maximum Mean Discrepancy distance during the training process, adding it to the loss function at a certain ratio to help the model better adapt to different data distributions and improve generalization. Through experiments on the Taillard and DMU datasets, this paper validated the effectiveness and feasibility of the DA-L2D model. The experimental results demonstrated that the domain adaptation method could effectively enhance the performance of scheduling models.

Acknowledgement. This research was supported by the National Natural Science Foundation of China (Nos. 62202395, 62176221, 62276215, 62276218), Sichuan Science and Technology Program (No. MZGC20230073).

References

1. Chen, R., Li, W., Yang, H.: A deep reinforcement learning framework based on an attention mechanism and disjunctive graph embedding for the job-shop scheduling problem. IEEE Trans. Ind. Inf. **19**(2), 1322–1331 (2022)
2. Cui, S., Wang, S., Zhuo, J., Su, C., Huang, Q., Tian, Q.: Gradually vanishing bridge for adversarial domain adaptation. In: Proceedings of the IEEE Conference on Computer Vision and Pattern Recognition, pp. 12455–12464 (2020)
3. Demirkol, E., Mehta, S., Uzsoy, R.: Benchmarks for shop scheduling problems. Eur. J. Oper. Res. **109**(1), 137–141 (1998)
4. Hinton, G., Vinyals, O., Dean, J.: Distilling the knowledge in a neural network. arXiv preprint arXiv:1503.02531 (2015)
5. Hochreiter, S., Schmidhuber, J.: Long short-term memory. Neural Comput. **9**(8), 1735–1780 (1997)
6. Iklassov, Z., Medvedev, D., Solozabal, R., Takac, M.: Learning to generalize dispatching rules on the job shop scheduling. arXiv preprint arXiv:2206.04423 (2022)
7. Khalil, E., Dai, H., Zhang, Y., Dilkina, B., Song, L.: Learning combinatorial optimization algorithms over graphs, vol. 30 (2017)
8. Liang, J., Hu, D., Feng, J.: Do we really need to access the source data? Source hypothesis transfer for unsupervised domain adaptation. In: International Conference on Machine Learning, pp. 6028–6039. PMLR (2020)
9. Liu, M.Y., Tuzel, O.: Coupled generative adversarial networks, vol. 29 (2016)

10. Long, M., Cao, Y., Wang, J., Jordan, M.: Learning transferable features with deep adaptation networks. In: International Conference on Machine Learning, pp. 97–105. PMLR (2015)
11. Long, M., Zhu, H., Wang, J., Jordan, M.I.: Deep transfer learning with joint adaptation networks. In: International Conference on Machine Learning, pp. 2208–2217. PMLR (2017)
12. Monaci, M., Agasucci, V., Grani, G.: An actor-critic algorithm with deep double recurrent agents to solve the job shop scheduling problem. arXiv e-prints. pp. arXiv–2110 (2021)
13. Ni, F., et al.: A multi-graph attributed reinforcement learning based optimization algorithm for large-scale hybrid flow shop scheduling problem. In: Proceedings of the 27th ACM SIGKDD Conference on Knowledge Discovery & Data Mining, pp. 3441–3451 (2021)
14. Park, J., Bakhtiyar, S., Park, J.: Schedulenet: learn to solve multi-agent scheduling problems with reinforcement learning. arXiv preprint arXiv:2106.03051 (2021)
15. Park, J., Chun, J., Kim, S.H., Kim, Y., Park, J.: Learning to schedule job-shop problems: representation and policy learning using graph neural network and reinforcement learning. Int. J. Prod. Res. **59**(11), 3360–3377 (2021)
16. Perron, L., Furnon, V.: Or-tools. https://developers.google.com/optimization/
17. de Puiseau, C.W., Tercan, H., Meisen, T.: Curriculum learning in job shop scheduling using reinforcement learning. arXiv preprint arXiv:2305.10192 (2023)
18. Schulman, J., Wolski, F., Dhariwal, P., Radford, A., Klimov, O.: Proximal policy optimization algorithms. arXiv preprint arXiv:1707.06347 (2017)
19. Taillard, E.: Benchmarks for basic scheduling problems. Eur. J. Oper. Res. **64**(2), 278–285 (1993)
20. Tassel, P., Gebser, M., Schekotihin, K.: A reinforcement learning environment for job-shop scheduling. arXiv preprint arXiv:2104.03760 (2021)
21. Tzeng, E., Hoffman, J., Darrell, T., Saenko, K.: Simultaneous deep transfer across domains and tasks. In: Proceedings of the IEEE International Conference on Computer Vision, pp. 4068–4076 (2015)
22. Tzeng, E., Hoffman, J., Saenko, K., Darrell, T.: Adversarial discriminative domain adaptation. In: Proceedings of the IEEE Conference on Computer Vision and Pattern Recognition, pp. 7167–7176 (2017)
23. Tzeng, E., Hoffman, J., Zhang, N., Saenko, K., Darrell, T.: Deep domain confusion: maximizing for domain invariance. arXiv preprint arXiv:1412.3474 (2014)
24. Xie, B., Li, S., Lv, F., Liu, C.H., Wang, G., Wu, D.: A collaborative alignment framework of transferable knowledge extraction for unsupervised domain adaptation. IEEE Trans. Knowl. Data Eng. (2022)
25. Xu, K., Hu, W., Leskovec, J., Jegelka, S.: How powerful are graph neural networks? (2019)
26. Yan, H., Ding, Y., Li, P., Wang, Q., Xu, Y., Zuo, W.: Mind the class weight bias: weighted maximum mean discrepancy for unsupervised domain adaptation. In: Proceedings of the IEEE Conference on Computer Vision and Pattern Recognition, pp. 2272–2281 (2017)
27. Yao, S., Kang, Q., Zhou, M., Rawa, M.J., Albeshri, A.: Discriminative manifold distribution alignment for domain adaptation. IEEE Trans. Syst. Man Cybern. Syst. **53**(2), 1183–1197 (2022)
28. Yosinski, J., Clune, J., Bengio, Y., Lipson, H.: How transferable are features in deep neural networks? vol. 27 (2014)

29. Zeng, Y., Liao, Z., Dai, Y., Wang, R., Li, X., Yuan, B.: Hybrid intelligence for dynamic job-shop scheduling with deep reinforcement learning and attention mechanism. arXiv preprint arXiv:2201.00548 (2022)
30. Zhang, C., Song, W., Cao, Z., Zhang, J., Tan, P.S., Chi, X.: Learning to dispatch for job shop scheduling via deep reinforcement learning, vol. 33, pp. 1621–1632 (2020)

Large-Scale Data Generation Using SWLBM on the Sunway TaihuLight Supercomputer and Subsequent Data Mining with Physics-Informed Neural Networks

Xuesen Chu[1,2,3(✉)] 🆔, Wei Guo[1,2], Tianqi Wu[1,2], Shengze Cai[4], and Guangwen Yang[3,5]

[1] China Ship Scientific Research Center, Wuxi 214082, China
chuxs@cssrc.com.cn
[2] Taihu Lake Laboratory of Deep Sea Technology and Science, Wuxi 214082, China
[3] Department of Computer Science and Technology, Tsinghua University, Beijing 100084, China
[4] College of Control Science and Engineering, Zhejiang University, Hangzhou 310027, China
[5] National Supercomputing Center in Wuxi, Wuxi 214072, China

Abstract. Generating large-scale datasets on supercomputers is a critical component of modern research, enabling sophisticated analysis and mining applications. In the context of SUBOFF model studies for advanced submarine development, we emphasize the significance of supercomputer-generated big data and its subsequent application. Utilizing the Sunway TaihuLight supercomputer and the SWLBM software, we employ the lattice Boltzmann method (LBM) to generate a comprehensive dataset of flows over a SUBOFF model. This dataset is then used to train a physics-informed neural network (PINN) model, which aims to reconstruct flow fields from sparse velocity measurements. The PINN is designed to perform super-resolution velocity reconstruction and concurrent pressure field inference. Our results show that the reconstructed flow fields, including pressure, are in good agreement with the full-resolution LBM references, highlighting the promise of this approach for complex flow motion reconstruction. This research demonstrates the potential of supercomputer-generated big data in enhancing simulations and experiments in the study of SUBOFF and beyond.

Keywords: Physics-informed neural networks · Deep learning · Lattice Boltzmann Method · Data mining · Computational fluid dynamics

1 Introduction

In recent years, the role of supercomputers in data generation and mining has become increasingly prominent, especially in the field of AI for Science research. These powerful computing devices not only tackle problems that are beyond the capabilities of traditional computer systems but also play a crucial role in data analysis and mining. For instance, AI supercomputers, through virtualization and distributed acceleration training, can effectively support the training of massive-scale AI models such as DALLE2 and CLIP, which involve multimodal heterogeneous data like images, text, and speech. These technologies have broad application prospects in fields such as meteorological forecasting, new material development, and molecular drug design. Furthermore, supercomputers are instrumental in addressing problems in scientific computing, such as solving partial differential equations, which assist in planetary simulations and genetic analysis. The high-performance computing capabilities of supercomputers are essential for meeting these computational demands. With the increase in data volume, future AI supercomputers are expected to solve previously unsolvable problems, providing strong support for scientific research and applications.

Computational Fluid Dynamics (CFD) has long been a pivotal application for supercomputers, particularly in the field of marine engineering. The use of supercomputers in CFD has enabled the simulation of complex fluid flow phenomena with unprecedented detail and accuracy, which is crucial for submarine design and performance prediction. Submarines, as a common marine equipment, have become promising in broad applications, including hydrographic and oceanographic exploration and military. In order to improve the capability of submarines, scientists and researchers mainly focus on the connection between the submarine geometry and the surrounding fluid dynamics, since the surrounding flow field is the dominant factor of the dynamic performance of submarines. As a typical geometry and the foundation for submarines, the SUBOFF model and its variants are well-recognized and frequently employed in the literature. Therefore, the accurate prediction of the flow fields over a SUBOFF model is generally the focus in the community, which can help to understand the hydrodynamics and optimize the parameters of SUBOFF model such as drag coefficient under different flow conditions [1–3].

During the past decades, CFD as been the go-to method for studying flow fields and analyzing SUBOFF motion mechanics. F The Reynolds-averaged Navier Stokes equations (RANS) are widely used for turbulence simulation due to their lower computational costs. However, these methods may fail to capture instantaneous turbulent fluctuations and can produce inaccurate results in the presence of no-slip boundaries and strong pressure gradients [4–6]; Large eddy simulation (LES) addresses some of these limitations by generating high-fidelity flow fields, but it remains computationally expensive and may not fully reflect real-world conditions [7–10]. Experimental technologies, such as flow visualization techniques, are essential for a more realistic understanding of submarine models, yet they often suffer from limited spatial resolution [11, 12].

In recent years, machine learning and deep learning have shown remarkable success in diverse physical applications, including fluid dynamics [13], ocean engineering [14], and building construction [15–17]. These advancements have inspired the use of neural networks and other machine learning methods to effectively solve complex problems.

Inspired by these advancements, this paper leverages the power of the Sunway TaihuLight supercomputer to conduct high-fidelity simulations of the flow over a SUBOFF model using the lattice Boltzmann method (LBM). The resulting dataset will be made public to serve as a benchmark for the community. We then apply a deep learning approach based on physics-informed neural networks (PINNs [18]) to reconstruct flow fields from sparse velocity measurements, mimicking PIV measurements from real experiments. Our objective is to infer fluid dynamics, such as velocity and pressure, from these low-resolution velocity fields, with the expectation that the PINNs approach will correct the pressure field.

The rest of the paper is organized as follows. The data used in the paper generated by SWLBM on Sunway TaihuLight Supercomputer is described in Sect. 2. The methodology based on PINNs for flow field reconstruction is explained in Sect. 3. Numerical experiments and results are given in Sect. 4, followed by a conclusion in Sect. 5.

2 Large-Scale Data Generation with SWLBM

2.1 SWLBM and Its Capacity for Generating Massive Data

SWLBM is a LBM based CFD framework designed for largescale flow simulation running on Sunway series supercomputer with Sunway manycore processors. The main motivation of this code developing is to produce an efficient tool to solve large-scale simulation problems within days in industrial area which acquire extreme super-size computing with Sunway Supercomputers. The scheme of LBM chosen in this work was the D3Q19 model which was introduced by Qian et al. [21].

Sunway TaihuLight supercomputer consiste with SW26010cmany-core processors. It has 260 heterogeneous cores divided by four core groups (CGs) which providing a peak performance of 3.06TFlops. As in Fig. 1 shows, each CG is composed of one management processing element (MPE) and 64 computing processing elements (CPEs).

The SWLBM code was optimized with multi-levels parallelization including MPI Athread and SIMD. The orchestrated strategy includes carefully designed domain decomposition consider with MPI load balance, data blocking in CG level with efficient data exchange scheme, thread-level data reuse, maximize the utilization of DMA (Direct Memory Access) bandwidth and 64k LDM, manual loop unroll and instructions reordering to exploit computational potential of the pipelines and 256-bit vectorization instructions of CPEs.

As the final result, SWLBM achieve high level parallel efficiency performance. The performance experiments were done up to the largest size with 5.6 trillion lattice cells running on 160,000 CGs and 10,400,000 cores, achieved a sustained performance 4.7 PFlops. Strong scaling and weak scaling were shown in Fig. 2

It achieves hundreds of times speedup compared to other CFD codes based on finite volume method such as OpenFOAM, allowing us to generate high-fidelity data with low computational cost. These factors enable SWLBM to possess ultra-high-resolution simulation capabilities, facilitating the generation of ultra-large-scale simulation data. Such as Wind field simulation with large areas as shown in Figs. 3.turbulence flow of submarine as shown in Fig. 4. More information about SWLBM could be found in reference [23, 24].

Different resolutions in simulations generate distinct data sets, with higher resolution capturing more physical information that is closer to reality. This is particularly crucial for leveraging generated data to supplement the physical world's data scarcity.

An example of a circular cylinder flow simulation at Re = 3900 illustrates the impact of resolution on the results. Three simulations were conducted for the same object, each

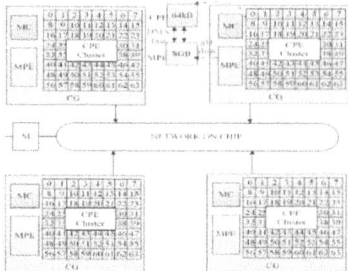

Fig. 1. Architecture of SW26010[22]

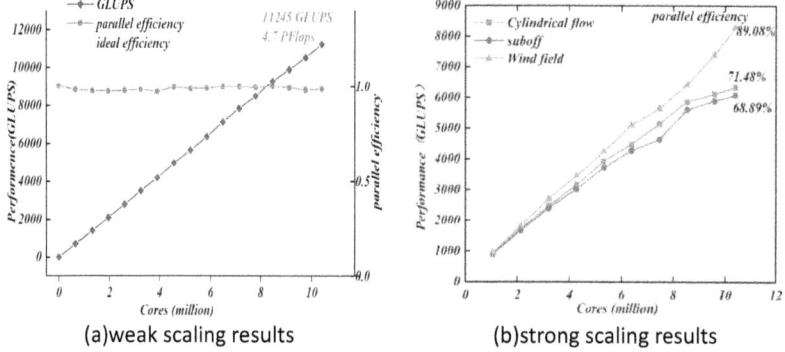

Fig. 2. Performance and parallel efficiency of weak scaling and strong scaling tests[20]

Fig. 3. Wind field simulation with large-scale region

employing different resolutions: $1000 \times 1000 \times 250$ and $2000 \times 2000 \times 500$. The final results are depicted in the figure below. The figure clearly shows that the high-resolution simulation results have more abundant vortex structure information. Data mining based on such high-fine data is more meaningful (Fig. 5).

Fig. 4. Turbulence flow of submarine

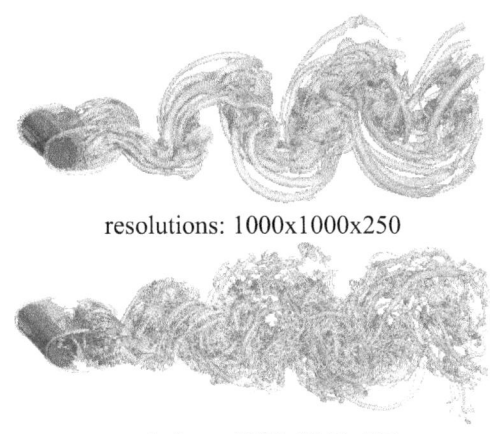

resolutions: 1000x1000x250

resolutions: 2000x2000x500

Fig. 5. Circular cylinder flow simulation at Re = 3900 with different resolutions

2.2 BigData Generated for Data Mining

For data mining study, the flow over a SUBOFF standard model placed at 1.25 m away from the inlet with a rectangular domain of 7.5 m × 1.5 m × 1.5 m with resolution of 2.5 mm was simulated, as shown in Fig. 1. The whole domain formed to 3000 × 600 × 600 consists of 1.08 billions lattices. The inlet and four lateral faces were defined with a free-stream velocity boundary condition of u = 3.23 m/s, v = 0, w = 0, with $Re_L = 1.4 \times 10^7$. The simulation was run with 1000 core groups of SW26010 processor on the Sunway Taihu Light supercomputer. The time step was set as 10^{-4} s. Each iteration costed about 0.06 s wall time (Fig. 6).

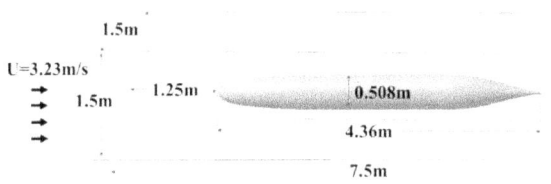

Fig. 6. Schematic diagram of simulation setting

The simulation results after 100,000 iterations where the flow converged was saved, the velocity and vorticity magnitudes in a 2D plane (the middle plane) are illustrated in Figs. 7(a) and 7(b), respectively, while the iso-surface of the Q-criterion (which is colored by the velocity magnitude) of the flow is demonstrated in Fig. 7(c). We note that the spatial resolution is high enough to resolve the small-scale structures of the flow. The details of the vortices structure can be observed in Fig. 7(c), indicating that the LBM simulation can provide high-fidelity flow fields. And performing 100,000 iterations takes only 1667 CG hours for calculation, with the cost being just a few hundred RMB. This demonstrates that SWLBM is a powerful tool for data generation.

But the challenge is that the data generated by LBM simulation occupy the computer memory of 64 Gb per snapshot, which is too large for post-processing treatment.

To address the challenge of processing large volumes of data, we first write the result data into a database, and then use SQL queries to extract the corresponding coarse-grained data from the database for further data minning.

In this study, we sample the flow fields with a spacing of 10 grids in the 3D domain from the original $3000 \times 600 \times 600$ data, resulting in a million-level dataset with a mesh of $300 \times 60 \times 60$. Such a dataset, which retains most of the details of the flow over the SUBOFF model, as shown in Fig. 8.

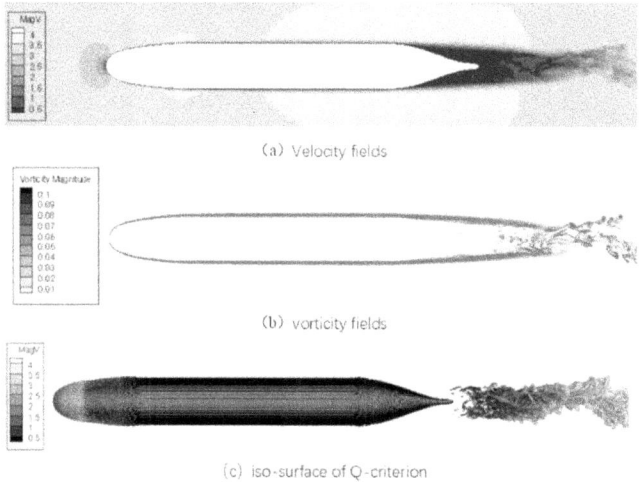

Fig. 7. The flow fields of LBM simulation after 100 000 iterations: (a) velocity magnitude, (b) vorticity magnitude, (c) iso-surface of the Q-criterion colored by the velocity magnitude.

To demonstrate the unsteady characteristic, we sample 100 snapshots with a time step of 10^{-3} s, corresponding to 10 iterations in simulation. The results of velocity and pressure are recorded at each snapshot, which are used to evaluate the reconstruction method in the following section.

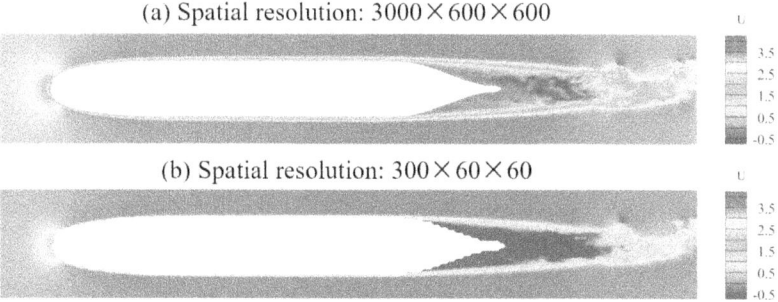

Fig. 8. The flow fields are down-sampled for testing the reconstruction method.

3 Flow Field Reconstruction Method with PINNs

In recent years, deep learning has been applied in fluid mechanic [25–27] for simulation [28, 29] and experiment [30–33]. In particular, the physicsi-informed neural networks (PINNs) have been proposed to solve partial differential equations (PDEs) [18, 34] and have become popular in many applications. By integrating physical governing equations in the loss function while training a neural network to approximate the solution of PDEs, PINNs are able to solve forward and inverse problems, especially to reconstruct the flow fields from partially known data [35, 36] and also to combine with LBM [37]. Inspired by the previous works, we introduce PINNs to reconstruction of the flow over the SUBOFF model. Following the context of PINNs [18], a fully-connected neural network is employed to approximate the solution of the NS equations, including velocity and pressure, by taking the spatial and temporal coordinates as inputs. For flow reconstruction problems, the output layer can be denoted to be the flow quantities (u, v, w, p), and the trainable parameters in the network are optimized by minimizing the loss function:

$$\mathcal{L}oss = \mathcal{L}_{NS} + \lambda \mathcal{L}_{data}, \quad (1)$$

with

$$\mathcal{L}_{NS} = \frac{1}{N_f} \sum_{i=1}^{N_f} \left| \frac{\partial u_i}{\partial t} + u_i \cdot \nabla u_i + \nabla p_i - \frac{1}{\text{Re}} \nabla^2 u_i \right|^2 + \frac{1}{N_f} \sum_{i=1}^{N_f} |\nabla \cdot u_i|^2 \quad (2)$$

and

$$\mathcal{L}_{data} = \frac{1}{N_d} \sum_{i=1}^{N_d} \left| u_i - u_i^{data} \right|^2, \quad (3)$$

where N_f and N_d represent the number of residual points and data points. $\mathcal{L}_{NS}$ penalizes the residual of the PDEs, while $\mathcal{L}_{data}$ represents the mismatch between the network output and the velocity measurement; λ is a weighting coefficient to balance different terms in the loss function and it can be unity. Notably, the automatic differentiation (AD) is directly applied to represent all the differential operators that exist in the PDEs.

As for the training process, we use the ADAM optimizer [38], which is an adaptive algorithm for gradient-based first-order optimization, to train the neural networks based on the above loss function. After training, the neural network can be used to predict the velocity and pressure at any given locations and time. We note that the PINNs method is different from those learning-based super-resolution methods [39], which are generally data driven, and requires that the input and output data are limited on a regular grid mesh. On the contrary, the PINNs method we introduce in this paper can provide the flow information in any location and any time by integrating the physical equations in training. It is also capable of learning the pressure field from sparse velocity vectors, which are significant for understanding the mechanism of the flow around the SUBOFF model and not introduced in existing super-resolution methods.

4 Results and Discussion

In order to validate the data mining effective, we train the neural network using the velocity vectors of the standard case, which are composed of 300 × 60 × 60 spatial grids and 100 snapshots. The neural network is composed of 10 hidden layers and 200 neurons per layer. The coefficient λ in the loss function (5) is 1. The network is trained with learning rate 0.001 for more than 100 000 iterations to converge. No pressure information is given for training. It is shown that the neural network of PINN can approximate the complex flow vector field with high accuracy: the root mean squared errors (RMSEs), expressed as $\sqrt{\frac{1}{n}\sum_{i=1}^{n}(y_i - \hat{y}_i)^2}$ with y is the quantity of interest, are less than 0.02 dimensionless unit for all three velocity components, whose maximum dimensionless value is 1. On the other hand, the velocity fields outputted by PINNs are consistent with the training data. Although there are errors behind the SUBOFF model due to the approximation error of deep neural networks, the magnitudes of the errors are acceptable, as shown in Fig. 9. The high reconstruction accuracy is also verified by the Q-criterion. The iso-surfaces of the Q-criterion from LBM and PINNs for one snapshot are illustrated in Fig. 10, where we can find that most of the vortices can be resolved by the PINN method.

More importantly, we should note that the pressure field generated by LBM contains disturbance (as shown in Fig. 9) since it is assumed as compressible flow in the LBM simulation. However, as we directly use the incompressible Navier-Stokes equations in PINNs, the pressure field reconstructed from PINNs is smoother and is expected to be more accurate. It is also worth noticing that there is no pressure information provided during the training of PINN. The spatio-temporal pressure fields are all inferred based on the velocity vectors and the equations in PINNs.

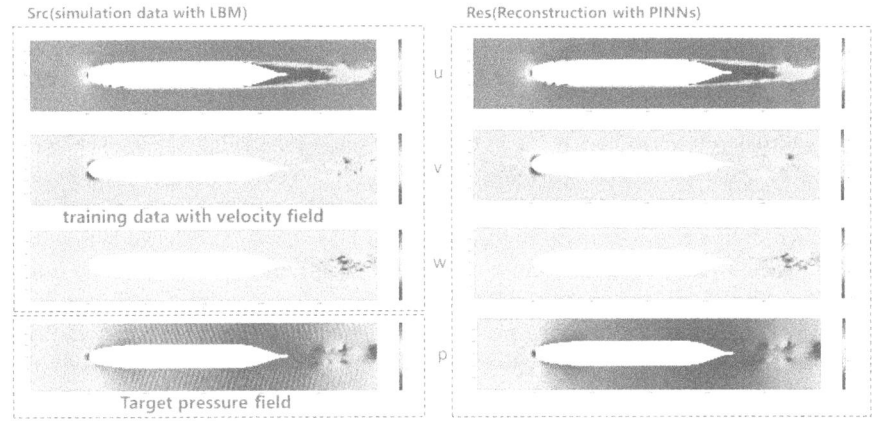

Fig. 9. Flow field reconstruction result. The velocity and pressure fields in the middle plane are shown.

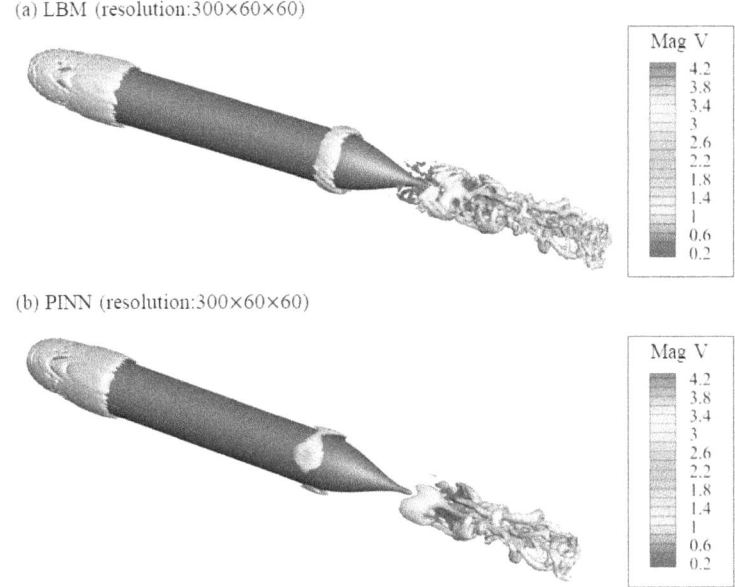

Fig. 10. Flow field reconstruction result. The iso-surface of the Q-criterion is demonstrated with color coded by the velocity magnitude.

5 Conclusion

In this paper, we have leveraged the Sunway TaihuLight supercomputer and SWLBM software to generate a large-scale dataset of flow fields over a SUBOFF model. We have employed physics-informed neural networks (PINNs) to reconstruct the flow fields, including velocity and pressure fields, from sparse velocity measurements. The results demonstrate that the PINNs model trained with sparse velocity data can effectively reconstruct high-resolution flow fields, and the reconstructed velocity and pressure fields are in good agreement with the LBM reference solutions, validating the potential of this approach in complex flow field reconstruction. The research in this paper indicates that large-scale datasets generated by supercomputers hold significant potential in enhancing simulations and experiments in the study of SUBOFF and beyond. We plan to make the generated dataset publicly available to facilitate research in related fields. Based on the experience of this work, we believe that future AI-HPC needs to balance both general computing and AI computing requirements. The bridge between them is the mining and processing of data generated by HPC, which should focus on large-scale simulation data. We should build a processing platform to create datasets and support data-driven AI model training.

Disclosure of Interests. The authors have no competing interests to declare that are relevant to the content of this article.

References

1. Liu, Y., Yu, Z., Zhang, L., et al.: A fine drag coefficient model for hull shape of underwater vehicles. Ocean Eng. **236**, 109361 (2021)
2. Li, D., Yand, Q., Zhai, L., et al.: Numerical investigation on the wave interferences of submerged bodies operating near the free surface. Int. J. Naval Arch. Ocean Eng. **13**, 65–74 (2021)
3. Meng, L., Yang, L., Su, T., et al.: Study on the influence of porous material on underwater vehicle's hydrodynamic characteristics. Ocean Eng. **191**, 106528 (2019)
4. Sezen, S., Dogrul, A., Delen, C., et al.: Investigation of self-propulsion of DARPA Suboff by RANS method. Ocean Eng. **150**, 258–271 (2018)
5. Liu, S., He, G., Wang, Z., et al.: Resistance and flow field of a submarine in a density stratified fluid. Ocean Eng. **217**, 107934 (2020)
6. Pan, Y., Zhang, H., Zhou, Q.: Numerical prediction of submarine hydrodynamic coefficients using CFD simulation. J. Hydrodyn. **24**(6), 840–847 (2012)
7. Rocca, A., Cianferra, M., Broglia, R., et al.: Computational hydroacoustic analysis of the BB2 submarine using the advective Ffowcs Williams and Hawkings equation with Wall-Modeled LES. Appl. Ocean Res. **129**, 103360 (2022)
8. Wang, S., Shi, B., Li, Y., et al.: A large eddy simulation of flows around an underwater vehicle model using an immersed boundary method. Theor. Appl. Mech. Lett. **6**(6), 302–305 (2016)
9. Posa, A., Broglia, R., Felli, M., et al.: Characterization of the wake of a submarine propeller via Large-Eddy Simulation. Comput. Fluids **184**, 138–152 (2019)
10. Posa, A., Balaras, E.: Large-Eddy Simulations of a notional submarine in towed and self-propelled configurations. Comput. Fluids **165**, 116–126 (2018)
11. Kareem, K.M., Korulla, M., Nagarajan, V., et al.: Steady velocity measurements in the stern wake of submarine hull form at high angles of incidence. Ocean Eng. **277**, 114281 (2023)
12. Ashok, A., Van-Buren, T., Smits, A.J.: The structure of the wake generated by a submarine model in yaw. Exp. Fluids **56**(6), 123 (2015)
13. Brunton, S.L., Noack, B.R., Koumoutsakos, P.: Machine learning for fluid mechanics. Annu. Rev. Fluid Mech. **52**, 477–508 (2020)
14. Panda, J.P.: Machine learning for naval architecture, ocean and marine engineering. J. Mar. Sci. Technol. **28**(1), 1–26 (2023)

15. Morasaei, A., Ghabussi, A., Aghlmand, S., et al.: Simulation of steel–concrete composite floor system behavior at elevated temperatures via multi-hybrid metaheuristic framework. Eng. Comput. **38**, 2567–2582 (2022)
16. Ma, X., Foong, L.K., Morasaei, A., et al.: Swarm-based hybridizations of neural network for predicting the concrete strength. Smart Struct. Syst. **26**(2), 241–251 (2020)
17. Ma, R., Karimzadeh, M., Ghabussi, A., et al.: Assessment of composite beam performance using GWO–ELM metaheuristic algorithm. Eng. Comput. **38**, 2083–2099 (2022)
18. Raissi, M., Perdikaris, P., Karniadakis, G.E.: Physics-informed neural networks: a deep learning framework for solving forward and inverse problems involving nonlinear partial differential equations. J. Comput. Phys. **378**, 686–707 (2019)
19. Chu, X., Liu, Z., Shi, S., Meng, H., Lv, X., Han, J: Development progress on SWLBM CFD software on sunway architecture. In: The tenth National Conference on Fluid Mechanics, HangZhou, China (2018)
20. Liu, Z., et al.: SunwayLB: enabling extreme-scale lattice boltzmann method based computing fluid dynamics simulations on sunway taihulight. IEEE International Parallel and Distributed Processing Symposium (IPDPS) (2019)
21. Qian, Y., d'Humières, D., Lallemand, P.: Lattice BGK models for Navier–Stokes equation. EPL (Europhys. Lett.) **17**(6), 479 (1992)
22. Lv, X., Liu, Z., Chu, X., Shi, S., Meng, H., Huang, Z.: Extreme-scale simulation based LBM computing fluid dynamics simulations. Comput. Sci. **47**(4), 13–17 (2020)
23. Xuesen, C., Xiang, H., Fang, L., Zhao, L., Guangwen, Y.: Development progress of SWLBM a Framework based on lattice Boltzmann method for fluid dynamics simulation. In: Gan, L., Wang, Y., Xue, W., Chau, T. (eds) Applied Reconfigurable Computing. Architectures, Tools, and Applications. ARC 2022. Lecture Notes in Computer Science, vol. 13569 (2022)
24. Chu, X., Liu, Y., Dong, Z., et al.: Direct simulation of flow field around SUBOFF in grid-generated turbulence with SWLBM. Comput. Fluids **265**, 106019 (2023)
25. Cai, S., Mao, Z., Wang, Z., et al.: Physics-informed neural networks (PINNs) for fluid mechanics: a review. Acta. Mech. Sin. **37**(12), 1727–1738 (2021)
26. Rabault, J., Ren, F., Zhang, W., et al.: Deep reinforcement learning in fluid mechanics: a promising method for both active flow control and shape optimization. J. Hydrodyn. **32**, 234–246 (2020)
27. Fan D., Jodin G., Consi T.R., et al.: A robotic intelligent towing tank for learning complex fluid-structure dynamics. Sci. Rob. **4**(36), eaay5063 (2019)
28. Kochkov, D., Smith, J.A., Alieva, A., et al.: Machine learning–accelerated computational fluid dynamics. Proc. Natl. Acad. Sci. **118**(21), e2101784118 (2021)
29. Ling, J., Kurzawski, A., Templeton, J.: Reynolds averaged turbulence modelling using deep neural networks with embedded invariance. J. Fluid Mech. **807**, 155–166 (2016)
30. Cai, S., Liang, J., Gao, Q., et al.: Particle image velocimetry based on a deep learning motion estimator. IEEE Trans. Instrum. Meas. **69**(6), 3538–3554 (2019)
31. Rabault, J., Kolaas, J., Jensen, A.: Performing particle image velocimetry using artificial neural networks: a proof-of-concept. Meas. Sci. Technol. **28**(12), 125301 (2017)
32. Liang, J., Xu, C., Cai, S.: Recurrent graph optimal transport for learning 3D flow motion in particle tracking. Nat. Mach. Intell. **5**(5), 505–517 (2023)
33. Yu, C., Luo, H., Bi, X., et al.: An effective convolutional neural network for liquid phase extraction in two-phase flow PIV experiment of an object entering water. Ocean Eng. **237**, 109502 (2021)
34. Lu, L., Meng, X., Mao, Z., et al.: DeepXDE: a deep learning library for solving differential equations. SIAM Rev. **63**(1), 208–228 (2021)
35. Clark Di Leoni, P., Agarwal, K., Zaki, T.A., et al.: Reconstructing turbulent velocity and pressure fields from under-resolved noisy particle tracks using physics-informed neural networks. Exp. Fluids **64**(5), 95 (2023)

36. Boster, K.A., Cai, S., Ladrón-de-Guevara, A., et al.: Artificial intelligence velocimetry reveals in vivo flow rates, pressure gradients, and shear stresses in murine perivascular flows. Proc. Natl. Acad. Sci. **120**(14), e2217744120 (2023)
37. Lou, Q., Meng, X., Karniadakis, G.E.: Physics-informed neural networks for solving forward and inverse flow problems via the Boltzmann-BGK formulation. J. Comput. Phys. **447**, 110676 (2021)
38. Kingma, D.P., Ba, J.: Adam: a method for stochastic optimization. Arxiv preprint arxiv:1412.6980 (2014)
39. Liu, B., Tang, J., Huang, H., et al.: Deep learning methods for super-resolution reconstruction of turbulent flows. Phys. Fluids **32**(2), 025105 (2020)

Enhancing Interaction Graph of Data Schema and Syntactic Structure with Pre-trained Language Model for Text-to-SQL

Wenbin Zhao[1]($^{\boxtimes}$), Long Zhao[1], Feng Wu[2], Zixuan Zheng[1], Haoxin Jin[1], and Bin Gu[1]

[1] School of Information Science and Technology, Shijiazhuang Tiedao University, Shijiazhuang, Hebei, China
zhaowb2013@stdu.edu.cn
[2] Hebei Science and Technology Information Processing Laboratory, Hebei Institute of Science and Technology Information, Shijiazhuang, Hebei, China

Abstract. Text-to-SQL generation is an important area of natural language processing. It can help non-specialists interact with databases using natural language, simplify the database query process, improve efficiency, enhance the user experience, etc. Existing work on Text-to-SQL mainly utilizes large-scale pre-trained models to improve the performance of model generation. Despite progress, Text-to-SQL still has some shortcomings in some respects, such as the discrepancy of words in natural languages, inaccurate scheme links, and inadequate domain generalization capabilities. In this paper, we present an SQL generating framework,which enhancing the interaction graph of data schema and syntactic structure with pre-trained language model for Text-to-SQL(SGIS), aimed at improving the domain generalization of models and the ability of models to deal with cross-cutting questions. Specifically, we first introduce a model linking method based on a pre-trained model, extracting input NL question and database scheme relationship structures to solve the scheme linking question between the NL questions in the model and database models. On this basis, the sentence in the input NL question is extracted from the reliant information and integrated into the well-structured chart data to solve the incomplete question of the relationship characteristics embedded in it. At the same time, in order to prevent the over-adaptation of embedded sides during the training process during the optimization process, we use a type-coding method to help the model effectively differentiate between the type of relationship that the sentence depends on when embedding sides, thereby reducing unnecessary entanglement. Numerous experiments have proven that SGIS's performance on both data sets Spider and Spider-SYN under standard settings is due to all comparative base lines.

Keywords: Text-to-SQL · Syntactic information · Data scheme

1 Introduction

Structured SQL is crucial for managing databases and essential for data analysis, but writing SQL statements can be difficult for both professionals and non-professionals. With advances in deep learning and natural language processing, generating SQL statements from natural language queries, known as Text-to-SQL, has become possible. In a Text-to-SQL task, the model converts natural language queries into structured query language (SQL), requiring an understanding of query intent and semantics to retrieve accurate database information. This technology lowers the barrier for using SQL, especially for complex queries, and reduces human input errors and ambiguity, enhancing query accuracy. Text-to-SQL tasks are widely used in natural language interfaces.

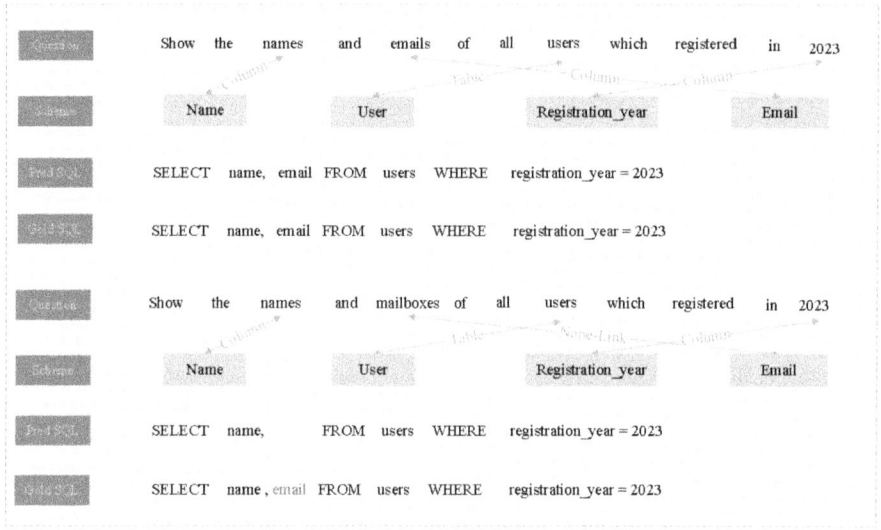

Fig. 1. After replacing the keywords mentioned in the NL question with synonyms, the comparison between the output results of the previous model and the target results is made.

For the challenges of the Text-to-SQL task at this stage, firstly, most existing models focus on single-domain datasets and lack the ability to generalize across domains, whereas the key to achieving domain generalization is schema linking——correctly aligning entities in NL questions to expected schema columns or tables. Second, these models typically ignore the syntactic structure of natural language questions, whereas syntactic information is critical for understanding question intent and generating accurate SQL queries. For example, as shown in Fig. 1, "emails" in an NL question can be matched with its corresponding "email" column, but when "emails" is replaced with its synonym mailboxes in an NL question, the model does not understand the semantics well and cannot find its corresponding schema item. This is also the question of domain generalization. Meanwhile, in Fig. 2, "names" and "emails" should belong to the juxtaposition relationship, but after changing the expression, the model does not fully understand the syntactic dependency information in the NL question, and does not encode the name column together with the "email" column, so there is a certain gap between the SQL statement predicted by the model and the target statement.

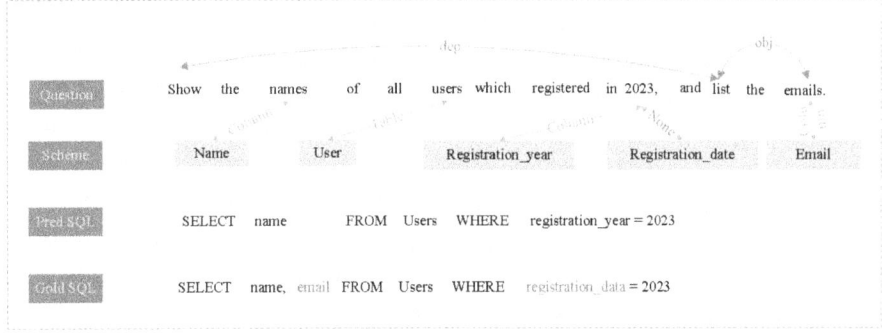

Fig. 2. When there is no syntactically dependent information in the NL question, the previous SQL generation model cannot generate the corresponding SQL statement.

In this paper, we mainly study the question of data scheme integration syntactic information-scheme interaction graph, which aims to improve the model's ability to link data schemes and text semantics as well as the accuracy and efficiency of the model for cross-domain data. This study intends to use a pre-trained language model to enhance the pattern link between data patterns and text semantics. At the same time, in order to reduce the impact of domain features, we integrate domain-independent syntactic structure information as additional embedded information. We conducted a large number of experiments on two data sets, and the experimental results are better than the baseline model. The main contributions of this paper are as follows :

- We study the influence of the correlation between heterogeneous data,and propose a SQL generation framework which enhancing the interaction graph of data schema and syntactic structure with a pre-trained language model
- In order to learn and identify rich syntactic dependency types, we introduce a type-embedding method to further improve performance.
- We conducted a large number of experiments to evaluate the effectiveness of our model. Experimental results on two datasets, Spider and Spider-SYN, show that SGIS outperforms all baseline models in terms of model accuracy.

We believe this work will inspire new methods for text data query and analysis, significantly advancing information retrieval, data mining, and intelligent question answering. We also explore future research directions and potential applications to further develop data query and analysis technology.

2 Related Work

Recently, meaningful advances in encoders [1–4], decoders [5,6], and table-based pre-training models [7–11] have appeared on the Spider benchmark [12]. In particular, Lei et al. [13] pointed out that the scheme linking module in the encoder is a key factor for successful prediction. Guo et al. [14] used heuristic rules to construct intermediate representations.Rui et al. [15] use a common attention mechanism to measure the similarity between NL tokens and schema tokens. A more recent approach, RAT-SQL [16], utilizes relational graph attention to deal with a variety of predefined relations and further considers local and non-local edge features. To address robustness in a more realistic setting, Gan et al. [17] proposed the use of different data augmentation techniques, including data annotation and adversarial training. Wang et al. [18] proposed a meta-learning training objective based on model agnosticity to improve the out-of-domain generalization of Text-to-SQL models. Scholak et al. [19] proposed PICARD, a method for constraining the autoregressive decoder of a language model by incremental parsing. Li et al. [20] proposed a Ranked Enhanced Coding and Skeleton Sensing Decoding Framework (RESDSQL) to decouple scheme linking and skeleton parsing to reduce the difficulty of Text-to-SQL tasks.

In addition, graph encoders have been widely used in cross-domain Text-to-SQL. Bogin et al. [3] were the first to use graph neural networks (GNN) to encode database schemas.Global-GNN [23] applies GNN to soft-select a subset of tables or columns for an output query.ShadowGNN [24] proposes a graph projection neural network to abstract the question and schema representations.LGESQL [25] utilizes line graphs to update edge features in Text-to-SQL heterogeneous graphs, further considering local and non-local, dynamic and static edge features.BRIDGE [26] serializes questions and schemes into tagged sequences and maximizes the use of BERT [27] and database content to capture question-schema links. SmBoP [28] provides the first semi-autoregressive bottom-up semantic parser for the decoding phase in Text-to-SQL.Cai et al. present SADGA, which proposes a structure-aware bi-graph aggregation network for solving two key questions in cross-domain Text-to-SQL tasks: question encoding methods and question-schema linking methods.

The success of PLMs has led to a large number of studies investigating and explaining the richness of knowledge implicitly learned by PLMs during pre-training [29–32] A typical approach is to probe PLMs using a small number of learnable parameters that take into account a variety of linguistic attributes, including morphology, lexical meaning, syntax [33], world knowledge [34]and semantics [35]. Unlike previous methods, our approach follows unsupervised techniques that ensure that all relational knowledge is extracted from the PLM. In addition to this, some work has attempted to apply pre-training objectives to textual table data.TAPAS [36] and TaBERT [37] utilize semi-structured table data to enhance the representational capabilities of language models. For Text-to-SQL, GraPPa [38] performs pre-training on synthetic data generated from synchronous context-independent grammars, while BLOG [39] utilizes a set of novel prediction tasks using a parallel text table corpus to help solve problematic scheme-linking challenges.GAP [40] explores the use of generators to generate pre-training data to augment the joint question and structured scheme encoding capabilities.

3 Method

In this section, we present and describe SGIS in detail.First, we present the relevant definitions and symbolic representations of the tasks, and then we present the architecture of the model, where we describe each component in detail.

3.1 Problem Definition

Given the natural language question Q and the database information S, the input X is input into the corresponding SQL statement Y through a series of neural network methods. More specifically, the natural language question $Q = \{q_1, q_2, \cdots, q_{|Q|}\}$, the database schema term $S = \langle T, C \rangle$, where $T = \{t_1, t_2, \cdots, t_{|T|}\}$ and $C = \{c_1, c_2, \cdots, c_{|C|}\}$ represent the tables and columns in the database, respectively.Formally, input $X = \langle Q, S \rangle$ integrates the natural language question Q and the database item S. In fact, we use the encoder-decoder framework to jointly map the input X into the embedding, and the decoder generates the AST of the SQL query by traversing the syntax tree.

3.2 Model Architecture

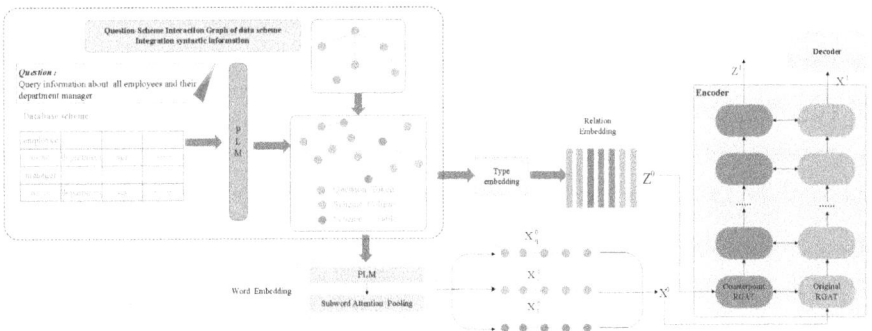

Fig. 3. Model architecture diagram

This section describes the architecture of the model. The overall architecture of SGIS is shown in Fig. 3. It consists of four main components: (i) Graph construction module, which uses pre-trained model to construct scheme link relationship and uses syntactic structure to enhance the semantic information in the graph; (ii) Graph Input module, which encodes the input node information and edge information into node matrix and edge matrix; (iii) Relational graph encoding module, the model uses the line Graph Enhanced Relational Graph Attention Network (RGAT) as the encoder, because its line graph network with dual structure can better capture node information and edge structure information, and obtain high-order node and edge type information; (iv) AST decoding module, the decoder decodes the vector output by the encoder into the form of AST syntax tree, the syntax tree can better deal with structured SQL statements and generate accurate SQL queries.

Graph Construction Module. In this section, the graph construction module aims to model the information in NL problems and database schemas as graph data. In the process of unifying NL sequences and structured database schemas into graph data, a pre-training model (PLM) is used to learn the scheme link structure between NL questions in the input and database schema items. At the same time, in order to reduce the impact of domain features on the performance of the model, domain-independent syntactic structure information is added as an additional input to fuse with the database information. Taking the integrated graph of syntactic dependency relations and scheme link relations as input can effectively improve the domain generalization ability of the model.

Scheme Link Based On Pre-training. For a given question Q and database schema term S, serialize into input X as follows: $X = (\langle s \rangle; q_1; \ldots; q_{|Q|}; \langle \backslash s \rangle; s_1 ; \langle \backslash s \rangle; \ldots; \langle \backslash s \rangle; s_{|T|+|C|})$, Where $|Q|$ denotes the number of words in the natural language question Q, T and C denote the number of data tables and columns, is a special tag for separating input tags. X is fed into the pre-training model for context learning.

We use a function $f(\cdot)$ to express the correlation between the question mark q_i and the database table or column s_j. In this regard, we use common mask markers to measure the relationship. Specifically, the input X is sent to the PLM. we use h_j^s to denote the contextualized representation of the j-th mode item s_j, where $1 \leq j \leq |T|+|C|$. Then, we replace the question tag with the mask tag [MASK], and feed the marked input to the PLM again, and we use $h_{j \backslash q_i}^s$ the new representation of the jth mode item when q_i is masked. Formally, we calculate the distance measure of and in the two processes to judge the correlation between the scheme item and the question mark. The specific distance measure is as follows :

$$f(q_i, s_j) = d_{Poin}\left(h_{j \backslash q_i}^s, h_j^s \right) \quad (1)$$

where $d_{Poin}(\cdot)$ is the distance metric used to measure the difference between two vectors in a hyperbolic space.

By repeating the learning process twice for each pair of s_j and q_i tokens, and computing the $f(q_i, s_j)$, we obtain the relationship matrix $R^{linking}$, denotes the question-scheme pair (q_i, s_j) relationships between them. We compute the relationship matrix $R^{linking}$ by:

$$R_{ij}^{linking} = \begin{cases} 0 & \text{if } f(q_i, s_j) < \tau \\ 1 & \text{if } f(q_i, s_j) > \tau \end{cases}, \quad (2)$$

where τ is a predefined threshold. When the correlation between the question and the scheme is less than the threshold, the corresponding link structure is not constructed.

Question-Schema Interaction Graph. The joint input questions and schema items can be viewed as a graph $G = (\mathcal{V}, R)$, where $\mathcal{V} = Q \cup T \cup C$, denotes that the set of nodes containing the input NL question tokens, and R represents the known relation between two elements in the input nodes. In this paper, we regard the syntactic information structure of natural language as a new edge type and integrate it with the constructed interaction graph. Specifically, for any two words v_i and v_j in the question, the first-order distance $D(v_i, v_j)$ of the attention score between them is calculated. If v_i and v_j have a certain dependency type, the distance is set to the first-order distance ; otherwise, set to 0. According to the calculated first-order distance D, the syntactic relationship between the question words is expressed as the relation matrix $R_{ij}^{question}$, which is as follows:

$$R_{ij}^{question} = \begin{cases} Dependence & if\ D(v_i, v_j) = 1 \\ NONE & if\ D(v_i, v_j) = 0 \end{cases} \quad (3)$$

Then the syntactic relationship is merged into the previous question-scheme interaction graph as an additional edge. This integration allows the model to consider both the structure of the question and the potential links in the syntactic information while ensuring the accuracy of the scheme link.

Graph Input Module

type embedding. In the node-based graph embedding, the edge embedding is often initialized according to the embedding vectors of the starting node x_i and the target node x_j, and does not consider the type relationship of the edges in the graph. Therefore, we adopt a type embedding method to make the model better understand the edge relationship type. Firstly, a unique encoding vector for each edge type is defined by a mapping function f. The mapping function maps the edge type t to the vector e_t in the embedding space. For each edge e_{ij} in the graph, in addition to its type t, its corresponding feature vector f_{ij}, the edge feature vector f_{ij} and its type coding vector e_t are spliced to form an embedded representation of the edge z_{ij}. As shown below :

$$z_{ij} = f_{ij} \oplus e_t \quad (4)$$

After obtaining the vector representation of each edge, they are sent to the corresponding BiLSTM according to their edge types, and concatenate the hidden state at both ends. The obtained edge representation is stacked to form an initial edge embedding matrix Z.

Dual Relation Graph Attention Network Module. After embedding the nodes and edges in the input graph, we model the graph G using the line graph Augmented Relational Graph Attention Network (RGAT) In this module, there are L stacked bi-layered relational graph attention network layers consisting of two RGAT modules per layer to capture the relational structure in the original and line graphs respectively, with the node embedding information in each graph characterizing the edges in their pairwise graphs.

For the RGAT layer of the original graph:

Given the node embedding matrix of layer $X^l \in \mathbb{R}^{V^n \times d}$ and the edge characterization matrix $[Z^l; Z_{\text{nlc}}] \in \mathbb{R}^{|V^e| \times d}$ (where Z^l are local edge features, and Z_{nlc} are non-local edge features), the output of the original graph is represented $\widetilde{x}_i^l$ computed by the following steps:

1. Calculation of attention weights $\widetilde{\alpha}_{ji}^h$:

$$\widetilde{\alpha}_{ji}^h = \frac{\left(\widetilde{x}_i^l W_q^h\right)\left(\widetilde{x}_j^l W_k^h + [\psi(r_{nj})]^h\right)^T}{\sqrt{d/H}} \qquad (5)$$

Here W_q^h and W_k^h are the trainable parameters, and H is the number of attention heads, and $\psi(r_{nj})$ is the edger r_{nj} of the feature vectors.
2. Apply softmax function to calculate normalized weights α_{ji}^h.
3. Updating the node representation $\widetilde{x}_i^{l+1}$

$$\widetilde{x}_i^{l+1} = \sum_{j \in N_i^n} \alpha_{ji}^h \left(\widetilde{x}_j^l W_v^h + [\psi(r_{nj})]^h\right) \qquad (6)$$

where N_i^n is the node v_j^n of the receiving field.
4. apply LayerNorm and Feedforward Neural Network (FFN) to get the node representation of the next layer x_i^{l+1}.

In this process, local edge features Z^l are dynamically provided and generated by node embedding in the line graph, while the non-local edge features Z_{nlc} are instead extracted directly from the parameter matrix. This distinction encourages the model to pay more attention to local edge features while maintaining information from multi-hop neighbors.

AST Decoder Module. For decoding, we use a syntax-based syntactic neural network decoder, which follows a tree-structured architecture to generate the Abstract Syntax Tree (AST) of the SQL query The decoder makes use of a predefined set of SQL syntax rules to select the rules to extend the current AST node by means of depth-first search, and selects the next action by means of action selection. Repeated application of rules and selection of actions gradually builds the AST for the complete SQL query, and once the AST is constructed, the decoder converts the AST into the final SQL query string.

4 Experiment

4.1 Datasets and Evaluation Metrics

We conduct experiments on Spider(Yu et al., 2018) and SpiderSyn (Gan et al., 2021). Spider is a large-scale, complex, cross-domain Text-to-SQL benchmark. The Spider dataset contains over 10,000 natural language questions and corresponding SQL queries, covering multiple database domains and complex queries The natural language questions in the Spider dataset cover a wide range of difficulty levels and query types, including aggregations, joins, conditional filters, and other SQL query operations. Spider-syn is an extended version of the Spider dataset that aims to further challenge model generalization and robustness. By introducing synthetic data instances to test the performance of the model when dealing with new domains and unseen query types, Spider-SYN adds diversity and complexity to the dataset. In the evaluation, we follow the official evaluation reporting exact matching accuracy.

4.2 Implementation Details

In the encoder, for PLM, the GNN hidden size is set to 512. The number of GNN layers is 8. The dropout is set to 0.1. The size of the inner layer of the positional feedforward network is 1024. In the decoder, the dimensions of the hidden state, action embedding, and node type embedding are set to 512, 128, and 128, respectively. The recurring dropout rate of the decoder LSTM is 0.21. In the encoder and decoder, the number of heads for multiple attention is 8 and the feature dropout rate is set to 0.2. The maximum gradient paradigm is set to 5. The batch size is 20 and the number of training epochs is 200.

4.3 Benchmark Model

We perform experiments on Spider and Spider-Syn and compare our approach to several baselines, including:

- RYANSQL (Choi et al., 2020) is a sketch-based slot-filling method that is proposed to synthesize each SELECT statement for its corresponding position.
- RATSQL (Wang et al., 2020) is a relationally aware scheme encoding model in which the question-scheme interaction graph is constructed from n-gram schemes.
- ShadowGNN (Chen et al., 2021) deals with schemas with domain-independent representations at the abstraction and semantic levels.
- BRIDGE (Lin et al., 2020) represents questions and schemes in a labeled sequence, where a subset of fields is expanded with cell values mentioned in the question.
- LGESQL (Cao et al., 2021) a line graph augmented Text-to-SQL model to mineunderlying relational features without the need to construct meta-paths.
- SADGA (cai et al.,2022) a structure-aware dual graph aggregation network is proposed, which uses graph structure to uniformly encode natural language questions and database architecture.

Table 1. Exact matching accuracy on Spider development and test sets.

Model	Dev	Test
GNN	40.7	39.4
IRNet v2	63.9	55.0
RATSQL+BERT	62.7	57.2
RYANSQL	70.0	60.6
ShadowGNN+RoBERTa	71.6	66.7
LGESQL+BERT-large	74.1	68.3
SGIS+BERT-large	**75.2**	**69.7**
w/o Te	74.3	–
w/o Syntax	73.7	–
S2SQL+ELECTRA	76.4	72.1
SGIS+ELECTRA	**77.2**	**73.0**
w/o Te	75.9	–
w/o Syntax	75.1	–

4.4 Overall Results

We first compare SGIS with other state-of-the-art models on Spider. As shown in Table 1, we can see that SGIS outperforms all existing models. Notably, SGIS + BERT-large has an accuracy of 69.7% test accuracy and 75.2% dev accuracy, which is 1.4% and 1.1% higher than the strong baseline LGESQL + BERT-large. Similarly, the SOTA model S2SQL + ELECTRA has an accuracy of 72.1% on the hidden test set and 76.4% on the development set, while SGIS + ELECTRA can achieve 73.0% test accuracy and 77.2% dev accuracy. Table 2 shows the performance of SGIS on the Spider-SYN dataset with an accuracy of 65.5%, which is 0.9% higher than the LGESQL + ELECTRA-large model.

Table 2. Exact matching accuracy on Spider-SYN.

Model	ACC
Global-GNN	23.6
IRNet	28.4
RATSQL	33.6
RATSQL+BERT	48.2
RATSQL+Grappa	49.1
LGESQL+ELECTRA-large	64.6
SGIS+ELECTRA-large	**65.5**

Table 3. Accuracy of ablation studies on the Spider Dev set by hardness levels.

Model	Easy	Medium	Hard	Extra hard
RATSQL	86.4	73.6	62.1	42.9
SADGA	**90.3**	72.4	63.8	49.4
SGIS	89.4	**74.5**	**68.1**	**52.6**
w/o Te	89.2	73.6	66.5	50.2
w/o Syntax	88.5	72.9	65.4	49.0

4.5 Ablation Experiments

In order to validate the effectiveness of each component of SGIS, the ablation study was conducted with different levels of parsing difficulty. The last two rows of Table 1 show that removing the syntactic information(i.e.Syntactic in the table) and scheme-linking(i.e.Te in the table) leads to a decrease in the model's performance on the Spider dev set by 1.5%, 0.9% and 2.1%, 1.3% respectively. In more detail, as shown in Table 3, we explore the accuracy of the results of the model in the Spider dataset at different difficulty levels. We find that in the easy case, the accuracy of our model is not as high as SADGA, because there are no very complex syntactic relationships and fuzzy links in complex sentences in simple questions, so SGIS is not effectively improved in this case. In the case of hard and extra hard, our model has greatly improved compared to the previous baseline model. At the same time, we found that in the case of removing scheme-linking and Syntax, the four difficulty levels decreased by 0.2%, 0.9%, 1.6%, 2.4% and 0.9 %, 1.6 %, 2.7%, 3.6% respectively. As the difficulty level increases, the results of the model prediction decrease more, which proves that our syntactic information and pre-trained scheme- linking modules have a great effect on the model improvement, especially in dealing with complex questions and data scenarios.

4.6 Case Studies

In Table 4, we compare the SQL queries generated by the model with the SQL queries created by the baseline model. We note that the performance of SGIS is better than the baseline model system, especially when the syntactic structure and data scheme links are needed. For example, in the first case, there are syntactic relationships (name, grades, CONJ), (order, names, NMOD), and the model captures the correspondence between grades words and math columns, but there is a column "grade" that does not exist in the table in the baseline model. The object of ORDER is also misunderstood as "grade". In the second and third cases, it is also reflected that the model may effectively improve the questions in the previous baseline model after data scheme linking and syntactic integration.

Table 4. Accuracy of ablation studies on the Spider Dev set by hardness levels.

Question	List the name and the grades of students whose math exceed 80, in alphabetical descending order for the names
Baseline	SELECT name, grade FROM stduent WHERE math>80 ORDER BY grade DESC
SGIS	SELECT name, math FROM stduent WHERE math>80 ORDER BY grade DESC
Gold	SELECT name, math FROM stduent WHERE math>80 ORDER BY name DESC
Question	What are the names of the singers who are not China?
Baseline	SELECT name FROM singer WHERE name!="China"
SGIS	SELECT name FROM singer WHERE citizenship="China"
Gold	SELECT name FROM singer WHERE citizenship!="China"
Question	Show the date with the least number of attendance, and list the id
Baseline	SELECT employee_data _____FORM employee WHERE MIN(employee_number)...
SGIS	SELECT employee_data , employees FORM employee WHERE MIN(employee_number)...
Gold	SELECT employee_data,employee_id FORM employee WHERE MIN(employee_number)...

5 Conclusion

We propose a enhanced interaction graph SQL generation framework (SGIS) of data schema and syntactic structure with Pre-trained Language Model. It can measure the correlation between the input text information and the data scheme information through the pre-trained model, help the model capture the link structure between the question keyword and the corresponding database scheme item, and construct the graph structure input on this basis. At the same time, we use the syntax analyzer to extract the syntactic structure in the text, and integrate these dependencies as the variable structure in the graph data into the input graph. The question-scheme interaction graph we constructed can effectively improve the data constraint of model input. However, due to the excessive types of syntactic dependencies in the integration process, the model is prone to over-fitting in the optimization process. We use a relational encoding method to process the edge embedding in the input graph, extra encode all the relationship types, and then splice the type information into the edge embedding. Finally, the model has made progress in the widely used Spider and Spider-SYN datasets and is superior to all comparable benchmark models.

Acknowledgments. This study was supported by the National Natural Science Foundation of China (61373160), Hebei Province Central Guidance Local Science and Technology Development Fund Project (No. 236Z0306G), the Natural Science Foundation of Hebei Province (F2021210003), Hebei Provincial Science and Technology Plan Project (22567636H).

References

1. Bogin, B., Berant, J., Gardner, M.: Representing schema structure with graph neural networks for Text-to-SQL parsing. In: ACL (2019)
2. Chen, Z., et al.: ShadowGNN: graph projection neural network for Text-to-SQL parser. In: NAACL (2021)
3. Hui, B., et al.: Dynamic hybrid relation exploration network for cross-domain context-dependent semantic parsing. In: AAAI (2021)
4. Hui, B., et al.: S2SQL: injecting syntax to question-schema interaction graph encoder for Text-to-SQL parsers. In: ACL (2022)
5. Choi, D., Shin, M., Kim, E., Shin, D.R.: RYANSQL: recursively applying sketch-based slot fillings for complex Text-to-SQL in cross-domain databases. ArXiv arxiv:2004.03125 (2020)
6. Hui, B., et al.: Improving text-to-sql with schema dependency learning. arXiv preprint arXiv:2103.04399 (2021)
7. Liu, Q., et al.: TAPEX: table pre-training via learning a neural SQL executor. In International Conference on Learning Representations (2022)
8. Qin, B., et al.: SDCUP: schema dependency-enhanced curriculum pre-training for table semantic parsing. arXiv preprint arXiv:2111.09486 (2021)
9. Shi, P., et al.: Learning contextual representations for semantic parsing with generation-augmented pre-training. In: AAAI (2021)
10. Yin, P., Neubig, G., Yih, W.T., Riedel, S.: TaBERT: pretraining for joint understanding of textual and tabular data. In: ACL (2020)
11. Yu, T., et al.: GraPPa: grammar- augmented pre-training for table semantic parsing. In: ICLR (2021)
12. Yu, T., et al.: Spider: a large-scale human-labeled dataset for complex and cross-domain semantic parsing and Text-to-SQL Task. In: EMNLP (2018)
13. Lei, W., et al.: Re-examining the role of schema linking in Text-to-SQL. In: EMNLP (2020)
14. Guo, J., et al.: Towards complex Text-to-SQL in cross-domain database with intermediate representation. In: ACL (2019)
15. Rui, Z., et al.: Editing-based SQL query generation for cross-domain context-dependent question. In: EMNLP (2019)
16. Wang, B., Shin, R., Liu, X., Polozov, O., Richardson, M.: RAT-SQL: relation-aware schema encoding and linking for Text-to-SQL parsers. In: ACL (2020)
17. Gan, Y., et al.: Towards robustness of Text-to-SQL models against synonym substitution. In: ACL (2021)
18. Wang, B., Lapata, M., Titov, I.: Meta-learning for domain generalization in semantic parsing. In NAACL (2020)
19. Scholak, T., Schucher, N., Bahdanau, D.: PICARD: parsing incrementally for constrained auto-regressive decoding from language models. In: EMNLP (2021)
20. Li, H., Zhang, J., Li, C., Chen, H.: Resdsql: decoupling schema linking and skeleton parsing for text-to-sql. In: Proceedings of the AAAI Conference on Artificial Intelligence, vol. 37, no. 11, pp. 13067–13075 (2023)
21. Bogin, B., Gardner, M., Berant, J.: Global reasoning over database structures for text-to-SQL parsing. In: Proceedings of the 2019 Conference on Empirical Methods in Natural Language Processing and the 9th International Joint Conference on Natural Language Processing (EMNLP-IJCNLP), Hong Kong, China, pp. 3659–3664. Association for Computational Linguistics (2019)

22. Chen, Z., et al.: Shad-owGNN: graph projection neural network for text-to-SQL parser. In: Proceedings of the 2021 Conference of the North American Chapter of the Association for Computational Linguistics: Human Language Technologies, pp. 5567–5577. Association for Computational Linguistics (2021)
23. Cao, R., Chen, L., Chen, Z., Zhao, Y., Zhu, S., Yu, K.: LGESQL: Line graph enhanced text-to-SQL model with mixed local and non-local relations. In: Proceedings of the 59th Annual Meeting of the Association for Computational Linguistics and the 11th International Joint Conference on Natural Language Processing, vol. 1: Long Papers, pp. 2541–2555. Association for Computational Linguistics (2021)
24. Lin, X.V., Socher, R., Xiong, C.: Bridging textual and tabular data for cross-domain text-to-SQL semantic parsing. In: Findings of the Association for Computational Linguistics: EMNLP 2020, pp. 4870–4888. Association for Computational Linguistics (2020)
25. Devlin, J., Chang, M.W., Lee, K., Toutanova, K.: BERT: pre-training of deep bidirectional transformers for language understanding. In Proceedings of the 2019 Conference of the North American Chapter of the Association for Computational Linguistics: Human Language Technologies, vol. 1 (Long and Short Papers), Minneapolis, Minnesota, pp. 4171–4186. Association for Computational Linguistics (2019)
26. Rubin, O., Berant, J.: SmBoP: semi-autoregressive bottom-up semantic parsing. In Proceedings of the 2021 Conference of the North American Chapter of the Association for Computational Linguistics: Human Language Technologies, pp. 311–324. Association for Computational Linguistics (2021)
27. He, W., et al.: Unified dialog model pre-training for task-oriented dialog understanding and generation. In: SIGIR (2022)
28. He, W., et al.: GALAXY: a generative pre-trained model for task-oriented dialog with semi-supervised learning and explicit policy injection. In: AAAI (2021)
29. Kovaleva, O., Romanov, A., Rogers, A., Rumshisky, A.: Revealing the dark secrets of BERT. In: EMNLP (2019)
30. Rogers, A., Kovaleva, O., Rumshisky, A.: A primer in bertology: what we know about how bert works. In: TACL (2020)
31. Belinkov, Y., Durrani, N., Dalvi, F., Sajjad, H., Glass, J.: What do neural machine translation models learn about morphology?. In: ACL (2017)
32. Reif, E., et al.: Visualizing and measuring the geometry of BERT. In: NeurIPS (2019)
33. Liu, N.F., Gardner, M., Belinkov, Y., Peters, M.E., Smith, N.A.: Linguistic knowledge and transferability of contextual representations. In: NAACL (2019)
34. Liu, Q., Yang, D., Zhang, J., Guo, J., Zhou, B., Lou, J.G.: Awakening latent grounding from pretrained language models for semantic parsing. In: ACL (2021)
35. Hewitt, J., Liang, P.: Designing and interpreting probes with control tasks. In: EMNLP (2019)
36. Herzig, J., Nowak, P.K., Müller, T., Piccinno, F., Eisenschlos, J.: TaPas: weakly supervised table parsing via pre-training. In: Proceedings of the 58th Annual Meeting of the Association for Computational Linguistics, pp. 4320–4333. Association for Computational Linguistics (2020)
37. Yin, P., Neubig, G., Yih, W., Riedel, S.: TaBERT: pretraining for joint understanding of textual and tabular data. In: Proceedings of the 58th Annual Meeting of the Association for Computational Linguistics, pp. 8413–8426. Association for Computational Linguistics (2020)
38. Yu, T., et al. Grappa: grammar-augmented pre-training for table semantic parsing. In International Conference on Learning Representations (2021)

39. Deng, X., et al.: Structure-grounded pretraining for text-to-SQL. In Proceedings of the 2021 Conference of the North American Chapter of the Association for Computational Linguistics: Human Language Technologies, pp. 1337–1350. Association for Computational Linguistics (2021)
40. Shi, P., et al.: Learning contextual representations for semantic parsing with generation-augmented pre-training. In: Proceedings of the AAAI Conference on Artificial Intelligence, pp. 13806–13814 (2021)

Spatial Evolution Analysis of the Level of Digital Economy Development in China

Jing Feng(✉) [ID], Tongle Han [ID], Ziqi Xia [ID], Xuehua Zhang [ID], and Yunhan Qu

School of Economics and Management, Tiangong University, Tianjin 300387, China
13182048817@163.com

Abstract. In rcent years, China's economic growth has been largely driven by advancements in the digital economy, facilitating high-quality development. To evaluate China's digital economy development, this study used a literature research approach to identify secondary indicators based on the five major new development concepts as primary benchmarks. Data from 31 provinces, autonomous regions, and municipalities directly under the central government were analyzed to assess China's digital economy development from 2013 to 2021. Using GeoDa software, this study conducted spatial autocorrelation analysis including both global and local spatial autocorrelation assessments. And the findings indicated significant agglomeration effects in China's digital economy development levels, characterized by higher levels observed in eastern coastal areas and lower levels in the northwest and northern regions. Besides, Temporal evolution characteristics revealed a gradual reduction in aggregation areas for regions with low levels of digital economy development in China, while an expansion was observed for regions with high levels. However, due to disparities in developmental foundations and speeds, there existed an imbalance in regional differences regarding the level of digital economy development.

Keywords: Level of digital economy development · Spatial autocorrelation analysis · Temporal · evolution characteristics · High-quality development

1 Introduction

China's 14th Five-Year Plan explicitly proposed developing the digital economy and deepening integration with the real economy. Analyzing the development of the digital economy across provinces and cities in China and studying their spatial evolution patterns are imperative for current societal and national requirements.

Additionally, they are beneficial for the establishment of a new development pattern with domestic circulation as the primary focus and dual circulation between domestic and international markets.

Many scholars have conducted extensive research on the digital economy. For example, Oloyede Abdulkarim A et al. (2023) [1] conducted a systematic literature review using the PRISMA model to investigate factors and indicators to measure the digital economy in developing countries, aiming to define and assess its impact on the national economy. In the construction and measurement of the evaluation system for the level of digital economy development, in Nakamura L's (2017) [2] study, the Dagum Gini coefficient was employed to measure the 'free' digital economy within the GDP and productivity accounts. Remlein Marzena et al. (2022) [3], based on principles from economic growth theory, computed coefficients of digital competitiveness as composite indicators comprising multiple global metrics. Their study examined how the levels of digitization in Poland and Ukraine influenced small and medium enterprises' contributions to the national Gross Domestic Product (GDP). The authors found that a unit change in the integrated coefficient of digital competitiveness is related to the greatest change in the contribution of SMEs to the country's GDP when the other factors in the model equation remain fixed. Besides, Maksim Vlasov et al. (2022) [4] explored the mutual impact between elements of the digital economy and social culture using methods such as the Pearson correlation coefficient, based on five-year regional data from Russia. They assessed various elements within social culture for their correlations with economic development. Ming Yang (2022) [5] employed principal component analysis to select five corresponding indicators from three dimensions-digital infrastructure level, digital industrial development level, and digital financial inclusive development-to construct an index system for measuring the level of digital economy development. Correlation and principal component analysis are employed by Giuseppe B et al. (2023) [6] to assess redundancies within the Digital Economy and Society Index (DESI). In the context of specific cases, a more practical model for assessing the digital economy divide between countries is established. Panel data from 30 provinces in China spanning from 2010 to 2019 was selected by Zhipeng Chen et al. (2022) [7] for their study. They used a coupling coordination degree model as well as a panel vector autoregressive model (PVAR), to identify relationships between various indicators, and conducted an analysis on each province's respective levels of Digital Economy Development.

In the research on the spatial characteristics of the level of digital economy development in China. The GIS spatial analysis method was utilized by Shengpeng Wang et al. (2022) [8] to analyze the spatiotemporal evolution characteristics of the level of digital economy development in 285 prefecture-level and above cities in mainland China, based on the measurement of the level of digital economy development.

The characteristics and effects of China's digital economy spatial correlation network results were studied by Zhentao Wang et al. (2022) [9] using a modified gravity model and social network analysis method. The spatial distribution pattern of digital economy development was described by Jiyin Wu et al. (2022) [10] through the natural break method and kernel density analysis. Furthermore, a spatial Durbin model was constructed to analyze the spatial spillover effects of the digital economy on industrial structure upgrading. The spatiotemporal evolution characteristics of the digital economy were analyzed by Jie Li et al. (2022) [11] utilizing kernel density, Dagum Gini coefficient, and ESDA. The regional development disparities of China's digital economy were analyzed by Luyang Tang et al. (2022) [12] using Theil index and kernel density estimation methods. Additionally, network methods were employed to analyze the correlation between the level of digital economy development in each province.

While significant progress has been made in researching digital economy development, there are still shortcomings that require attention. Differences in indicators exist due to the incomplete evaluation index system and lack of unified standards when measuring the level of digital economy development. This makes it difficult to compare the level of digital economy development in different regions with different indicator standards. Thus, This study aimed to enhance the current evaluation index system for assessing the level of digital economy development, emphasizing the "new development philosophy" with innovation, coordination, green development, openness, and sharing as it's five primary indicators. Several secondary indicators are constructed within each of these five dimensions to assess the level of digital economy development in 31 provinces and cities in China. Subsequently, the Moran spatial analysis method is utilized to investigate the spatial evolution patterns of China's digital economy development from 2013 to 2021. Finally, the paper examines and evaluates the digital economy development in different regions based on these analyses.

2 Construction of the Evaluation Index System for the Level of Digital Economy Development

2.1 Development of the Index System

Based on the "new development philosophy", this paper constructed a table to evaluate the level of digital economy development in each province based on primary indicators including innovation, green development, coordination, openness, and sharing. The innovation indicator considered the driving force behind digital economy development, while the coordination indicator focuses on regional balance. The green indicator assessed environmental friendliness and sustainability, the openness indicator addresses internal and external linkage issues, and the sharing indicator emphasized the range and depth of digital economy radiation to citizens. 16 sub-indicators under these five primary indicators were constructed in Table 1:

Table 1. Evaluation System for Digital Economy Development Level

Primary Indicators	Secondary Indicators	Data Sources
Innovation	R&D expenditure of industrial enterprises above a certain scale (in ten thousand yuan)	National Bureau of Statistics--《China Statistical Yearbook》(https://www.stats.gov.cn/sj/ndsj/)
	Number of R&D projects in industrial enterprises above a certain scale	
	Number of legal entities in the digital industrialization and industrial digitalization industries	

(*continued*)

Table 1. (*continued*)

Primary Indicators	Secondary Indicators	Data Sources
Coordination	Total profits of the manufacturing sector in the electronic information industry	
	The proportion of the value added of the tertiary industry to the regional GDP	
	Average number of employees in high-tech industries	
	The proportion of technology market transactions to regional GDP	
Openness	Total import and export value of foreign-invested enterprises	
	Total import and export value of goods	
	E-commerce sales volume	
	Exports of the electronic information industry	
Sharing	Internet broadband access ports (ten thousand)	National Bureau of Statistics-- 《China Statistical Yearbook》 (https://www.stats.gov.cn/sj/ndsj/)
	Number of Internet broadband users	
	Total volume of telecommunications services	
Green	The ratio of interest expenses of the six major high-energy-consuming industries to the total interest expenses of the industrial sector (negative indicator)	China National Institute of Statistics- 《China Industrial Statistical Yearbook》 (https://www.stats.gov.cn/fw/zgtjzlg/)
	Per capita GDP electricity consumption (negative indicator)	National Bureau of Statistics- 《China Statistical Yearbook》 (https://www.stats.gov.cn/sj/ndsj/)

2.2 Research Methodology

Data Standardization. Due to the dimensional differences among different indicators, direct comparison and analysis are not feasible. Standardizing the data is necessary in order to unify the indicator dimensions. This paper employs the method of range standardization, which linearly transforms the original data to map the result values between 0 and 1. The specific formula is divided into two types:

For positive indicators:

$$X_{ij}^* = \frac{X_{ij} - X_{min}}{X_{max} - X_{min}}, X_{min} \leq X_{ij} \leq X_{max} \tag{1}$$

For negative indicators:

$$X_{ij}^* = \frac{X_{max} - X_{ij}}{X_{max} - X_{min}}, X_{min} \leq X_{ij} \leq X_{max} \tag{2}$$

where X_{ij} is the dimensionless processed indicator value, X_{max} is the maximum value of the data under this indicator dimension, and X_{min} is the minimum value of the data under this indicator dimension.

Entropy Weighting Method. The entropy weighting method is an objective approach to assign weights. In the specific application process, the entropy of each indicator is calculated based on its data dispersion using information entropy. Subsequently, the entropy weights of each indicator are adjusted to obtain a more objective weight for each indicator. This paper utilizes the entropy weighting method to calculate the weights of each sub-indicator. The standardized data of indicators are then weighted to derive comprehensive scores for evaluating the level of digital economy development in 31 provinces and cities in China.

Firstly, calculate the information entropy Hi of each indicator Xi:

$$H_i = -\sum_{j=1}^{m_i} P_{ij} \log P_{ij} \tag{3}$$

Secondly, calculate the weight Wi of each indicator:

$$W_i = -\frac{1 - H_i}{\sum_{j=1}^{n}(1 - H_i)} \tag{4}$$

In Eqs. (3) and (4), mi is the number of values for indicator Xi, and Pij is the probability of indicator Xi being in the j-th value.

Global Spatial Autocorrelation Analysis. Global spatial autocorrelation is utilized to characterize the spatial attributes of attribute values across an entire region and indicate the similarity of attribute values in adjacent areas.

(1) Definition of Spatial Weight Matrix

The spatial weight matrix reflects the inter-dependency among individuals in a given space. As illustrated in the figure, a spatial weight matrix W is defined to describe the spatial interdependence among n individuals. Here, W_{ij} represents the extent of influence that individual i has on individual j within the given space. It is evident that the spatial weight matrix quantifies the level of mutual influence between individuals in a spatial context.

(2) Moran's Index

Moran's Index is divided into global Moran's Index and local Moran's Index. In a narrow sense, the value of Moran's Index is a rational number. After variance normalization, its value will be standardized to fall within the range of −1 to 1. A Moran's value equal to 0 indicates that the spatial distribution of digital economy development in China is random. A Moran's value between 0 and 1 indicates positive spatial autocorrelation, while a Moran's value between −1 and 0 indicates negative spatial autocorrelation.

(3) Global Correlation Analysis

In order to investigate the potential correlation in the spatial distribution of digital economy development levels among different provinces in China, this paper conducts a spatial autocorrelation analysis. Moran's I statistic, z-score, and p-value are calculated to assess the significance of the observed correlation pattern.

First, the hypothesis is proposed:

H_0: There is no spatial correlation in the level of digital economy development among provinces in China.

H_1: There is spatial correlation in the level of digital economy development among provinces in China.

The formula for Moran's I statistic is:

$$I = \frac{n}{S_0} \frac{\sum_{i=1}^{n} \sum_{j=1}^{n} w_{i,j} z_i z_j}{\sum_{i=1}^{n} z_i^2} \tag{5}$$

In this formula, I is the Moran's Index statistic, which measures the strength and direction of spatial autocorrelation. n represents the total number of study areas, which is 31 provinces in China (excluding Hong Kong, Macau, and Taiwan). The attribute value of element i is denoted as xi, which represents the level of digital economy development in region i. z_i is the deviation of the attribute of element i from its attribute mean. $w_{i,j}$ represents the spatial weight between observation points i and j. S_0 is the sum of all spatial weights [13].

Local Spatial Autocorrelation Analysis. Local spatial autocorrelation analysis calculates and analyzes the spatial correlation between the attribute values of individual spatial objects and the neighboring areas within a specific region. It reflects local spatial heterogeneity and instability by utilizing Moran's I index, LISA cluster maps, and LISA significance maps to analyze differences in local features' spatial distribution. The establishment of a distribution weight matrix for all provinces in the country allows for the use of Geoda software to conduct univariate local Moran's I analysis [14].

3 The Measurement of China's Digital Economic Development Level

The data on the 16 selected sub-indicators were used to evaluate the level of digital economic development. These data were sourced from the "China Statistical Yearbook," "China Information Industry Yearbook," and "China Industrial Statistical Yearbook," as well as provincial statistical yearbooks spanning the period from 2013 to 2021 across 31 Chinese provinces, municipalities, and autonomous regions (excluding Hong Kong, Macao, and Taiwan). In cases where data were missing, interpolation was applied for sporadic gaps, while consecutive gaps were forecasted using Excel's forecast function with a linear trend. Instances where the linear forecast yielded negative values were uniformly adjusted to zero.

Following data standardization, weights for each indicator were determined using the entropy weighting method with Eqs. (3) and (4), and the data were subsequently weighted. Analysis using SPSS software revealed that the index of digital economic development levels ranged from 0.0049 to 0.8823 during the study period. These levels exhibited dynamic temporal changes across regions, with significant disparities observed in measurement values between regions. The subsequent chapter provided a detailed spatial analysis of the evolution patterns based on the findings from Table 2.

Table 2. Digital Economy Development Index of 31 Provinces and Municipalities

Provincial Level	2013	2014	2015	2016	2017	2018	2019	2020	2021
Beijing Municipality	0.1767	0.2867	0.1728	0.1519	0.1754	0.1960	0.1732	0.2048	0.2001
Tianjin Municipality	0.1161	0.0858	0.0960	0.0901	0.0898	0.0912	0.0959	0.0965	0.0943
Hebei Province	0.0885	0.0558	0.0832	0.0689	0.0813	0.0935	0.1027	0.1120	0.1021
Shanxi Province	0.0695	0.0365	0.0509	0.0421	0.0509	0.0572	0.0645	0.0653	0.0620
Inner Mongolia Autonomous Region	0.0483	0.0241	0.0340	0.0267	0.0312	0.0333	0.0360	0.0362	0.0351
Liaoning Province	0.1080	0.0750	0.0909	0.0726	0.0838	0.0975	0.1006	0.1017	0.0998
Jilin Province	0.0729	0.0333	0.0454	0.0365	0.0403	0.0404	0.0479	0.0436	0.0438
Heilongjiang Province	0.0709	0.0338	0.0445	0.0342	0.0404	0.0408	0.0473	0.0437	0.0438
Shanghai Municipality	0.2378	0.2378	0.2019	0.2123	0.2347	0.2634	0.2809	0.2758	0.2830
Jiangsu Province	0.3962	0.3476	0.3931	0.4191	0.4173	0.5437	0.5549	0.5958	0.5943
Zhejiang Province	0.2055	0.1654	0.2336	0.2136	0.2460	0.2954	0.2923	0.3636	0.3564
Anhui Province	0.1001	0.0673	0.0987	0.1193	0.1161	0.1161	0.1293	0.1401	0.1283

(*continued*)

Table 2. (*continued*)

Provincial Level	2013	2014	2015	2016	2017	2018	2019	2020	2021
Fujian Province	0.1305	0.0922	0.1198	0.1138	0.1288	0.1521	0.1591	0.1739	0.1616
Jiangxi Province	0.0772	0.0520	0.0681	0.0596	0.0702	0.0887	0.1033	0.1120	0.1012
Shandong Province	0.2052	0.1740	0.2312	0.1961	0.2204	0.2476	0.2282	0.2713	0.2489
Henan Province	0.1254	0.0956	0.1367	0.1131	0.1297	0.1603	0.1741	0.1818	0.1720
Hubei Province	0.1054	0.0787	0.1090	0.0975	0.1037	0.1199	0.1311	0.1389	0.1299
Hunan Province	0.0942	0.0595	0.0842	0.0765	0.0874	0.1044	0.1200	0.1343	0.1195
Guangdong Province	0.5548	0.5046	0.4804	0.5188	0.6032	0.7517	0.7725	0.8823	0.8020
Guangxi Zhuang Autonomous Region	0.0666	0.0351	0.0501	0.0454	0.0544	0.0627	0.0670	0.0722	0.0673
Hainan Province	0.0553	0.0220	0.0262	0.0183	0.0228	0.0227	0.0261	0.0234	0.0241
Chongqing Municipality	0.0981	0.0704	0.0776	0.0765	0.0887	0.1067	0.1156	0.1173	0.1132
Sichuan Province	0.1166	0.0816	0.1025	0.0896	0.2110	0.1536	0.1748	0.1811	0.1698
Guizhou Province	0.0433	0.0199	0.0315	0.0249	0.0331	0.0403	0.0467	0.0479	0.0450
Yunnan Province	0.0481	0.0234	0.0353	0.0289	0.0377	0.0463	0.0519	0.0569	0.0517
Tibet Autonomous Region	0.0511	0.0127	0.0114	0.0081	0.0134	0.0125	0.0170	0.0136	0.0144
Shaanxi Province	0.0868	0.0623	0.0756	0.0666	0.0788	0.0978	0.1063	0.1095	0.1045
Gansu Province	0.0345	0.0148	0.0238	0.0188	0.0254	0.0259	0.0439	0.0307	0.0335

(*continued*)

Table 2. (*continued*)

Provincial Level	2013	2014	2015	2016	2017	2018	2019	2020	2021
Qinghai Province	0.0050	0.0049	0.0080	0.0070	0.0109	0.0105	0.0216	0.0116	0.0146
Ningxia Hui Autonomous Region	0.0206	0.0104	0.0166	0.0120	0.0160	0.0158	0.0233	0.0178	0.0190
Xinjiang Uyghur Autonomous Region	0.0449	0.0196	0.0279	0.0221	0.0270	0.0295	0.0280	0.0356	0.0310

4 Analysis of the Spatial Evolution Patterns of Digital Economy Development Level in China

Spatial geographic factors were taken into consideration when measuring the level of digital economy development. GeoDa software was utilized to import the provincial distribution maps of 31 provinces, municipalities, and autonomous regions in China (excluding Hong Kong, Macau, and Taiwan). Through global spatial autocorrelation analysis and local spatial autocorrelation analysis, the spatial characteristics and evolutionary patterns of digital economy development level in China were studied from macro to micro perspectives.

4.1 Global Spatial Autocorrelation Analysis

After importing the data from the table into GeoDa software and conducting spatial correlation analysis, the global spatial autocorrelation data of digital economy development in China from 2013 to 2021 were presented in Table 3:

Table 3. Global Spatial Autocorrelation Attribute Table

	year								
index	2013	2014	2015	2016	2017	2018	2019	2020	2021
Moran's Index	0.2160	0.1900	0.2860	0.2750	0.1990	0.2120	0.2100	0.2070	0.2080
Z-value	2.5324	2.2014	3.0714	3.1271	2.3486	2.5778	2.5867	2.5643	2.1933
P-value	0.0190	0.0270	0.0050	0.0090	0.0260	0.0220	0.0200	0.0200	0.0380

Based on Table 3, it is evident that, in the year 2013 to 2021, the Moran's I values of digital economy development level in China ranged from 0.190 to 0.286. The significance level (P-value) obtained after 999 randomizations varied from 0.005 to 0.045, all of which are less than 0.05. Furthermore, all z-values exceed the threshold of 1.99. These findings suggest that there was significant positive spatial autocorrelation in the overall digital economy development level in China during the specified time period.

4.2 Local Spatial Autocorrelation Analysis

Based on confirming the positive spatial effect of China's digital economy development during the study period, this paper conducted a local spatial autocorrelation analysis to calculate the spatial correlation of attribute values between individual spatial objects within a specified area and neighboring areas. By analyzing the local feature differences in spatial distribution, it reflected the spatial heterogeneity and instability within local regions. Based on establishing a distribution weight matrix for provinces nationwide, the Geoda software was utilized for univariate local Moran's I index analysis. Taking the local spatial analysis of China's digital economy development level in 2020 as an example,the results of the local correlation analysis were presented in Fig. 1:

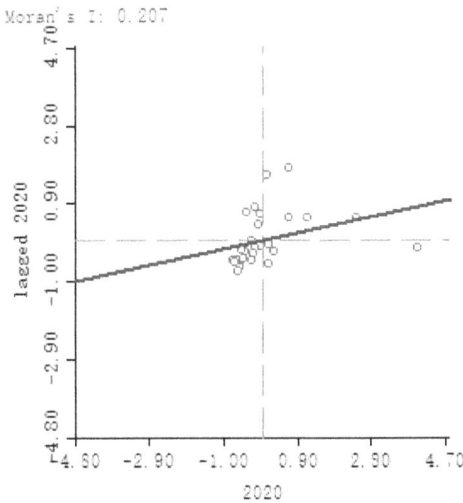

Fig. 1. Local Moran scatter plots of provincial digital economy development levels for selected years in China.

Based on the Moran scatter plots in Fig. 1,we can see that the majority of scatter points in the Moran scatter plot are located in the fourth quadrant,indicating a pronounced positive spatial autocorrelation pattern within the study area. In other words,regions with similar levels of digital economic development tend to cluster together,demonstrating a distinct spatial clustering effect. This study concluded that there was significant spatial autocorrelation in the digital economy development levels across multiple provinces in China. These levels could be classified into four categories based on their own and neighboring areas' digital economy development attributes: High-High (HH) for high levels surrounded by high levels, Low-Low (LL) for low levels surrounded by low levels,

Low-High (LH) for low levels surrounded by high levels,and High-Low (HL) for high levels surrounded by low levels.

Based on computed Moran's I values and Digital Economy Development Index, this study conducted univariate local Moran analysis using GeoDa software, with a significance level set at 95%, to automatically classify spatial autocorrelation patterns across regions. The significant clustering regions were listed in Table 4:

Table 4. Significance Table for 2020

year	HH	HL	LL	LH
2020	Fujian Province 、Shanghai Municipality	Sichuan Province	Xinjiang Uyghur Autonomous Region 、Gansu Province 、Inner Mongolia Autonomous Region	Jiangxi Province

In 2020, there was a spatial clustering effect observed in the development of China's digital economy. The eastern coastal areas of China, including Shanghai and Fujian Province, along with neighboring provinces and municipalities (Guangdong, Jiangxi, Zhejiang, Jiangsu), formed a significantly high-level cluster area of digital economy development ($p \leq 0.05$). However, in some regions of northwest and north China, there appeared significantly low-level clusters of digital economy development ($p \leq 0.05$). Specifically, these include regions centered around Xinjiang, Gansu, and Inner Mongolia, along with their adjacent provinces.

Furthermore, Jiangxi Province was categorized as an LH region, indicating that the province itself had a relatively low level of digital economy development. In contrast, the overall level of neighboring provinces and municipalities (Hubei, Hunan, Anhui, Zhejiang, Fujian, Guangdong) was relatively high. Similarly, Sichuan Province exhibited a relatively high level of digital economy development compared to its neighboring provinces and municipalities (Qinghai, Tibet, Yunnan, Guizhou, Shaanxi, Gansu), which have a relatively low overall level.

Following the steps,the paper analyzed the spatial correlation of digital economy development levels in various provinces and regions of China from 2013 to 2021. The results are listed in Table 5:

Based on Table 5 and in conjunction with the digital economy development index of each province and municipality,it is evident that there had been a significant decrease in the number of LL characteristic areas. Tibet, Qinghai, and Sichuan had gradually escaped from the LL characteristics (low digital economy level both within and neighboring regions), indicating a reduction in the area of low digital economy level clustering in China. These three regions had shown significant improvement in their digital economy levels either internally or in collaboration with neighboring provinces and municipalities.

Sichuan Province had been designated as a high-level characteristic area since 2017, signifying its rapid development in the digital economy and creating a significant gap with neighboring regions such as Qinghai, Tibet, Yunnan, Guizhou, Shaanxi, and Gansu.

Table 5. Agglomeration Table of China's Digital Economy Level from 2013 to 2021

year	HH	HL	LL	LH
2013	Fujian Province		Xinjiang Uyghur Autonomous Region, Tibet Autonomous Region, Qinghai Province, Sichuan Province, Gansu Province, Inner Mongolia Autonomous Region	
2014			Xinjiang Uyghur Autonomous Region, Tibet Autonomous Region, Qinghai Province, Sichuan Province, Gansu Province, Inner Mongolia Autonomous Region	Fujian Province
2015	Fujian Province, Jiangsu Province, Shanghai Municipality		Xinjiang Uyghur Autonomous Region, Tibet Autonomous Region, Qinghai Province, Sichuan Province, Gansu Province, Inner Mongolia Autonomous Region	Jiangxi Province, Anhui Province
2016	Fujian Province, Zhejiang Province, Anhui Province, Shanghai Municipality		Xinjiang Uyghur Autonomous Region, Tibet Autonomous Region, Qinghai Province, Sichuan Province, Gansu Province, Inner Mongolia Autonomous Region	Jiangxi Province
2017	Fujian Province, Shanghai Municipality	Sichuan Province	Inner Mongolia Autonomous Region, Xinjiang Uyghur Autonomous Region	Jiangxi Province

(*continued*)

Table 5. (*continued*)

year	HH	HL	LL	LH
2018	Fujian Province, Shanghai Municipality	Sichuan Province	Inner Mongolia Autonomous Region, Xinjiang Uyghur Autonomous Region, Gansu Province	Jiangxi Province, Anhui Province
2019	Fujian Province, Shanghai Municipality	Sichuan Province	Inner Mongolia Autonomous Region, Xinjiang Uyghur Autonomous Region, Gansu Province	Jiangxi Province
2020	Fujian Province, Shanghai Municipality	Sichuan Province	Xinjiang Uyghur Autonomous Region, Gansu Province, Inner Mongolia Autonomous Region	Jiangxi Province
2021	Fujian Province, Shanghai Municipality	Sichuan Province	Xinjiang Uyghur Autonomous Region, Gansu Province, Inner Mongolia Autonomous Region	Jiangxi Province

The digital economy development index of Sichuan had exhibited a growth rate of 45.62% from 2013 to 2021, ranking third among the 31 provinces and municipalities, indicating rapid progress.

After analyzing the evolution of HH characteristic areas, it was found that Shanghai became an HH characteristic area in 2015. In 2013, Shanghai ranked third among the 31 provinces and municipalities in terms of digital economy development index. However, adjacent provinces Jiangsu and Zhejiang ranked second and fourth, respectively, but their calculated significance p-values were greater than 0.05. This indicates that Shanghai was not identified as an HH characteristic area that year. Shanghai's digital economy level increased by 19.00% from 2013 to 2021. Jiangsu and Zhejiang provinces showed growth rates of 50.00% and 73.43%, respectively, ranking third and second among the 31 provinces and municipalities, demonstrating rapid growth. Fujian Province and its neighboring areas have maintained a strong development momentum, consistently maintaining a high level of digital economy development.

From 2013 to 2021, Anhui and Jiangxi provinces had been identified as areas with characteristics similar to the Lianhua (LH) region at certain times. During this period, Anhui Province ranked eighth and Jiangxi Province ranked seventh in terms of the growth rate of the digital economy level among the 31 provinces and municipalities, indicating rapid development. Anhui Province had gradually shed the LH characteristic area label, narrowing the gap with its surrounding regions. Jiangxi Province's digital economy development index was 0.1012 in 2021, ranking fourteenth among the 31 provinces and municipalities. Despite the progress made by Jiangxi Province, it still fell within the LH characteristic area. This indicated a limited influence from highly developed digital economy areas and emphasizes the significant gap in digital economy development between LL and HH characteristic areas.

5 Conclusion

This study had developed an evaluation system to assess the level of development in the digital economy. The system consisted of primary indicators such as innovation, environmental sustainability, coordination, openness, and sharing, along with 16 secondary indicators. The entropy weight method was used to measure the digital economy's development level in 31 provinces, municipalities, and autonomous regions in China from 2013 to 2021. The study also examined the spatial evolution characteristics of the digital economy's development level in these regions through global spatial autocorrelation analysis and local spatial autocorrelation analysis. The conclusions were as follows:

(1) Over the study period, the digital economy development level indices of various regions ranged from 0.0049 to 0.8823, with dynamic changes observed over time and significant differences between regions.
(2) China's digital economy development level exhibited significant spatial autocorrelation, with clustering effects mainly observed in high-level digital economy development regions composed of some cities in the eastern coastal areas and low-level digital economy development regions composed of cities in the northwest and north.
(3) The clustering area of low-level digital economy development regions in China was gradually decreasing, while the clustering area of high-level digital economy development was expanding rapidly, especially in some coastal provinces and municipalities. However, despite the overall progress in China's digital economy, regional disparities in the level of digital economy development remain significant due to differences in development foundations and speeds.

Disclosure of Interests. The authors have no competing interests to declare that are relevant to the content of this article.

References

1. Oloyede, A.A., Faruk, N., Noma, N., Tebepah, E., Nwaulune, A.K.: Measuring the impact of the digital economy in developing countries: a systematic review and meta-analysis. Heliyon **9**(7), e17654 (2023)
2. Nakamura, L., Sameuls, J., Solveichik, R.H.: Measuring the "free" digital economy within the GDP and productivity accounts. In: Bea Working Papers (2017)
3. Marzena, R., Svitlana, C., Grażyna, D., et al.: Challenges of the MSE sector in the digital economy in Poland and Ukraine: comparative and statistical analysis. Central Eur. Manag. J. **32**(1), 134–151 (2024)
4. Vlasov, M., Polbitsyn, N.S., Olumekor, M., et al.: Socio-cultural factors and components of the digital economy in ethnic minority regions. Sustainability **16**(9), 3825 (2024)

5. Yang, M.: Research on the impact of industrial investment on economic growth in Jilin Province. Mod. Ind. Econ. Informationizat. **12**(10), 1–2+5 (2022)
6. Giuseppe, B., Antonio, D., Carmela, P., et al.: A reduced composite indicator for digital divide measurement at the regional level: an application to the digital economy and society index (DESI). Technol. Forecast. Social Change **190**, 122461 (2023)
7. Chen, Z., Ke, R., Li, H.: A study on the development of China's digital economy based on provincial panel data. Acad. J. Bus. Manag. **4**(10), 19–25 (2022)
8. Wang, S., Tengtang, W., Xia, Q., Bao, H.: The spatio-temporal characteristics of china's digital economy development leveland its driving mechanism of innovation. J. Econ. Geogr. **42**(07), 33–43 (2022)
9. Wang, Z., Xu, Y.: Analysis of the network structure characteristics and its effectson china's digital economy space-based on social network analysis. Finan. Econ. (10), 29–42 (2022)
10. Reinsdorf, M., Schreyer, P.: Measuring consumer inflation in a digital economy. In: Measuring Economic Growth and Productivity, pp. 339-362 (2020)
11. Li, J., Wang, M.: Measurement and spatial-temporal evolution of digital economy development level. Stat. Decis. **38**(24), 73–78 (2022)
12. Luyang, T., Bangke, L., Tianhai, T.: Spatial correlation network and regional differences for the development of digital economy in China. Entropy **23**(12), 1575 (2021)
13. Zhang, L., Li, X., Wang, Z., Wei, L., Zhiqing, X.: Spatial autocorrelation analysis of PM2.5 in Beijing-Tianjin-Hebei region based on Moran's I Index. Sichuan Environ. **40**(02), 53 (2021)
14. Qiu, M.: China's Regional Disparity and Spatial Analysis Based on New HDI, pp. 35–38. Southwest University of Finance and Economics (2019)

Predicting Calibrated Conversion Rate of Online Advertising Using a Multi-task Mixture-of-Experts Calibration Model

Xinyue Zhang[1,2,3], Yuyao Guo[2,3], and Xiang Ao[2,3(✉)]

[1] School of Information Engineering, China University of Geosciences, Beijing, China
`zhangxinyue244@mails.ucas.ac.cn`
[2] Key Lab of AI Safety, Chinese Academy of Sciences(CAS), Beijing, China
[3] Key Lab of Intelligent Information Processing, Institute of Computing Technology CAS, Beijing, China
`aoxiang@ict.ac.cn`

Abstract. Accurately predicting conversion rate (CVR) is paramount in online advertising. However, traditional models may face problems such as delayed feedback, where there is a delay of an indeterminate amount of time between click and conversion. Calibration is an effective way to optimize conversion rate estimates in online advertising. Unlike conversion delays, post-click user behaviors occur rapidly and are informative to conversion rate prediction.

Our proposed solution, the Multi-Task Mixture-of-Experts Calibration (MTMEC) framework, integrates multi-task learning and mixture-of-experts models. It modifies CVR prediction using post-click user behavior data, utilizing streaming learning for real-time data access. Each task is weighted by a gating network, enabling adaptive loss functions through multi-task learning. Parametric scaling further minimizes calibration errors, enhancing prediction accuracy without excessive parameters.

Experiments on real-world datasets validate the effectiveness of MTMEC. It improves model prediction and reduces calibration errors. This framework offers a robust solution for online advertising, bridging the gap between calibration accuracy and system responsiveness.

Keywords: Online advertising · Conversion rate prediction · Calibration · Multi-task learning · Mixture-of-experts

1 Introduction

The evolution of bidding methods underscores the growing importance of precise conversion rate estimation. This aspect of prediction carries significant weight, as it directly influences the optimization of product detail page displays and ultimately maximizes benefits for merchants. As shown in Fig. 1, a CVR model scores all candidates. The displayed ads and user feedback are then recorded, with which we continuously train new models. Advertising systems expose ads to users and collect user behaviors such as clicking, favoriting, and adding items to the shopping cart.

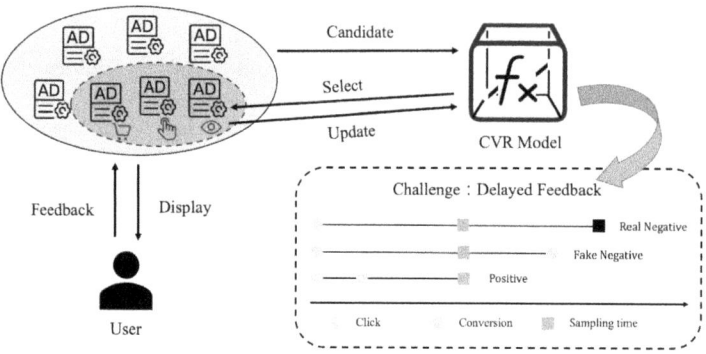

Fig. 1. Illustration of the CVR model's serving and updating process in online ads.

One of the key challenges to CVR prediction for online advertising is the problem of *delayed feedback* [1,2]. This delayed feedback can lead to some positive samples being incorrectly labeled as negative due to the narrow sampling time window, referred to as fake negatives [3]. The presence of these fake negative samples poses challenges to conversion modeling. Biased training data will significantly impact the model's accuracy and reliability in predicting conversions. For the delayed feedback problem, calibration is an effective solution that we can use to further optimize the model structure. Recent calibration methods are relatively simplistic, while post-click behaviors have been proven effective and can be leveraged to further optimize the model structure. In real-world advertising and recommendation systems, there are numerous events related to conversions beyond just the conversion label [4]. For instance, after clicking on an item, users might favorite it or add it to cart. Research by Wen et al. indicates that approximately 12% of items in the shopping cart are eventually purchased, while less than 2% do not make it into the cart [5].

Although there is a strong correlation between post-click behaviors and final conversions, it brings forth new challenges: (1) The correlation between post-click behavior data and conversions within a short timeframe is complex. Moreover, user behaviors may involve multiple dimensions (such as favoriting, adding to cart, etc.), making it difficult for existing calibration methods to handle such intricate relationships effectively. (2) Advertising systems require high real-time performance. Incorporating post-click behavior data, which often involves multiple dimensions and large data volumes, poses challenges. Prolonged model training periods can adversely impact the system's timeliness and responsiveness.

To tackle these challenges, this paper introduces a methodology for calibrating CVR prediction models utilizing a multi-task mixture-of-experts calibration framework. This approach involves capturing user preferences and assembling experts for various tasks through gating networks to personalize learning weight assignments. The primary contributions of this paper are as follows:

- We propose MTMEC, a multi-task mixture-of-experts framework designed to address delayed feedback in online advertising. This framework models various user behavior patterns through experts, each with dedicated gating and activation units for weight assignment across tasks.
- To calibrate the conversion model further, we incorporate a scaling layer and freeze the previously trained parameters to fine-tune the existing prediction model, ensuring more precise calibration of the CVR.
- We validate the effectiveness of our model using two real-world datasets. The experimental results show that our model outperforms existing methods.

2 Related Work

Delayed Feedback Models. To address the delayed feedback issue, extending the time window with a waiting period is common for collecting more realistic data samples. However, this approach does not fully eliminate biases, and training on such biased data can significantly harm performance [6]. Recent stream learning approaches either eliminate the waiting window [3] or use a short waiting window [6,7].

DFM was the first study to emphasize the delayed feedback issue [1], modeling the time delay between ad exposure and eventual conversion, but it is challenging to apply this method to online training. FSIW [8], ES-DFM [9], DEFER [10], and DEFUSE [11] redesign the data pipeline and utilize the importance sampling method to estimate the actual distribution of CVR. Ktena et al. used a fake negative data pipeline in online streaming and recommended the use of fake negative weighted loss (FNW) and fake negative calibration (FNC) [3]. The FTP model method constructs an ideal data subset and trains a Prophet model as the current time's Oracle model to optimize the conversion rate before the maximum feedback window [2].

Calibration Methods. Calibration methods proposed for deep neural networks can be classified into three categories based on the stage at which they are applied: regularization during the training phase, uncertainty estimation methods for neural networks, and post-processing calibration methods.

Regularization methods applied in the training phase optimize the size of the prediction value end-to-end by modifying the prediction target, regularization process, etc., to construct essentially calibrated deep neural networks. Neural network uncertainty estimation based on Bayesian methods [12,13] or deeply integrated [14,15] methods, attempts to invert the factors that lead to bias in model prediction into the feature processing or loss function. Post-processing methods are more flexible and are currently the most widely studied and applied form [16–18]. That is, a calibration module is strung on top of the underlying prognostic model, usually obtaining a mapping function on an independent validation set, which converts the raw output into calibration results.

Bin-based post-processing calibration methods enhance predictive model reliability by aligning predicted probabilities with observed outcomes. Histogram Binning (HB) divides the range of predicted probabilities into fixed-width bins and adjusts the predictions within each bin to match the observed frequencies. [17]. Bayesian Binning into Quantiles (BBQ) uses equal-frequency binning integrated by the BDeu score [19]. Isotonic Regression (IR) learns a non-decreasing function for calibration [20]. Mix-n-Match (MNM) [21] uses the weighted average output of different calibration models as the calibration output. Smoothed Isotonic Regression (SIR) [22] obtains post-calibration predictions for each sample by linear interpolation on top of the order-preserving regression. MBCT proposed Feature-aware Binning [23]. Neural Calibration integrates parameterization techniques with neural networks and uses neural networks for calibration [24].

3 Calibration Metrics

3.1 LogLoss

The logarithmic loss function, also known as cross-entropy loss, is commonly used to evaluate calibration error [16].

$$\text{LogLoss} = -\frac{1}{|\mathcal{D}|} \sum_{i=1}^{|\mathcal{D}|} [y_i \log p_i + (1 - y_i) \log (1 - p_i)] . \tag{1}$$

3.2 Brier Score

The Brier Score measures the mean square error between the algorithm's predictions and the true values, denoted as:

$$\text{BrierScore} = \frac{1}{|\mathcal{D}|} \sum_{i=1}^{|\mathcal{D}|} (y_i - p_i)^2 , \tag{2}$$

where the $\mathcal{D}$ is the number of samples, i denotes the ith sample, p_i is the model predicted label, and y_i is the true label.

3.3 ECE

Expected Calibration Error (ECE) measures the average difference between model accuracy and confidence. It is calculated by dividing the sample into K buckets and is expressed as:

$$\text{ECE} = \frac{1}{|\mathcal{D}|} \sum_{k=1}^{K} \left| \sum_{i=1}^{|\mathcal{D}|} (y_i - p_i) \, \mathbf{1}_{[a_k, b_k)} (p_i) \right| , \tag{3}$$

where $\mathbf{1}_{[a_k, b_k)}(p_i)$ is an indicator with a value of 1 if the predicted probability p_i located in the $[a_k, b_k)$, and otherwise 0.

3.4 AUC

Area under the Curve of ROC (AUC) quantifies the probability that a randomly chosen positive sample and a randomly chosen negative sample from the test set will be correctly classified, with the positive sample's predicted value exceeding that of the negative sample.

$$\text{AUC} = \frac{\sum_{i \in \mathcal{X}^+} rank_i - \frac{|\mathcal{X}^+| \times (|\mathcal{X}^+|+1)}{2}}{|\mathcal{X}^+| \times |\mathcal{X}^-|}, \quad (4)$$

where $\mathcal{X}^+$ and $\mathcal{X}^-$ represent the positive and negative classes in the samples, and $rank_i$ denotes the ranking of the predicted scores for sample i.

4 Methodology

4.1 Calibrated Conversion Model

In the context of online learning, predicting the CVR without calibration can be considered as a binary classification task. To address delayed feedback in advertising systems, we introduce waiting window w_w for capturing prompt conversions and attribution window w_a for setting the maximum delay in sample acceptance.

The training set $\mathcal{D}_{\text{train}}$ of the converted model at the current timestamp of τ is denoted as:

$$\mathcal{D}_{\text{train}} = \{(s_i, t_i, x_i, y_{i,\tau})\}_{i=1}^{N_\tau - w_w} \mid s_i < \tau - w_w, \quad (5)$$

x represents a specific item sample, s represents the time of the click when the state of the item sample is recorded, t represents the time when the conversion of the item label occurs, and $y_{i,\tau}$ is the label of the sample at time τ. If the label is not converted at the time τ, we record $t_i = \infty$ and $y_{i,\tau} = 0$. If no conversion record is obtained before $s_i + w_a$, the sample is considered a negative sample, and the label y_i is denoted as follows:

$$y_i = \begin{cases} 1, & t_i \leq s_i + w_a \\ 0, & t_i > s_i + w_a \end{cases} \quad (6)$$

The development set $\mathcal{D}_{\text{dev}}$ is divided based on user behavior occurring within a specific time window w_b following a click, ensuring w_b does not exceed the waiting window w_w.

$$\mathcal{D}_{\text{dev}} = \{(s_i, t_i, x_i, y_{i,j,\tau})\}_{i=1}^{N_\tau - w_w + w_b} \mid t_{i,j} <= \tau - w_w + w_b, \quad (7)$$

where $t_{i,j}$ denotes the time when the jth behavior of sample i was logged.

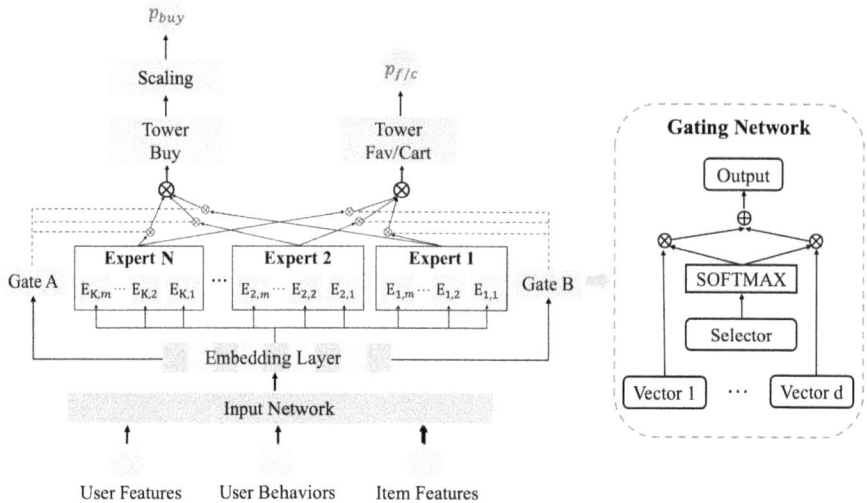

Fig. 2. Multi-Task Mixture-of-Experts Calibration (MTMEC) Model

4.2 Multi-task Mixture-of-Experts Calibration Model

The model structure of MTMEC is illustrated in Fig. 2. The model consists of three primary components: input network, mixture-of-experts network, and multi-gating network.

Input Network. The training data encompasses user features, user behaviors, and item features, which are input into the input network for processing. The training set $\mathcal{D}_{\text{train}}$ and devlopment set $\mathcal{D}_{\text{dev}}$ are divided based on the w_w, w_a and w_b. Given the presence of numerous sparse features in the input, both $\mathcal{D}_{\text{train}}$ and $\mathcal{D}_{\text{dev}}$ are channeled into the embedding layer. Here, the original input x of a sample is represented as $x = [x_1, x_2, ..., x_d]$, where d denotes the number of distinct features. The embedding layer is defined as:

$$\hat{e}_{im} = w_{im} e_{im} + w_{if} \phi_i(x_i) + w_{iu} \psi_i(x_i)$$
$$\hat{e}_m = [\hat{e}_{1\,m}, \hat{e}_{2\,m}, ...\hat{e}_{Nm}] \,, \tag{8}$$

where $\hat{e}_{im}$ is the embedding of the ith feature, w_{im}, w_{if}, w_{iu} are trainable linear transformation weights. $\phi_i(x_i)$ and $\psi_i(x_i)$ are the transformation functions of the embedding layer for embedding, respectively. Next, the embedding $\hat{e}_m$ is input to the expert network and gating network in the MTMEC model.

Mixture-of-Experts Network. The modules controlling the experts are shared among all the tasks at the bottom of the multi-task model, for which the user's personality behaviors are captured. Experts are introduced as each sub-problem needs to be solved by a different expert. Each expert is an independent entity with weights and parameters, and consists of multiple self-networks.

All experts amalgamate to form a mixture-of-experts layer, receiving input from the embedding layer and transmitting it to the subsequent layer's tower network for end-to-end model training. The mixture-of-experts is delineated as:

$$y = \sum_{i=1}^{N} g(x)_i f_i(x) , \qquad (9)$$

$\sum_{i=1}^{N} g(x)_i = 1$, the ith logit of the $g(x)_i$ output represents the probability of the expert f_i. Here, f_i, i = 1, ..., N are N expert networks. Subsequently, experts are mobilized for diverse tasks through a gating network. Tailored to the number of tasks, the gating network activates the expert system independently, assigning distinct weights. When addressing task k, the selected experts are denoted by the $E^k(x)$ matrix.

$$E^k(\mathbf{x}) = \left[E_{(K,m)}^T, \ldots, E_{(K,1)}^T, \ldots, E_{(2,n)}^T, \ldots, E_{(2,1)}^T, \ldots, E_{(1,r)}^T, \ldots, E_{(1,1)}^T \right] . \qquad (10)$$

Multi-Gating Network. The user behavior sequences are fed into the expert and gating networks, with one gating unit and one activation unit for each task. For the kth task, the individual gating network added is g_k, as in Fig. 2. The gating network is based on a single-layer feed-forward network, For each gating network use an input linear transformation with a SoftMax layer:

$$g^k(\mathbf{x}) = \text{softmax}(M_{gk} x) , \qquad (11)$$

where $M_{gk} \in \mathbb{R}^{N \times d}$ is a trainable matrix. The predicted value of the final k^{th} task is:

$$p_k = t^k \sum_{i=1}^{N} g^k(x)_i f_i(x) , \qquad (12)$$

where t^k denotes the tower network of task k.

Uncertainty-Based Optimization. Multi-task learning involves optimizing all losses together, but model performance can be influenced by manually adjusting weights, which is a tedious task. Specifically, different behaviors' contributions and correlations with conversions vary and may change dynamically. Here, we've adopted an uncertainty-based approach [25] to automate weight acquisition, continuously adjusting loss weights for tasks during multi-task training. The general formulation of the multi-task joint loss function $\mathcal{L}$ is as follows:

$$\mathcal{L} = \frac{1}{S} \sum_{i=1}^{S} \sum_{j=1}^{K} \frac{1}{exp(\sigma_j)} \cdot ((y_{i,j} - p_{i,j})^2 + \sigma_j) , \qquad (13)$$

where S represents the number of samples, and the trainable parameters σ_j signifies the variance of the jth output variable.

Parameterized Scaling. After training the mixture-of-experts model, the parameters are then fixed and undergo parametric scaling to further optimize the calibration effect. Specifically, a *sigmoid* function is applied after the tower network output l_{tower} for the CVR calibration task, defined as follows:

$$q(y=1) = \frac{1}{1+\exp\left(-(al_{\text{tower}})+b\right)} \ . \tag{14}$$

The model optimizes the parameters a and b using maximum likelihood estimation from the development set $(l_{\text{tower, i}}, y_i)$, which is trained using the $\mathcal{D}_{\text{train}}$.

5 Experiments

5.1 Datasets

We evaluate MTMEC on two real-world datasets on E-commerce, including the **CIKM2019 EComm AI**[1] and **Taobao UserBehavior**[2] dataset. The CIKM 2019 E-Commerce dataset, provided by the CIKM2019 EComm AI challenge, contains 16 days of data. It includes user characteristics (i.e., ID, age, purchasing power), user behaviors (i.e., click, purchase, favorite, add-to-cart), and item category IDs. There are 47,867,752 samples in total, tested online over 10 days of data. The Taobao UserBehavior dataset, provided by Alibaba, contains 82,219,000 samples. It also includes user IDs, item IDs, item categories, and user behaviors. We conduct online testing on the last 5 days of data from the Taobao dataset, excluding conversions with delays exceeding 3 days. The statistical information of the two processed datasets is shown in Table 1, where r_{buy} signifies the average buy rate, r_{fav} signifies the average favorite rate, and r_{cart} signifies the average add-to-cart rate.

Table 1. Statistics of CIKM2019 and Taobao dataset.

Dataset	#features	#samples	#log period	#w_a	#r_{buy}	#r_{fav}	#r_{cart}
CIKM2019	13	47,867,752	16 days	10 days	0.0347	0.0168	0.0055
Taobao	7	82,219,000	9 days	5 days	0.0198	0.0139	0.0053

5.2 Competitors

We compare MTMEC against baseline models to evaluate its performance.

Base Model (Base) is trained on the training set $\mathcal{D}_{\text{train}}$.

Valid model (Valid) is trained on the $\mathcal{D}_{\text{train}}$ and the behavioral set $\mathcal{D}_{\text{dev}}$.

---

[1] https://tianchi.aliyun.com/competition/entrance/231721/introduction.
[2] https://tianchi.aliyun.com/dataset/649.

Histogram Binning Uniform Mass (HBUM) is a data binning technique that divides data into equally sized bins and uses the average predicted values to estimate bias for each bin.

Bayesian Binning into Quantiles (BBQ) uses equal-frequency binning models with varying bin counts, integrated by the BDeu score.

Isotonic Regression (IR) is a non-parametric regression method designed to learn a segmented non-decreasing constant function to calibrate the data.

Platt Scaling (PS) ifits a logistic regression model.

Temperature Scaling (TS) adjusts the classifier's output by introducing an additional learnable parameter, temperature T.

Scaling Binning (SB) bins the data after transforming or normalizing the features or predictions using a scaling function.

Beta Calibration (Beta) assumes data follows a beta distribution and learns location parameters during training alongside regular model parameters.

Neural Calibration (Nerual) combines ILPS and neural networks. **KD Calibration (KD)** uses knowledge distillation to train a teacher model on full feedback data and a student model on online samples.

5.3 Experimental Settings

Online Simulation. In the experiments, we split the dataset into daily intervals based on behavior type and timestamp to simulate online training. For the CIKM dataset, we set $w_a = 5$ days and $w_w = 1$ day, and for the Taobao dataset, $w_a = 3$ days and $w_w = 1$ day. During calibration, we use post-click user behavior data as the development set $\mathcal{D}_{\text{dev}}$, training models, and the scaling layer with $w_w = 1$ day. To ensure fair comparisons, we train existing calibration methods and retrain a validation model using $\mathcal{D}_{\text{dev}}$. Model evaluation occurs on the day following τ, repeating this process.

Parameter Settings. The base CVR calibration model uses a DNN with fixed hidden size [128,128] and utilizes the Adam optimizer with a learning rate of 0.001 for updates. The MTMEC model consists of 12 expert networks, each with a single fully connected layer of size 4. The tower network has a single-layer structure with a size of 8.

5.4 Results

Primary Analysis. (1) The experiments in Table 2 demonstrate that MTMEC outperforms all baseline models significantly. MTMEC model results also illustrate the validity of calibration using post-click behaviors. The selection of post-click behaviors is reflected in the division of the development set. It is unrelated to the model structure, thus not leading to the complexity of parameters and the model. MTMEC model employs shared experts, with each individual task controlled by a separate gating network. This approach allows for the capture

Table 2. Performance comparisons of the proposed model with baseline methods.(LL: Log Loss, BS: Brier Score)

Method	CIKM				Taobao			
	AUC↑	LL↓	BS↓	ECE↓	AUC↑	LL↓	BS↓	ECE↓
Base	0.6885	0.1713	0.0350	0.0267	0.6899	0.1257	0.0177	0.0166
Valid	0.6518	0.1722	0.0355	0.0258	0.6356	0.1054	0.0188	0.0121
HBUM	0.6873	0.1483	0.0351	**0.0040**	0.6908	0.0854	0.0173	**0.0017**
BBQ	0.6839	0.1484	0.0342	0.0042	0.6512	0.0865	0.0174	0.0019
IR	0.6898	0.1484	0.0342	0.0040	0.6928	0.0853	0.0173	0.0017
PS	0.6891	0.1482	0.0342	0.0041	0.6930	0.0853	0.0173	0.0018
TS	0.6891	0.1482	0.0342	0.0052	0.6924	0.0861	0.0174	0.0017
SB	0.6872	0.1483	0.0342	0.0040	0.6911	0.0854	0.0173	0.0017
Beta	0.6892	0.1480	0.0342	0.0040	0.6930	0.0853	0.0173	0.0018
Neural	0.6901	0.1703	0.0350	0.0159	0.6864	0.1308	0.0177	0.0132
KD-(0.2)	0.6864	0.1897	0.0326	0.0264	0.6864	0.1308	0.0177	0.0168
MTMEC	0.6951	0.1869	0.0326	0.0042	0.6970	0.1264	0.0177	0.0018
MTMEC-S	**0.6965**	**0.1378**	**0.0315**	0.0041	**0.6994**	**0.0850**	**0.0173**	0.0018

of differences between tasks and the rational utilization of experts without the need for additional model parameters, thereby maintaining model simplicity.

(2) By comparing the experimental results of MTMEC with various post-processing calibration methods, we conclude that the scaling layer is very effective, and the specific data are shown in Table 2. In our experiments, we find that the scaling layer can flexibly adjust the prediction results according to the characteristics of the samples and the distribution of the data, which reduces the degree of overconfidence in the model and improves the accuracy and stability of the prediction. This finding is significant for improving the accuracy and practicality of conversion rate prediction in online advertising systems.

Parameter Tuning Analysis. (1) Figure 3 visualizes how the model's accuracy varies with the number of experts and the duration of the different waiting windows. As shown in Fig. 3(a), varying the waiting window from 1 to 4 days, the model's accuracy was best for CVR calibration when the waiting window $w_w = 1$ day. It illustrates that to solve the problem of delayed advertisement feedback, the key challenge is to find the right balance to ensure the samples' freshness and the labels' accuracy.

(2) As shown in Fig. 3(b), optimal CVR calibration involves 12 experts. Increasing the number of experts could improve model performance through enhanced information diversity, boosting generalization and accuracy. However, excessive experts may lead to overfitting and increased computational costs. Therefore, careful evaluation and experimentation are crucial when adjusting the number of experts.

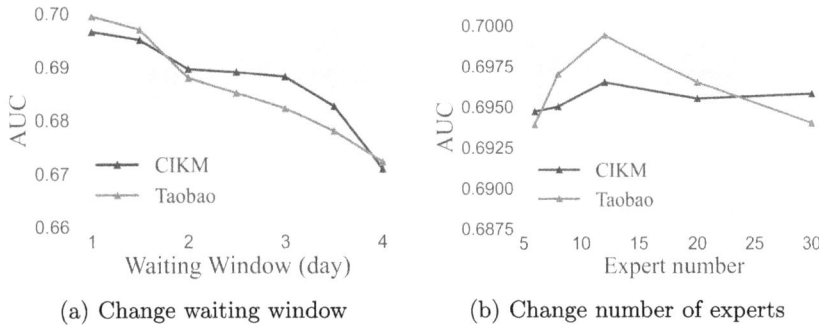

Fig. 3. Impact of waiting window and number of experts to models under two datasets

6 Conclusion and Future Work

This paper introduces a novel CVR calibration model named the Multi-Task Mixture-of-Experts Calibration Model designed to enhance information sharing and joint representation learning effectively. Mixture-of-experts networks are employed to capture diverse feature interaction patterns, with separate gating networks established for each task. These gating networks, comprising gating and activation units, facilitate refined weight allocation, improving prediction accuracy and the model's robustness and generalization, particularly for long-tailed user behaviors. Importantly, MTMEC incorporates parameter sharing and simplified network structures, ensuring that the model does not significantly increase parameter count or deployment complexity. Online simulation experiments conducted on two public datasets demonstrate a significant enhancement in MTMEC performance compared to baseline methods. Future research will focus on identifying commonalities and distinctions among user behaviors for further utilization.

Acknowledgments. The research work is supported by the National Key R&D Plan No. 2022YFC3303302. Xiang Ao is also supported by the Project of Youth Innovation Promotion Association CAS, and Beijing Nova Program 20230484430.

References

1. Chapelle, O.: Modeling delayed feedback in display advertising. In: Proceedings of the 20th ACM SIGKDD International Conference on Knowledge Discovery and Data Mining, pp. 1097–1105 (2014)
2. Li, H., et al.: Follow the prophet: accurate online conversion rate prediction in the face of delayed feedback. In: Proceedings of the 44th International ACM SIGIR Conference on Research and Development in Information Retrieval, pp. 1915–1919 (2021)
3. Ktena, S.I., et al.: Addressing delayed feedback for continuous training with neural networks in ctr prediction. In: Proceedings of the 13th ACM Conference on Recommender Systems, pp. 187–195 (2019)

4. Wen, H., Zhang, J., Lv, F., Bao, W., Wang, T., Chen, Z.: Hierarchically modeling micro and macro behaviors via multi-task learning for conversion rate prediction. In: Proceedings of the 44th International ACM SIGIR Conference on Research and Development in Information Retrieval, pp. 2187–2191 (2021)
5. Wen, H., et al.: Entire space multi-task modeling via post-click behavior decomposition for conversion rate prediction. In: Proceedings of the 43rd International ACM SIGIR Conference on Research and Development in Information Retrieval, pp. 2377–2386 (2020)
6. Yang, J.Q., et al.: Capturing delayed feedback in conversion rate prediction via elapsed-time sampling. In: Proceedings of the AAAI Conference on Artificial Intelligence, vol. 35, pp. 4582–4589 (2021)
7. Gu, S., Sheng, X.R., Fan, Y., Zhou, G., Zhu, X.: Real negatives matter: continuous training with real negatives for delayed feedback modeling. In: Proceedings of the 27th ACM SIGKDD Conference on Knowledge Discovery & Data Mining, pp. 2890–2898 (2021)
8. Zhou, G., et al.: Deep interest network for click-through rate prediction. In: Proceedings of the 24th ACM SIGKDD International Conference on Knowledge Discovery & Data Mining, pp. 1059–1068 (2018)
9. Wang, R., Fu, B., Fu, G., Wang, M.: Deep & cross network for ad click predictions. In: Proceedings of the ADKDD 2017, pp. 1–7 (2017)
10. Gawlikowski, J., et al.: A survey of uncertainty in deep neural networks. Artif. Intell. Rev. **56**(Suppl 1), 1513–1589 (2023)
11. Chen, Y., et al.: Asymptotically unbiased estimation for delayed feedback modeling via label correction. In: Proceedings of the ACM Web Conference 2022, pp. 369–379 (2022)
12. Zhang, Z., Dalca, A.V., Sabuncu, M.R.: Confidence calibration for convolutional neural networks using structured dropout. arXiv preprint arXiv:1906.09551 (2019)
13. Wilson, A.G., Izmailov, P.: Bayesian deep learning and a probabilistic perspective of generalization. Adv. Neural. Inf. Process. Syst. **33**, 4697–4708 (2020)
14. Mehrtash, A., Wells, W.M., Tempany, C.M., Abolmaesumi, P., Kapur, T.: Confidence calibration and predictive uncertainty estimation for deep medical image segmentation. IEEE Trans. Med. Imaging **39**(12), 3868–3878 (2020)
15. Lakshminarayanan, B., Pritzel, A., Blundell, C.: Simple and scalable predictive uncertainty estimation using deep ensembles. Adv. neural Inf. Process. Syst. **30** (2017)
16. Guo, C., Pleiss, G., Sun, Y., Weinberger, K.Q.: On calibration of modern neural networks. In: International Conference on Machine Learning, pp. 1321–1330. PMLR (2017)
17. Zadrozny, B., Elkan, C.: Obtaining calibrated probability estimates from decision trees and naive bayesian classifiers. In: ICML, vol. 1, pp. 609–616 (2001)
18. Gupta, C., Podkopaev, A., Ramdas, A.: Distribution-free binary classification: prediction sets, confidence intervals and calibration. Adv. Neural. Inf. Process. Syst. **33**, 3711–3723 (2020)
19. Naeini, M.P., Cooper, G., Hauskrecht, M.: Obtaining well calibrated probabilities using bayesian binning. In: Proceedings of the AAAI Conference on Artificial Intelligence, vol. 29 (2015)
20. Zadrozny, B., Elkan, C.: Transforming classifier scores into accurate multiclass probability estimates. In: Proceedings of the Eighth ACM SIGKDD International Conference on Knowledge Discovery and Data Mining, pp. 694–699 (2002)

21. Zhang, J., Kailkhura, B., Han, T.Y.J.: Mix-n-match: ensemble and compositional methods for uncertainty calibration in deep learning. In: International Conference on Machine Learning, pp. 11117–11128. PMLR (2020)
22. Jiang, X., Osl, M., Kim, J., Ohno-Machado, L.: Smooth isotonic regression: a new method to calibrate predictive models. In: AMIA Summits on Translational Science Proceedings 2011, vol. 16 (2011)
23. Huang, S., et al.: Mbct: tree-based feature-aware binning for individual uncertainty calibration. In: Proceedings of the ACM Web Conference 2022, pp. 2236–2246 (2022)
24. Pan, F., et al.: Field-aware calibration: a simple and empirically strong method for reliable probabilistic predictions. In: Proceedings of The Web Conference 2020, pp. 729–739 (2020)
25. Kendall, A., Gal, Y., Cipolla, R.: Multi-task learning using uncertainty to weigh losses for scene geometry and semantics. In: Proceedings of the IEEE Conference on Computer Vision and Pattern Recognition, pp. 7482–7491 (2018)

A Transformer-Based Spatio-Temporal Graph Neural Network for Anomaly Detection on Dynamic Graphs

Yuanjun Gao, Quntao Zhu, Xuanhua Shi(✉), and Hai Jin

National Engineering Research Center for Big Data Technology and System, Services Computing Technology and System Lab, Cluster and Grid Computing Lab, School of Computer Science and Technology, Huazhong University of Science and Technology, Wuhan 430074, China
{gaoyj,quntaozhu,xhshi,hjin}@hust.edu.cn

Abstract. Dynamic graphs have emerged as a pivotal data structure underpinning real-world network applications. Against this backdrop, detecting anomalies in dynamic graphs has become particularly important, serving as a foundation for multiple downstream tasks. However, existing research has been limited by only considering the simplistic spatial features of the target edge and being confined to local spatial structures. This confinement imposes significant limitations on model performance. Furthermore, current models have not adequately addressed the issue of missing node attributes, relying solely on one-hot encoding to represent nodes, which restricts the expressive capacity of spatial features. In response to these challenges, we propose an end-to-end transformer-based spatio-temporal graph neural network model called **TransSTGNN** for detecting anomalous edges on dynamic graphs. This model innovatively integrates the graph convolution network with the transformer model to extract multi-dimensional spatial features of the target edge. Simultaneously, TransSTGNN incorporates a novel node enhancement module that reinforces the representation capability of spatial features by encoding the spatial attributes of nodes, effectively tackling the problem of missing node attributes. Extensive experiments confirm that TransSTGNN surpasses the state-of-the-art methods in anomaly detection on six benchmark datasets and effectively resolves the challenges mentioned above.

Keywords: Dynamic graphs · Anomaly detection · Graph convolution networks · Transformer · Node enhancement

1 Introduction

With the increasing development of network applications, graphs, as core data structures, have extended beyond static entities and interactions. To better reflect temporal changes in entities and relationships, dynamic graphs have been widely adopted. These graphs effectively integrate temporal features with the topological structure, adding a temporal dimension to network interactions. This integration allows for deeper analysis that leverages temporal features, revealing time-related events that are difficult to detect in static graphs. For instance, in recommendation systems [7], an in-depth analysis of users' behavioural patterns over time in shopping networks enables precise predictions about future interests and needs. Specifically, the system can track and analyze multi-dimensional data such as users' purchase history, browsing habits, and social interactions, utilizing these temporal sequences to accurately predict new products or services that may pique users' interest.

In the era of big data, global interconnectivity has spurred the rapid expansion of the scale of graph data. This massive dataset not only encompasses a wealth of information but may also conceal potential anomalous behaviours that could threaten the healthy development of networks. Consequently, anomaly detection on dynamic graphs has emerged as a crucial task, generating widespread research interest. For example, analyzing and identifying anomalous behaviour patterns on social media platforms can effectively curb the spread of false information [15,20]. In the context of shopping network comment analysis, it is essential to identify misleading comments to protect consumers from harmful information [14,24]. In the financial sector, the timely detection of abnormal trading patterns aids in preventing financial fraud [6,23].

It's particularly challenging to detect anomalies in dynamic graphs. The first challenge is acquiring rich spatial features to enhance the model's sensitivity to changes in spatial structures. Current efforts have not given adequate attention to spatial features, resulting in overly simplistic extraction of such features. Specifically, existing studies [1,25] primarily utilize one-hot encoding to represent node features and employ the Graph Convolution Network (GCN) to extract adjacency relationships between nodes. However, GCN tends to generate simplistic adjacency representations, making it difficult to comprehensively capture the structural characteristics of the target edge and its neighboring nodes. As the variety of anomalous data types increases, relying solely on GCN becomes insufficient for detecting complex anomaly patterns. The second challenge involves capturing the spatial structure of the target edge from multiple dimensions. Existing research [1,12,25] predominantly focuses on the local spatial structure of the target edge. This limitation arises because aggregating distant neighbor information through GCN requires deeper network structures, leading to the issue of over-smoothing. Over-smoothing occurs when node features gradually homogenize and lose their distinctiveness as the number of network layers increases. Consequently, they often limit the number of GCN layers, restricting the model's ability to capture the global spatial structure.

In response to the outlined challenges, we propose an end-to-end **Trans**former-based **S**patio-**T**emporal **G**raph **N**eural **N**etwork model, named **TransSTGNN**, which specializes in identifying abnormal edges within dynamic graphs. TransSTGNN comprises two primary components: a dynamic graph encoder and an anomaly detector. The dynamic graph encoder includes subgraph sampling, attention sampling, node enhancement, and spatio-temporal feature encoding modules. The subgraph sampling module captures the local spatial structure of the target edge by sampling h-hop subgraphs centered around it; the attention sampling module focuses on the global spatial structure of the target edge by sampling the Global Relevance Node Set (GRS) based on a relevance matrix; the node enhancement module employs a novel encoding method to enhance the spatial attributes of subgraph nodes and those in the GRS; the spatio-temporal feature encoding module innovatively combines the transformer model with the GCN to extract multidimensional spatial features. It utilizes the transformer model to aggregate spatial attributes of nodes in the GRS, effectively avoiding the over-smoothing issues commonly associated with deep GCN. Meanwhile, the local spatial structures of the target edge are extracted by the GCN, which addresses the limitations of the transformer model in this regard. Subsequently, the features are fused to obtain enhanced spatial features of the target edge. Finally, the module employs the Long Short-Term Memory (LSTM) network to model the evolution patterns of spatial features, thereby obtaining the spatio-temporal features of the target edge. Moreover, we utilize a multilayer perceptron as the anomaly detector, leveraging the spatio-temporal features to calculate the anomaly probability of the target edge.

We summarize the principal contributions of this paper as follows:

(1) We integrate the Transformer model with the GCN to extract multidimensional spatial features for the first time. This combination significantly enhances the model's ability to perceive global spatial structures and effectively mitigates the over-smoothing issue commonly encountered in GCN.
(2) We design a novel node enhancement module that encodes the statistical characteristics of nodes from multiple aspects, endowing them with enhanced spatial attributes. This module significantly enhances the representational capability of spatial features.
(3) We introduce a transformer-based spatio-temporal graph neural network model (**TransSTGNN**), specializing in identifying abnormal edges within dynamic graphs.
(4) TransSTGNN surpasses state-of-the-art methods on six datasets from real-world dynamic graphs, demonstrating its superior ability to capture rich and diverse spatial features that significantly enhance model performance.

2 Related Work

Early research primarily relied on traditional data processing methods. CM-Sketch [18] comprehensively considers local structural features and historical behaviors to identify anomalous edge characteristics. Node2vec [5] generates random walk paths for nodes by combining breadth-first and depth-first search strategies. Simultaneously, this algorithm employs Skip-gram techniques to learn node embeddings for anomaly detection. DeepWalk [17] derives node embeddings based on the context within multiple short random walks to detect anomalies.

However, the effectiveness of these approaches is limited due to their reliance on shallow mechanisms for anomaly detection.

Advances in deep learning technology have significantly improved the performance of models. NetWalk [21] employs an autoencoder for node representation learning and projects edges into a common latent space through node embeddings. The anomaly probability of an edge is calculated based on its distance from the nearest edge cluster center. As the graph evolves, the embedding representations of edges are updated accordingly. However, NetWalk is more sensitive to short-term anomalies while overlooking the long-term evolutionary patterns of edges. AddGraph [25] combines the GCN with the Gated Recurrent Units (GRU) equipped with a self-attention mechanism [2]. This model extracts spatial features of edges using the GCN and models their evolution with the GRU. It also aggregates historical information through the self-attention mechanism, fed into the GRU as a long-term feature. Inspired by AddGraph, StrGNN [1] similarly employs the GCN and GRU to model dynamic graphs but limits anomaly detection to the h-hop subgraph of the target edge, significantly enhancing detection efficiency. Nevertheless, both AddGraph and StrGNN place insufficient emphasis on spatial features, as they both use one-hot encoding as the sole attribute of nodes and extract simple spatial features through the GCN. TADDY [12] designs a node encoding method aimed at fusing temporal information and spatial structures of nodes. Subsequently, it adopts the transformer to aggregate the encodings of the endpoints of the target edge and their neighboring nodes for anomaly detection. However, TADDY fails to comprehensively consider the spatial structure of the target edge from multiple dimensions and does not focus on the evolutionary characteristics of dynamic graphs.

3 Proposed TransSTGNN Framework

In this section, we detail the structure of TransSTGNN. As illustrated in Fig. 1, TransSTGNN primarily consists of two main components: a dynamic graph encoder and an anomaly detector. The dynamic graph encoder comprises four core modules: subgraph sampling, attention sampling, node enhancement, and spatio-temporal feature encoding. Additionally, we utilize a multilayer perceptron as the anomaly detector.

3.1 Problem Definition

Definition 1 (Dynamic graph). *A dynamic graph is a collection of snapshots arranged in temporal order, where each snapshot represents the state of the dynamic graph at a specific time. Given a dynamic graph with a maximum timestamp T, it can be represented as $\mathbb{G} = \{\mathcal{G}^1, \ldots, \mathcal{G}^T\}$. Each $\mathcal{G}^t$ represents the snapshot of the dynamic graph at timestamp t, specifically in the form $\mathcal{G}^t = (\mathcal{V}^t, \mathcal{E}^t)$, where $\mathcal{V}^t$ and $\mathcal{E}^t$ denote the sets of nodes and edges at that time. Furthermore, let n^t and m^t denote the number of nodes and edges, respectively, in the snapshot $\mathcal{G}^t$, that is $n^t = |\mathcal{V}^t|$, $m^t = |\mathcal{E}^t|$.*

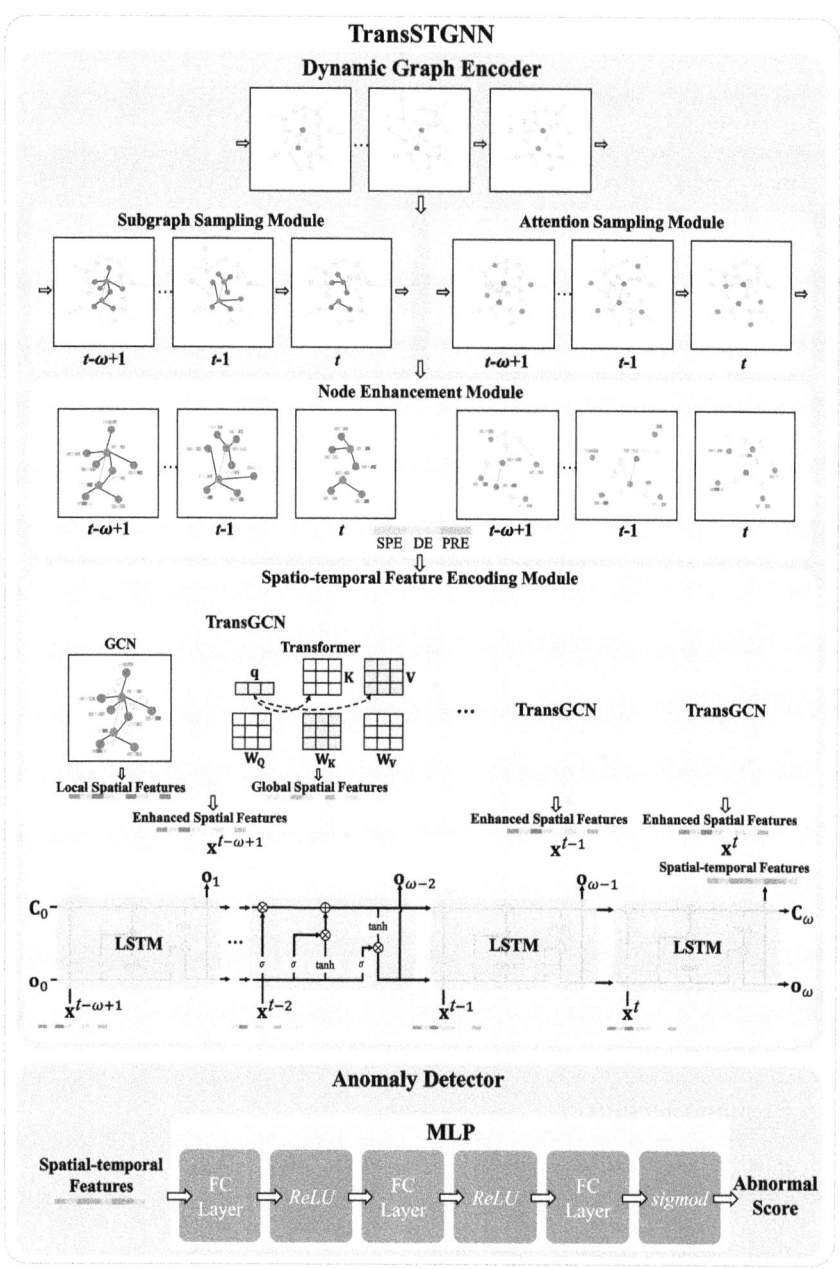

Fig. 1. The overall framework of TransSTGNN.

Definition 2 (Target edge). *The target edge in a dynamic graph anomaly detection task refers to an edge that the model classifies. At timestamp t, the target edge with nodes j and k as endpoints is defined as $e_{j,k}^t = (v_j^t, v_k^t)$, where v_j^t and v_k^t respectively represent node j and node k, and $v_j^t, v_k^t \in \mathcal{V}^t$, $e_{j,k}^t \in \mathcal{E}^t$.*

Definition 3 (h-hop subgraph). *Consider the target edge $e_{j,k}^t$ at timestamp t, with nodes j and k as endpoints. The h-hop subgraph centered at $e_{j,k}^t$ at timestamp t is defined as $\mathcal{G}_{e_{j,k}^t}^t = (\mathcal{V}_{e_{j,k}^t}^t, \mathcal{E}_{e_{j,k}^t}^t)$, where $\mathcal{V}_{e_{j,k}^t}^t$ denotes the set of nodes that are at a distance less than or equal to h from node j or node k, and $\mathcal{E}_{e_{j,k}^t}^t$ denotes the corresponding set of edges. Further, the set of h-hop subgraphs within the time window is defined as $\mathbb{G}_h = \{\mathcal{G}_{e_{j,k}^t}^i\}_{i=t-\omega+1}^{t}$, where ω represents the length of the time window.*

Definition 4 (Abnormal edge detection on dynamic graphs). *Given the dynamic graph $\mathbb{G}$, for a target edge $e_{j,k}^t$, the task of the abnormal edge detection involves determining the anomaly score. This score quantifies the deviation of the target edge from the normal pattern; an elevated score indicates a higher probability that the edge exhibits anomalous characteristics.*

3.2 Subgraph Sampling Module

Several research studies [1,11,12] have explicitly stated that anomalies tend to occur within the local substructures of graphs. For the target edge $e_{j,k}^t$ with nodes j and k as endpoints, we first sample the h-hop subgraph centered at $e_{j,k}^t$ at timestamp t. This process can be described as follows:

$$\mathcal{G}_{e_{j,k}^t}^t = \left(\mathcal{V}_{e_{j,k}^t}^t, \mathcal{E}_{e_{j,k}^t}^t\right) \quad (1)$$

$$\mathcal{V}_{e_{j,k}^t}^t = \left\{v \mid d\left(v, v_j^t\right) \leq h \cup d\left(v, v_k^t\right) \leq h, v \in \mathcal{V}^t\right\} \quad (2)$$

$$\mathcal{E}_{e_{j,k}^t}^t = \left\{(v_x, v_y) \mid v_x, v_y \in \mathcal{V}_{e_{j,k}^t}^t, (v_x, v_y) \in \mathcal{E}^t\right\} \quad (3)$$

where $d(\cdot)$ represents the shortest path function. Given the time window length ω, the set of h-hop subgraphs within this period is defined as:

$$\mathbb{G}_h = \left\{\mathcal{G}_{e_{j,k}^t}^i\right\}_{i=t-\omega+1}^{t} \quad (4)$$

3.3 Attention Sampling Module

For the target edge, $e_{j,k}^t$, aggregating information from nodes across the entire graph to obtain its global spatial features is computationally expensive. A more efficient approach focuses on the nodes most relevant to it within the graph. This strategy enhances computational efficiency and minimizes interference from irrelevant information. To this end, we employ the PageRank algorithm [13] to compute a graph diffusion matrix [4,8]. Through this matrix, we can sample a fixed-size set of nodes from the entire graph related to the target edge based on their importance. This set is called the Global Relevant Node Set, denoted as $GRS(e^t_{j,k})$. Equation 5 defines the computational method for the graph diffusion matrix $\mathbf{S_{diff}}$:

$$\mathbf{S_{diff}} = \beta(\mathbf{I}_n - (1-\beta)\mathbf{D}^{-\frac{1}{2}}\mathbf{A}\mathbf{D}^{-\frac{1}{2}})^{-1} \quad (5)$$

where $\mathbf{A} \in \mathbb{R}^{n \times n}$ represents the adjacency matrix of the snapshot, $\mathbf{D} \in \mathbb{R}^{n \times n}$ represents the diagonal degree matrix, $\mathbf{I}_n \in \mathbb{R}^{n \times n}$ denotes the identity matrix, and $\beta \in (0,1)$ is the teleportation probability. Given the graph diffusion matrix $\mathbf{S_{diff}}$, the i-th row $\mathbf{s}_i$ represents the relevance of the i-th node to other nodes. At timestamp t, the relevance vector for the target edge $e^t_{j,k}$, which relates to all nodes in the graph, is obtained by summing the relevance vectors $\mathbf{s}^t_j$ and $\mathbf{s}^t_k$ of nodes j and k. This vector is defined as:

$$\mathbf{s}^t_{e^t_{j,k}} = \mathbf{s}^t_j + \mathbf{s}^t_k \quad (6)$$

Subsequently, based on $\mathbf{s}^t_{e^t_{j,k}}$, the top N_k nodes with the highest relevance are selected as the global relevant node set $GRS(e^t_{j,k})$ for the target edge $e^t_{j,k}$ at timestamp t.

3.4 Node Enhancement Module

Given that the dataset used in this study lacks node attribute features, we adopt one-hot encoding as node attributes. Furthermore, we employ a node augmentation module to enhance the spatial features to endow nodes with spatial attributes. The node enhancement module employs a novel encoding method to augment the spatial attributes of nodes. This encoding includes the shortest path, node degree, and global importance. For each node within the subgraph and the $GRS(e^t_{j,k})$, spatial attributes are assigned as follows.

Position Encoding Based on Shortest Path (SPE): The topological distance is a crucial structural information in graphs and is significant for describing the topological similarity between nodes. Specifically, the shortest path function $d(\cdot)$ can be utilized to calculate the shortest path of a node to the endpoints of the target edge, choosing the smaller value as the node's shortest path distance. For node v, its position encoding based on shortest path can be defined as:

$$SPE(v) = linear(min\{d(v, v^t_j), d(v, v^t_k)\}) \quad (7)$$

where $linear(\cdot)$ represents a linear layer, and $min\{\cdot\}$ denotes the minimum value function.

Position Encoding Based on Node Degree (DE): In graphs, the degree of a node is an important indicator that reflects the complexity of its local spatial structure. Generally, nodes with higher degrees tend to exhibit more complex local structures. For node v, with its degree denoted by $deg(v)$, its position encoding based on node degree can be defined as:

$$DE(v) = linear(deg(v)) \tag{8}$$

Position Encoding Based on Global Importance (PRE): The PageRank algorithm evaluates a node's importance by considering its structural properties within the graph. It then assigns a corresponding rank, where higher rank values are associated with greater importance. For node v, with its importance in relation to the target edge $e_{j,k}^t$ denoted as $rank(v)$, its position encoding based on global importance can be defined as:

$$PRE(v) = linear(rank(v)) \tag{9}$$

Finally, the three types of encodings are concatenated to obtain the composite encoding for node v:

$$\mathbf{enc}_v^t = Combine(SPE(v), DE(v), PRE(v)) \tag{10}$$

where $\mathbf{enc}_v^t$ denotes the composite encoding of node v at timestamp t, and $Combine(\cdot)$ represents the vector concatenation function.

3.5 Spatio-Temporal Feature Encoding Module

The spatio-temporal feature encoding module comprises a transformer, a GCN, and an LSTM. Specifically, the encoding vectors $\mathbf{enc}_v^t$ of all nodes in the $GRS(e_{j,k}^t)$ are stacked into a matrix $\mathbf{H}_{GRS} \in \mathbb{R}^{N_k \times d_{node}}$, where N_k represents the size of $GRS(e_{j,k}^t)$ and d_{node} represents the dimension of the encoding vectors $\mathbf{enc}_v^t$. This matrix serves as the input to the transformer model, which aggregates the encodings of nodes within $GRS(e_{j,k}^t)$ to extract the global spatial features of the target edge $e_{j,k}^t$. This process can be formalized as follows:

$$\begin{aligned} \mathbf{Q}^l &= \mathbf{H}_{GRS}^{l-1} \mathbf{W}_\mathbf{Q}^l \\ \mathbf{K}^l &= \mathbf{H}_{GRS}^{l-1} \mathbf{W}_\mathbf{K}^l \\ \mathbf{V}^l &= \mathbf{H}_{GRS}^{l-1} \mathbf{W}_\mathbf{V}^l \end{aligned} \tag{11}$$

$$\mathbf{H}_{smp}^l = Softmax\left(\frac{\mathbf{Q}^l \mathbf{K}^{l\,T}}{\sqrt{d_{node}}}\right) \mathbf{V}^l \tag{12}$$

$$\mathbf{global}^t = \mathbf{h}_j^t + \mathbf{h}_k^t \tag{13}$$

where $\mathbf{W}_\mathbf{Q}, \mathbf{W}_\mathbf{K}, \mathbf{W}_\mathbf{V} \in \mathbb{R}^{d_{node} \times d_{node}}$, $\mathbf{H}_{GRS}^l$ denotes the output of the l-th layer of the tranformer, $\mathbf{h}_j^t$ and $\mathbf{h}_k^t$ respectively represent the j-th and k-th rows of

$\mathbf{H}_{GRS}^{l}$, and $\mathbf{global}^{t}$ denotes the global spatial features of the target edge $e_{j,k}^{t}$ at timestamp t.

Subsequently, a GCN with L layers is utilized to extract the local spatial structure of the subgraph $\mathcal{G}_{e_{j,k}^{t}}^{t}$. The input is the encoding matrix of the subgraph node set at timestamp t, $\mathbf{H}_{Sub}^{t} \in \mathbb{R}^{N_s \times d_{node}}$, where N_s represents the size of the subgraph node set, as defined below:

$$\mathbf{C}^0 = \mathbf{H}_{Sub}^{t}$$
$$\mathbf{C}^l = ReLU\left(\hat{\mathbf{A}}^t \mathbf{C}^{l-1} \mathbf{W}^{l-1}\right) \quad (14)$$
$$\mathbf{C}^L = ReLU\left(\hat{\mathbf{A}}^t \mathbf{C}^{L-1} \mathbf{W}^{L-1}\right)$$

where $\hat{\mathbf{A}}^t$ denotes the normalized adjacency matrix, $l \in [1, L-1]$, and $ReLU(\cdot)$ represents the activation function. The computation process for the normalized adjacency matrix $\hat{\mathbf{A}}^t$ is defined as follows:

$$\tilde{\mathbf{A}}^t = \mathbf{A}^t + I_n$$
$$\tilde{\mathbf{D}} = \sum_y \tilde{\mathbf{A}}_{xy}^t \quad (15)$$
$$\hat{\mathbf{A}}^t = \tilde{\mathbf{D}}^{-\frac{1}{2}} \tilde{\mathbf{A}}^t \tilde{\mathbf{D}}^{-\frac{1}{2}}$$

The local spatial features of the target edge $e_{j,k}^{t}$ at timestamp t are defined as follows:

$$\mathbf{local}^t = \mathbf{c}_j^t + \mathbf{c}_k^t \quad (16)$$

where $\mathbf{c}_j^t$ and $\mathbf{c}_k^t$ denote the j-th and k-th rows of $\mathbf{C}^L$, respectively, with $\mathbf{c}_j^t, \mathbf{c}_k^t \in \mathbb{R}^{d_{gcn}}$, where d_{gcn} represents the output dimension of the GCN. The enhanced spatial features are obtained by concatenating the local spatial features $\mathbf{local}^t$ and the global spatial features $\mathbf{global}^t$ of the target edge $e_{j,k}^{t}$:

$$\mathbf{x}^t = Combine(\mathbf{local}^t, \mathbf{global}^t) \quad (17)$$

where $\mathbf{x}^t$ represents the enhanced spatial features of the target edge at timestamp t, the process is repeated to extract these features at each timestamp within the time window. This results in a sequence of enhanced spatial features, $\mathcal{X} = (\mathbf{x}^{t-\omega+1}, \ldots, \mathbf{x}^t)$, with each $\mathbf{x}^i$ belongs to $\mathbb{R}^{d_{trans}+d_{gcn}}$. Finally, the evolution pattern of these spatial features is modeled using the LSTM, which can be defined as follows:

$$\mathbf{o}^t = LSTM(\mathbf{o}^{t-1}, \mathbf{x}^t) \quad (18)$$

where $\mathbf{o}^{t-1}$ represents the output of the LSTM at timestamp $t-1$ and $\mathbf{x}^t$ is the input to the LSTM at timestamp t. The output $\mathbf{o}^t$ of the LSTM at timestamp t is taken as the spatio-temporal feature $\mathbf{st}$ of the target edge, defined as:

$$\mathbf{st} = \mathbf{o}^t \quad (19)$$

3.6 Anomaly Detector

The anomaly detector utilizes a multilayer perceptron (MLP) architecture that comprises three fully connected layers. $ReLU$ activation functions are employed between each layer due to their ability to maintain non-saturating gradients in the positive range, effectively mitigating the vanishing gradient problem. This characteristic is pivotal for enabling efficient and effective learning of data features during the model's training phase. The final layer of the architecture employs a sigmoid activation function to transform the last fully connected layer's output into an anomaly score.

$$score = MLP\left(\mathbf{st}\right) \tag{20}$$

We employed the Binary Cross-Entropy Loss function to guide the learning process of the model. For a given graph snapshot $\mathcal{G}^t = (\mathcal{V}^t, \mathcal{E}^t)$, where m^t represents the number of edges, the loss function is defined as follows:

$$\mathcal{L} = -\sum_{i=1}^{m^t}((1-y)\log(1-score^i) + y\log(score^i)) \tag{21}$$

4 Experiments

4.1 Experimental Design

Datasets. TransSTGNN is evaluated on six real-world dynamic graph datasets previously utilized in related research [1,12,25], including Bitcoin-OTC [9], UCI Messages [16], Digg [3], As-Topology [22], Email-DNC [19], and Bitcoin-Alpha [10]. Table 1 shows the statistics of the six datasets. Bitcoin-OTC signifies a network of user trust within a Bitcoin trading platform, where nodes represent users and edges indicate trust evaluations among them. UCI Messages contains network data from the University of California, Irvine's social platform, where nodes represent users, and connections reflect exchanges of information between them. Email-DNC is a network dataset documenting email communications, where email accounts are treated as network nodes and email interactions form the edges of the network. Digg originates from a news platform, with nodes symbolizing the website's users and lines illustrating the comment interactions between them. Bitcoin-Alpha is also a user trust network dataset collected from a Bitcoin trading platform. AS-Topology represents the linkage among autonomous systems on the Internet, where each node denotes an autonomous system, and each edge signifies the link between systems.

Baselines. TransSTGNN was subjected to an in-depth performance comparison with six state-of-the-art graph anomaly detection methods: DeepWalk [17], Node2vec [5], NetWalk [21], StrGNN [1], AddGraph [25], and TADDY [12].

Table 1. The statistics of six benchmark datasets.

Dataset	Sum. Nodes	Sum. Edges	Avg. Degree	Size. Snapshot
Bitcoin-OTC	5,881	35,588	12.10	1000
UCI Messages	1,899	13,838	14.57	1000
Digg	30,360	85,155	5.61	4000
AS-Topology	171,420	34,761	9.86	4000
Email-DNC	1,866	39,264	42.08	500
Bitcoin-Alpha	3777	24,173	12.80	1000

Implementations. In configuring the TransSTGNN model, we assign a value of 100 to both the size of the GRS, N_k, and the subgraph node set, N_s, for the AS-Topology dataset. For the Digg dataset, these values are set at 80. For the remaining datasets, these values are set at 20. The subgraph hop count, h, is fixed at 1, and the time window length, ω, is set to 2. Additionally, both the node encoding dimension, d_{node}, and the output dimension of GCN, d_{gcn}, are established at 32. The GCN comprises two layers. For training, TransSTGNN utilizes the Adam optimizer with a learning rate of 0.001 across 30 epochs. The snapshot sizes for the datasets are as shown in Fig. 1. Following previous works [1, 12,21,25], all edges in the training data are considered positive, with duplicates removed. Anomalous edges are inserted into the test set for each dataset at anomaly percentages of 1%, 5%, and 10%. We assess model performance by the area under the curve (AUC) of the receiver operating characteristic (ROC), with higher values indicating better discrimination between normal and anomalous edges. The experimental platform consists of four Tesla V100 SXM2 32GB GPUs.

4.2 Effectiveness Evaluation

As shown in Table 2, the TransSTGNN model exhibits significant advantages over other baseline models. TransSTGNN achieves substantial performance improvements across all datasets compared to the state-of-the-art TADDY model. For instance, TransSTGNN improves the average AUC on the Bitcoin-OTC dataset by 0.016, consistently exceeding 0.95 across all anomaly ratios. For the Email-DNC datasets, the increases in average AUC are notable at 0.021, with peaks of up to 0.9553. The significant performance enhancements of TransST-GNN can be primarily attributed to two factors: firstly, TransSTGNN comprehensive capture of the spatial structure of target edges from multiple dimensions. Compared to previous studies, TransSTGNN elevates the focus on spatial features to new heights by not only enriching the node's spatial attributes through the introduction of the node enhancement module but also by integrating the transformer into the GCN to extract both local and global spatial features of the target edge. These rich and multidimensional spatial features significantly increase the model's sensitivity to spatial structural changes. Secondly, TransST-

GNN thoroughly considers the evolutionary characteristics of dynamic graphs and employs LSTM to model the evolution of the enhanced spatial features.

In addition, we have discovered through comparative analysis that the TransSTGNN model demonstrates remarkable stability in experimental results across various anomaly ratios on all datasets compared to previous models. For instance, on the UCI Messages dataset, the maximum difference in experimental outcomes for TransSTGNN across different anomaly ratios is only 0.028, while for the TADDY model, it reaches 0.054. These findings provide compelling evidence of the strong robustness and generalization capability of the TransSTGNN model. The primary reason for this stability is that detailed and comprehensive spatial features significantly enhance the model's performance threshold, allowing it to maintain stable performance across various anomaly ratios. Additionally, we have observed that the StrGNN model performs poorest at a 1% anomaly ratio, in contrast to its performance at 5% and 10% ratios. This observation underscores the crucial role of spatial features in anomaly detection.

Table 2. Experimental results of TransSTGNN for anomalous edge detection on six real-world dynamic graph datasets.

Method	Bitcoin-OTC			UCI Messages			Digg		
	1%	5%	10%	1%	5%	10%	1%	5%	10%
DeepWalk	0.7423	0.7356	0.7287	0.7514	0.7391	0.6979	0.7080	0.6881	0.6396
Node2vec	0.6951	0.6883	0.6745	0.7371	0.7433	0.6960	0.7364	0.7081	0.6508
NetWalk	0.7785	0.7694	0.7534	0.7758	0.7647	0.7226	0.7563	0.7176	0.6837
StrGNN	0.9012	0.8775	0.8836	0.8179	0.8252	0.7959	0.8162	0.8254	0.8272
AddGraph	0.8352	0.8455	0.8592	0.8083	0.8090	0.7688	0.8341	0.8470	0.8369
TADDY	0.9455	0.9340	0.9425	0.8912	0.8398	0.8370	0.8617	0.8545	0.8440
Ours	**0.9674**	**0.9506**	**0.9529**	**0.9449**	**0.9203**	**0.9170**	**0.9256**	**0.9107**	**0.9141**

Method	AS-Topology			Email-DNC			Bitcoin-Alpha		
	1%	5%	10%	1%	5%	10%	1%	5%	10%
DeepWalk	0.6844	0.6793	0.6682	0.7481	0.7303	0.7197	0.6985	0.6874	0.6793
Node2vec	0.6821	0.6752	0.6668	0.7391	0.7284	0.7103	0.6910	0.6802	0.6785
NetWalk	0.8018	0.8066	0.8058	0.8105	0.8371	0.8305	0.8385	0.8357	0.8350
StrGNN	0.8553	0.8352	0.8271	0.8775	0.9103	0.9080	0.8574	0.8667	0.8627
AddGraph	0.8080	0.8004	0.7926	0.8393	0.8627	0.8773	0.8665	0.8403	0.8498
TADDY	0.8953	0.8952	0.8934	0.9348	0.9257	0.9210	0.9451	0.9341	0.9423
Ours	**0.9246**	**0.9109**	**0.9168**	**0.9423**	**0.9468**	**0.9553**	**0.9691**	**0.9583**	**0.9567**

4.3 Ablation Study

Node Enhancement Module. This section investigates the specific impacts of various types of positional encoding in the node enhancement module on the model's performance. Experimental results are presented in Fig. 2. The experimental results indicate that as the anomaly ratio increases, the complexity and diversity of anomaly patterns also increase, leading to a greater impact on model performance when encodings are missing. A further comparison of the impact of different encodings revealed significant findings: the absence of SPE resulted in an average AUC decline of 0.15; the absence of PRE led to a decrease of 0.10; and the absence of DE caused a decline of 0.04. These results underscore the crucial roles of SPE and PRE in enhancing model performance. Specifically, SPE captures the topological similarity between nodes, reflecting the local spatial structure of the target edge. Conversely, PRE highlights the nodes' importance to the target edge, representing its global spatial structure. DE is also significant, as it reflects the complexity of the local spatial structure, thereby further refining the model's local spatial features.

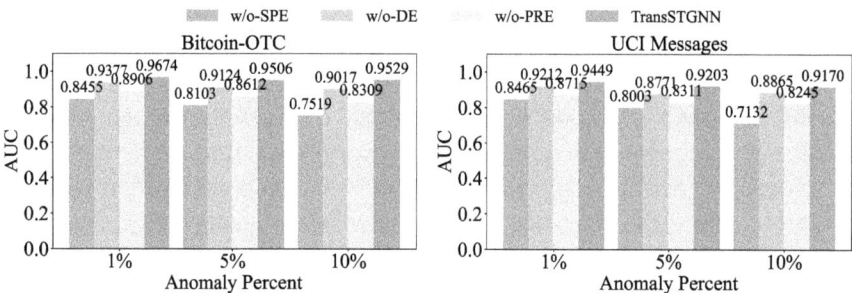

Fig. 2. Results of the ablation experiments for the node enhancement module. "w/o" denotes the removal of the corresponding encoding.

TransGCN. This section investigates the specific impact of the transformer model within the TransGCN module on the performance of the TransSTGNN. As illustrated in Fig. 3, experimental results show that TransSTGNN significantly outperforms the variant without the transformer across all datasets. By integrating the transformer, the model effectively aggregates information from globally relevant nodes, overcoming the limitations posed by local substructures. This capability allows it to capture and extract spatial features from multiple dimensions comprehensively, an aspect not emphasized in previous studies. Furthermore, the importance of the transformer model becomes more pronounced as the anomaly ratio increases. For instance, when the anomaly ratio escalated from 1% to 5% and then to 10%, the average performance improvements across all

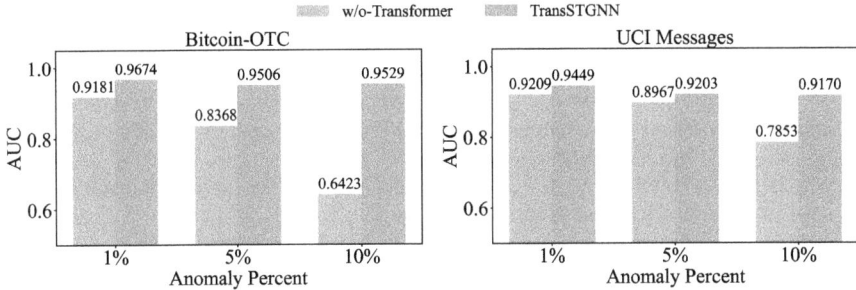

Fig. 3. Results of the ablation experiments for TransGCN. "w/o" denotes the removal of the transformer.

datasets were 0.044, 0.11, and 0.25, respectively. These results indicate that relying solely on local spatial features is insufficient to capture complex and diverse anomaly patterns, especially in datasets with a higher proportion of anomalies, thereby necessitating a combination of global information.

5 Conclusion

In this paper, we introduce a novel end-to-end transformer-based spatio-temporal graph neural network model, termed TransSTGNN, specializing in identifying abnormal edges within dynamic graphs. TransSTGNN consists of a dynamic graph encoder and an anomaly detector. The dynamic graph encoder includes the subgraph sampling module, attention sampling module, node enhancement module, and spatio-temporal feature encoding module. The subgraph sampling and attention sampling modules focus on the target edge's local and global spatial structures, respectively. The node enhancement module assigns spatial attributes to nodes. Additionally, the spatio-temporal feature encoding module integrates the Transformer model with GCN and employs LSTM to model the evolution of features, thereby capturing the spatio-temporal characteristics of the target edge. A multilayer perceptron serves as the anomaly detector, calculating the anomaly score of the target edge. Extensive experiments demonstrate that the TransSTGNN model surpasses state-of-the-art models on six datasets from real-world dynamic graphs. Additionally, ablation studies confirm that the node enhancement and TransGCN modules effectively enrich spatial features and enhance the model's capability to capture anomaly patterns.

Acknowledgments. This work was supported in part by the National Key R&D Program of China under Grant 2020AAA0108501.

References

1. Cai, L., et al.: Structural temporal graph neural networks for anomaly detection in dynamic graphs. In: Proceedings of the 30th ACM International Conference on Information & Knowledge Management, pp. 3747–3756 (2021)

2. Cui, Q., Wu, S., Huang, Y., Wang, L.: A hierarchical contextual attention-based network for sequential recommendation. Neurocomputing **358**, 141–149 (2019)
3. De Choudhury, M., Sundaram, H., John, A., Seligmann, D.D.: Social synchrony: predicting mimicry of user actions in online social media. In: Proceedings of the 2009 International Conference on Computational Science and Engineering, vol. 4, pp. 151–158. IEEE (2009)
4. Gasteiger, J., Weißenberger, S., Günnemann, S.: Diffusion improves graph learning. Adv. Neural Inf. Process. Syst. **32** (2019)
5. Grover, A., Leskovec, J.: node2vec: scalable feature learning for networks. In: Proceedings of the 22nd ACM SIGKDD International Conference on Knowledge Discovery and Data Mining, pp. 855–864 (2016)
6. Guo, Q., Li, Z., An, B., Hui, P., Huang, J., Zhang, L., Zhao, M.: Securing the deep fraud detector in large-scale e-commerce platform via adversarial machine learning approach. In: Proceedings of the world Wide Web Conference, pp. 616–626 (2019)
7. Guo, Q., et al.: A survey on knowledge graph-based recommender systems. IEEE Trans. Knowl. Data Eng. **34**(8), 3549–3568 (2020)
8. Hassani, K., Khasahmadi, A.H.: Contrastive multi-view representation learning on graphs. In: International Conference on Machine Learning, pp. 4116–4126. PMLR (2020)
9. Kumar, S., Hooi, B., Makhija, D., Kumar, M., Faloutsos, C., Subrahmanian, V.: Rev2: fraudulent user prediction in rating platforms. In: Proceedings of the Eleventh ACM International Conference on Web Search and Data Mining, pp. 333–341 (2018)
10. Kumar, S., Spezzano, F., Subrahmanian, V., Faloutsos, C.: Edge weight prediction in weighted signed networks. In: Proceedings of the IEEE 16th International Conference on Data Mining (ICDM), pp. 221–230. IEEE (2016)
11. Liu, Y., Li, Z., Pan, S., Gong, C., Zhou, C., Karypis, G.: Anomaly detection on attributed networks via contrastive self-supervised learning. IEEE Trans. Neural Netw. Learn. Syst. **33**(6), 2378–2392 (2021)
12. Liu, Y., et al.: Anomaly detection in dynamic graphs via transformer. IEEE Trans. Knowl. Data Eng. **35**(12), 12081–12094 (2021)
13. Ma, N., Guan, J., Zhao, Y.: Bringing pagerank to the citation analysis. Inf. Process. Manag. **44**(2), 800–810 (2008)
14. Miao, K., Shi, X., Zhang, W.A.: Attack signal estimation for intrusion detection in industrial control system. Comput. Secur. **96**, 101926 (2020)
15. Nguyen, V.H., Sugiyama, K., Nakov, P., Kan, M.Y.: Fang: leveraging social context for fake news detection using graph representation. In: Proceedings of the 29th ACM International Conference on Information & Knowledge Management, pp. 1165–1174 (2020)
16. Opsahl, T., Panzarasa, P.: Clustering in weighted networks. Social Netw. **31**(2), 155–163 (2009)
17. Perozzi, B., Al-Rfou, R., Skiena, S.: Deepwalk: online learning of social representations. In: Proceedings of the 20th ACM SIGKDD International Conference on Knowledge Discovery and Data Mining, pp. 701–710 (2014)
18. Ranshous, S., Harenberg, S., Sharma, K., Samatova, N.F.: A scalable approach for outlier detection in edge streams using sketch-based approximations. In: Proceedings of the 2016 SIAM International Conference on Data Mining, pp. 189–197. SIAM (2016)
19. Rossi, R., Ahmed, N.: The network data repository with interactive graph analytics and visualization. In: Proceedings of the AAAI Conference on Artificial Intelligence, vol. 29 (2015)

20. Song, C., Shu, K., Wu, B.: Temporally evolving graph neural network for fake news detection. Inf. Process. Manag. **58**(6), 102712 (2021)
21. Yu, W., Cheng, W., Aggarwal, C.C., Zhang, K., Chen, H., Wang, W.: Netwalk: a flexible deep embedding approach for anomaly detection in dynamic networks. In: Proceedings of the 24th ACM SIGKDD International Conference on Knowledge Discovery & Data Mining, pp. 2672–2681 (2018)
22. Zhang, B., Liu, R., Massey, D., Zhang, L.: Collecting the internet as-level topology. ACM SIGCOMM Comput. Commun. Rev. **35**(1), 53–61 (2005)
23. Zhang, G., et al.: efraudcom: an e-commerce fraud detection system via competitive graph neural networks. ACM Trans. Inf. Syst. (TOIS) **40**(3), 1–29 (2022)
24. Zhang, X., Zitnik, M.: Gnnguard: defending graph neural networks against adversarial attacks. Adv. Neural. Inf. Process. Syst. **33**, 9263–9275 (2020)
25. Zheng, L., Li, Z., Li, J., Li, Z., Gao, J.: Addgraph: anomaly detection in dynamic graph using attention-based temporal gcn. In: Proceedings of the 28th International Joint Conference on Artificial Intelligence, pp. 4419–4425 (2019)

Morphological Semantic Ensemble Filtering of Massive Sentence Pairs for Neural Machine Translation

Lin Wang[1] and Wuying Liu[2(✉)]

[1] Xianda College of Economics and Humanities, Shanghai International Studies University, Shanghai 200083, China
lwang@xdsisu.edu.cn
[2] Shandong Key Laboratory of Language Resources Development and Application, Ludong University, Yantai 264025, China
wyliu@ldu.edu.cn

Abstract. The dazzling brilliance of deep learning makes it a consensus that "give me large-scale high-quality sentence pairs to train on, I can obtain a high-performance neural machine translation model." The current ever-increasing productivity of human language data greatly reduces the scarcity of sentence pairs, but the explosion of language data also brings potential problems such as uneven translation quality of sentences. We address the issue of sentence pair filtering based on translation quality, explore a new method of morphological semantic ensemble filtering, which can fully leverage the efficiency of morphological measurement derived from Levenshtein editing distance and the accuracy of semantic measurement derived from pretrained models, and achieve an efficient estimation of sentence pair translation quality taking into account both morphology and semantics. We first conduct morphological filtering, semantic filtering, and morphological semantic ensemble filtering experiments on the datasets of 17 language pairs respectively, and then use the filtered sentence pairs to enhance the retraining of the 17 neural machine translation models. Experimental results show that all three filtered results can significantly improve the performance of neural machine translation models, among which, the morphological semantic ensemble filtering has the best effect, improving 2–3 BLEU points compared with the other two methods. Experimental results clarify that our new method can use morphological computing to strengthen perfect homonym data, while can use semantic computing to strengthen shaped synonym data. The corresponding implementation is an industrial-level straightforward and efficient algorithm.

Keywords: Ensemble Filtering · Morphological Similarity · Semantic Similarity · Pretrained Model · Neural Machine Translation

1 Introduction

The remarkable progress made by deep learning in the field of machine translation (MT) is mainly attributed to the supercomputing power of parallel vector computing units, the ultra-deep neural network algorithm supporting massive fine-grained features, and the large-scale computing materials of sentence pairs provided by Internet language big data [1]. The current scientific research and industrial applications of MT are affected to varying degrees by the convergence of computing power and algorithms, while the scale and quality of sentence pairs have gradually become the most critical new production factors that determine the performance of neural machine translation (NMT).

Mobile smart terminals and wireless network technology have further improved the productivity of human language data, and the accumulated Internet language big data has become the main source of large-scale sentence pairs. The Internet language big data is usually unstructured, nonstandard, and even contains fallacies, which makes it particularly important to automatically evaluate and filter large-scale sentence pairs [2]. Automatic sentence pair filtering can usually be regarded as a binary classification problem of sentence pairs, that is, for a sentence pair <XSen, YSen>, the filter estimates a confidence score belonging to [0, 1] interval according to the translation quality between sentence XSen and sentence YSen, and then makes high-quality or low-quality judgments based on the preset confidence threshold, finally eliminates low-quality sentence pairs and retains high-quality sentence pairs.

Automatic sentence pair filtering is a special classification problem due to the processing of two sentences in different languages. Back translation technology, which was proposed in the early days to solve the problem of low-resource MT, that is, a process of re-translating a translation back to its original language [3], can also be used for translation quality evaluation and automatic filtering of sentence pairs. Later, some studies integrate the advantages of different translation directions and different MT systems and use morphological similarity to filter sentence pairs [4]. With the development of deep learning and pretrained models, the effectiveness and efficiency of sentence embedding vector representation and sentence semantic similarity calculation have been significantly improved [5]. Some studies have used Levenshtein morphological similarity and pretrained model embedding semantic similarity respectively to compare the quality of artificial translations [6]. The morphology and semantics of a sentence can be regarded as the two sides of the same coin, but what is the quantitative relationship between the morphological similarity and the semantic similarity? Is it possible to develop a new method for better morphological semantic ensemble filtering based on this correlation? With these considerations in mind, we use detailed statistics analysis to show the correlation between morphological and semantic similarity.

2 Correlation Analysis

Correlation analysis usually refers to the numerical statistics of two or more interrelated variables, so as to measure the closeness of the correlation between variables. Among the many correlation analysis methods, the Pearson correlation coefficient method is widely used to measure the degree of linear correlation between two variables. We use the Pearson correlation coefficient method to quantitatively analyze the correlation between morphological similarity and semantic similarity.

2.1 Correlation Analysis Framework

Our proposed framework for correlation analysis is shown in Fig. 1. It mainly includes a machine translator (XChi Machine Translator) that supports automatic translation from language X to Chinese, a Levenshtein similarity scorer (Leven Similarity Scorer), three similarity scorers based on pretrained models (Roberta Similarity Scorer, Mpnet Similarity Scorer, MiniLM Similarity Scorer) and a Pearson Correlation Analyzer.

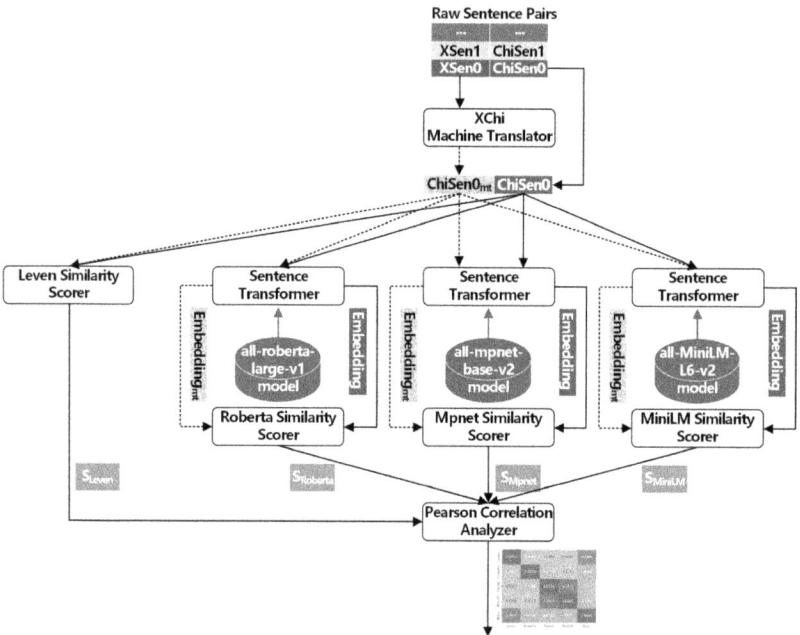

Fig. 1. Correlation Analysis Framework.

When a set of raw sentence pairs arrives, each pair of sentences <XSen, ChiSen> is taken out in turn for processing. First, the sentence XSen is sent to the XChi Machine Translator, and the translated output ChiSen$_{mt}$. Then, the Levenshtein similarity scorer and three similarity scorers based on the pretrained model receive ChiSen$_{mt}$ and ChiSen at the same time and calculate the similarity scores S_{Leven}, $S_{Roberta}$, S_{Mpnet}, and S_{MiniLM} that output the [0, 1] interval respectively. Finally, the Pearson Correlation Analyzer receives the above similarity scores and calculates the Pearson correlation coefficient between each two similarities using the formula (1).

$$r_{xy} = \frac{\sum_{i=0}^{n-1}(x_i - \bar{x})(y_i - \bar{y})}{\sqrt{\sum_{i=0}^{n-1}(x_i - \bar{x})^2}\sqrt{\sum_{i=0}^{n-1}(y_i - \bar{y})^2}} \qquad (1)$$

The value range of the Pearson correlation coefficient r_{xy} is [−1, 1], where a negative value indicates a negative correlation, that is, as the value of one variable increases, the other decreases; a positive value indicates a positive correlation, that is, as the value of one variable increases, the other also increases; 0 indicates no correlation between the two variables. The strength of the linear correlation is determined by the value of $|r_{xy}|$. The usual 5 grades include: extremely weak correlation ($|r_{xy}| \in [0.00, 0.19]$), weak correlation ($|r_{xy}| \in [0.20, 0.39]$), moderate correlation ($|r_{xy}| \in [0.40, 0.59]$), strong correlation ($|r_{xy}| \in [0.60, 0.79]$), extremely strong correlation ($|r_{xy}| \in [0.80, 1.00]$). There are also 3 grades including: weak correlation ($|r_{xy}| \in [0.10, 0.30]$), moderate correlation ($|r_{xy}| \in (0.30, 0.50)$), strong correlation ($|r_{xy}| \in [0.50, 1.00]$).

When using the formula (1) for correlation analysis, we also simultaneously calculated the classical BLEU4 score S_{BLEU} in MT evaluation and recorded the variables corresponding to the similarity scores S_{Leven}, $S_{Roberta}$, S_{Mpnet}, S_{MiniLM}, and S_{BLEU} as Leven, Roberta, Mpnet, MiniLM, and BLEU. Therefore, the variables x and y in the formula (1) belong to the set {Leven, Roberta, Mpnet, MiniLM, BLEU}, where $\bar{x}$ and $\bar{y}$ represent the mean of variables x and y, respectively.

2.2 Morphological Similarity

We used the Levenshtein morphological similarity score S_{Leven} in the correlation analysis framework to assess the morphological similarity between two sentences. The score is based on the edit distance proposed by Soviet mathematician Vladimir Levenshtein in 1965. The edit distance is the minimum number of editing operations required to convert one string to another, including replacements, insertions, and deletions. The calculation of S_{Leven} is shown in Formula (2), where $LevenDis(ChiSen, ChiSen_{mt})$ represents the Levenshtein edit distance between the human-translated sentence (ChiSen) and the machine-translated sentence ($ChiSen_{mt}$), $Len(ChiSen)$ represents the length of the reference translation, and $Len(ChiSen_{mt})$ represents the length of the candidate translation. The S_{Leven} value ranges from 0 to 1, with the higher the value, the more similar the morphology of the two sentences.

$$S_{Leven} = 1 - \frac{LevenDis(ChiSen, ChiSen_{mt})}{Max(Len(ChiSen), Len(ChiSen_{mt}))} \quad (2)$$

Since the reference sentences and candidate sentences in this paper are all Chinese, a Chinese-character-level 1-g token is used to calculate the S_{Leven} value. Therefore, the finest granular unit of edit distance is the Chinese-character-level 1-g token, and the sentence length is the total number of Chinese-character-level 1-g tokens.

2.3 Semantic Similarity

In order to calculate semantic similarity in the correlation analysis framework, we first use Sentence Transformer[1] to map human-translated sentence (ChiSen) and machine-translated sentence ($ChiSen_{mt}$) to three pretrained models (Model) and obtain corresponding embedding vector representations Embedding(Model) and

---

[1] https://www.sbert.net.

Embedding$_{mt}$(Model) [7]. Then the cosine similarity of each pair of sentence embedding vectors <Embedding(Model), Embedding$_{mt}$(Model)> is calculated in three semantic similarity scorers (Roberta Similarity Scorer, Mpnet Similarity Scorer, and MiniLM Similarity Scorer), and then normalized to the [0, 1] interval according to the formula (3) as the semantic similarity score S$_{(Model)}$ of the original sentence pair.

$$S_{(Model)} = \frac{CosSim(Embedding(Model), Embedding_{mt}(Model)) + 1}{2} \quad (3)$$

Any Sentence-BERT series of pretrained models[2] that support sentence embedding can be used for semantic similarity calculation in the correlation analysis framework. These serial models use siamese and triplet network structures to derive semantically meaningful sentence embeddings. This representation easily supports cosine similarity calculation and greatly reduces computational overhead while maintaining the accuracy from BERT. Three pretrained models Roberta, Mpnet, and MiniLM are used in this paper, and the detailed parameters of their corresponding and highly representative multilingual engineering implementation versions are shown in Table 1. The three models in Table 1 are all trained on a large and diverse dataset of over 1 billion training pairs, and they are all all-round models tuned for many use cases, which can be directly used for the correlation analysis framework.

Table 1. Model Overview.

Model (Base Model)	Max Sequence Length	Dimensions	Model Size (MB)	Speed	Sentence Embedding Performance	Semantic Search Performance	Avg Performance
all-roberta-large-v1 (roberta-large)	256	**1,024**	**1,360**	800	**70.23**	53.05	61.64
all-mpnet-base-v2 (microsoft/mpnet-base)	**384**	768	420	2,800	69.57	**57.02**	**63.30**
all-MiniLM-L6-v2 (nreimers/MiniLM-L6-H384-uncased)	256	384	80	**14,200**	68.06	49.54	58.80

In Roberta, Mpnet and MiniLM pretrained models, the improved Roberta model proposed by the replication study of BERT hyperparameter selection achieves state-of-the-art results on GLUE, RACE and SQuAD [8]; The Mpnet model that can see a full sentence leverages the dependency among predicted tokens through permuted language modeling, and takes auxiliary position information as input, which can reduce the position discrepancy, and outperform masked language modeling and permuted language modeling by a large margin [9]; The deep self-attention distillation model (MiniLM) proposed to address the challenges for fine-tuning and online serving in real-life applications due to latency and capacity constraints adopts the formula (4) to minimize the KL-divergence between the self-attention distributions of the teacher and student, and

---

[2] https://huggingface.co/models.

can simply and effectively compress large Transformer based pretrained models [10].

$$L_{AT} = \frac{1}{A_h |x|} \sum_{a=1}^{A_h} \sum_{t=1}^{|x|} D_{KL}\left(A_{L,a,t}^T || A_{M,a,t}^S\right) \tag{4}$$

Under the different parameter sizes of the student model, the monolingual MiniLM model is better than the optimal baseline. The data in Table 1 also shows that MiniLM is the fastest, 5 times faster than Mpnet, and more than 17 times faster than Roberta. Under the premise that the average performance is not much different, the extremely high processing speed is also very critical for big data applications.

2.4 Correlation Result

For each language X in 17 different languages (Arabic (Ara), Czech (Ces), Filipino (Fil), Indonesian (Ind), Italian (Ita), Kazakh (Kaz), Khmer (Khm), Kyrgyz (Kir), Lao, Malay (Msa), Myanmar (Mya), Polish (Pol), Russian (Rus), Slovak (Slk), Thai (Tha), Ukrainian (Ukr) and Vietnamese (Vie)), we have respectively selected 100,000 pairs of parallel sentences between language X and Chinese and called the multiloop incremental bootstrapping framework based low-resource MT interface [11] to translate the above 1.7 million language X sentences into Chinese. According to the morphological and semantic similarity between the Chinese MT sentence and the Chinese sentence of the sentence pair, the Pearson correlation coefficient between any two variables in the {Leven, Roberta, Mpnet, MiniLM, BLEU} set is calculated, and the corresponding heatmap is finally drawn as shown in Fig. 2.

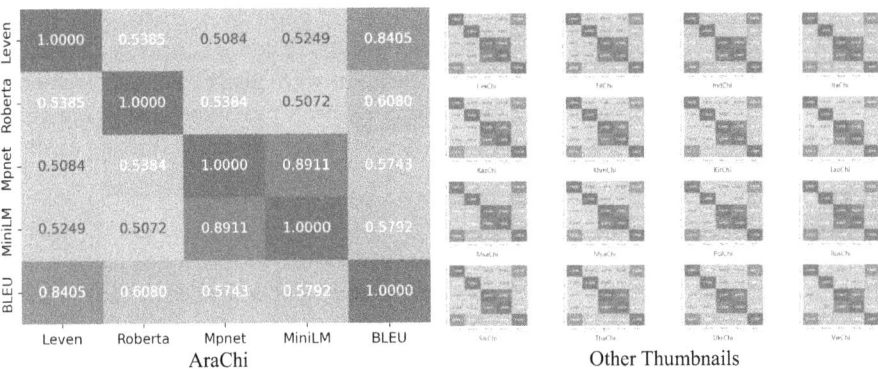

Fig. 2. Pearson Correlation Coefficient Heatmap.

The five variables in Fig. 2 all reflect the performance of the X-Chinese MT model from different perspectives. Since the performance of the low-resource MT model based on the multiloop incremental bootstrapping framework is stable, the Pearson correlation coefficient between each two variables is more focused on reflecting a certain essential attribute of the Chinese language itself. From Fig. 2, we can see that the statistical results on the set of 17 kinds of sentence pairs are basically the same. Without losing generality, we selected the Pearson correlation coefficient heatmap of Arabic-Chinese (AraChi) from the thumbnails for detailed analysis. We found that all the correlation coefficients are greater than 0.5, which shows that there is a strong linear correlation between the five variables. Among them, the correlation coefficient of semantic similarity Mpnet and MiniLM is as high as 0.8911, and the correlation coefficient of morphological similarity Leven and BLEU is as high as 0.8405, indicating that there is a strong linear correlation within the two groups. In particular, the strong linear correlation between Leven and BLEU also shows that the BLEU standard focuses on morphological evaluation. After comparing Leven, Roberta, Mpnet, and MiniLM, it is found that although morphological similarity and semantic similarity are related, there are also big differences, which also shows that morphology and semantics are two attributes of one body and two sides, which are not exactly the same.

Table 2. Average Pearson Correlation Coefficient.

	Leven	Roberta	Mpnet	MiniLM	BLEU
Leven	1.0000	0.5304	0.5014	0.5228	0.8368
Roberta	0.5304	1.0000	0.5265	0.4989	0.6020
Mpnet	0.5014	0.5265	1.0000	0.8910	0.5725
MiniLM	0.5228	0.4989	0.8910	1.0000	0.5817
BLEU	0.8368	0.6020	0.5725	0.5817	1.0000

We also calculate the arithmetic mean of Pearson correlation coefficients on the set of 17 kinds of sentence pairs. As shown in Table 2, the average Pearson correlation coefficient is almost the same as the Pearson correlation coefficient in each set of sentence pairs. This also further confirms that this is the attribute of the Chinese language itself. This stable property provides scientific data support for our ensemble filtering method.

3 Ensemble Filtering

There is a correlation between morphological similarity and semantic similarity in Chinese, but they cannot completely replace each other. The Levenshtein edit distance can straightforwardly and efficiently implement morphological similarity judgment, and according to the corresponding relationship of bilingual sentences, it can better support the training of NMT models to deal with homonym phenomena. The pretrained model can enhance the semantic representation of sentences through sentence embedding, incorporate more world knowledge, and support the cosine similarity calculation of vectors straightforwardly and efficiently, which can better support the training of NMT models to deal with synonym phenomena. The achievements of ensemble learning in text categorization [12] and natural language processing [13] have inspired us to propose an ensemble filtering idea based on morphological semantic similarity.

3.1 Ensemble Filtering Framework

Our proposed ensemble filtering framework is shown in Fig. 3. It mainly includes a machine translator (XChi Machine Translator) that supports automatic translation from language X to Chinese, a pretrained model (Pretrained Model) that supports multiple languages and a corresponding sentence embedding processing unit (Sentence Transformer), a Morphological Similarity Scorer, a Semantic Similarity Scorer, three filters (Morphological Filter, Semantic Filter, Ensemble Filter) using different filtering strategies, and an MT model retrainer (XChi MT Model Retrainer) with three copies.

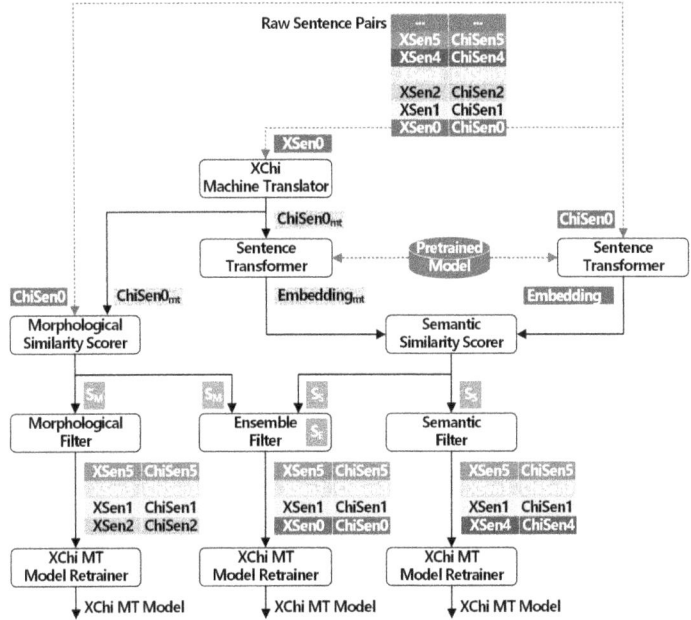

Fig. 3. Ensemble Filtering Framework.

When a set of raw sentence pairs arrives, each pair of sentences <XSen, ChiSen> is taken out for processing. Firstly, the **XChi Machine Translator** receives the sentence XSen and translates it to ChiSen$_{mt}$. Secondly, the **Morphological Similarity Scorer** receives ChiSen$_{mt}$ and ChiSen at the same time, calculates and outputs a morphological similarity score S_M belonging to the [0, 1] interval. Simultaneously, the **Sentence Transformer** represents ChiSen$_{mt}$ and ChiSen into corresponding embedding vectors Embedding$_{mt}$ and Embedding respectively according to the **Pretrained Model**. The **Semantic Similarity Scorer** receives Embedding$_{mt}$ and Embedding vectors at the same time and calculates a semantic similarity score S_S belonging to the [0, 1] interval. Thirdly, the **Ensemble Filter** linearly combines the morphological similarity score S_M and the semantic similarity score S_S according to formula (5) to calculate an ensemble similarity score S_E belonging to the [0, 1] interval.

$$S_E = \alpha S_M + (1 - \alpha) S_S \quad \alpha \in [0, 1] \tag{5}$$

Finally, according to the preset truncation rate of t%, the three filters respectively filter out high-quality sentence pairs with top (100-t)% similarity score from the set of raw sentence pairs and then run the **XChi MT Model Retrainer** to obtain their respective XChi MT Model.

The above morphological filtering process mainly includes four steps: MT to obtain candidate translations, Chinese sentence tokenization, morphological similarity calculation, and filtering high-quality sentence pairs according to the preset truncation rate. The most critical morphological similarity score S_M in the process is calculated using S_{Leven} in formula (2), that is, $S_M = S_{Leven}$. The above semantic filtering process includes four steps: MT, pretrained representation of sentences, semantic similarity calculation, and filtering according to the preset truncation rate. To address the challenge of dealing with synonyms in the most critical semantic similarity calculation step, we introduce a Sentence-BERT pretrained model suitable for large-scale industrial application scenarios. The semantic similarity score S_S is calculated using $S_{(Model)}$ in formula (3), that is, $S_S = S_{(Model)}$.

3.2 Ensemble Filtering Algorithm

The ensemble filtering algorithm that linearly combines morphological similarity and semantic similarity according to the ensemble filtering framework is shown in Fig. 4. The input of the algorithm is the list of raw sentence pairs *rsp*, the pretrained model *m*, the morphological weight α, and the truncation rate *t*. The output of the algorithm is a filtered list of sentence pairs *fsp*.

---
1. // **Ensemble Filtering Algorithm**
2. **Input**:   List<String, String> *rsp*; // raw sentence pairs
3.   Model *m*; // pretrained model
4.   Float α; // morphological weight, here α=0.5
5.   Int *t*; // truncation rate, here *t*=40(%)
6. **Output**: List<String, String> *fsp*; // filtered sentence pairs
7. List<Float> *ses*; // ensemble similarity scores
8. **For** <String, String> *sp* **In** *rsp*:
9.   String *csmt*←*MachineTranslator*(*sp*[0]);
10.   Float *sm*←*MorphologicalSimilarityScorer*(*sp*[1], *csmt*);
11.   Vector *e*←*SentenceTransformer*(*m*, *sp*[1]);
12.   Vector *emt*←*SentenceTransformer*(*m*, *csmt*);
13.   Float *ss*←*SemanticSimilarityScorer*(*e*, *emt*);
14.   *ses*.add(α×*sm*+(1-α)×*ss*);
15. **End For**
16. *fsp*←*Sorter*(*rsp*, *ses*); // sort raw sentence pairs in descending order of ensemble similarity score
17. *fsp*←*Truncator*(*fsp*, *t*); // truncate the top sentence pairs based on truncation rate
18. **Return** *fsp*.

Fig. 4. Ensemble Filtering Algorithm.

The main loop of the ensemble filtering algorithm is to take each group of sentence pairs in the list of sentence pairs *rsp* in turn, and translate the foreign sentence *sp*[0] in it into a Chinese sentence *csmt*; Then, two Chinese sentences *sp*[1] and *csmt* are represented as embedding vectors *e* and *emt* according to the pretrained model *m*;

The similarity scores *sm* and *ss* are calculated in ***MorphologicalSimilarityScorer*** and ***SemanticSimilarityScorer*** according to S_{Leven} and $S_{(Model)}$ calculation formulas; The two similarity scores are linearly combined to form a final ensemble similarity score. At the end of the loop, the sentence pairs in the list of raw sentence pairs are arranged in descending order according to the ensemble similarity score, and finally, the Top(100-*t*)% high-quality sentence pairs are filtered according to the truncation rate. When the size of the list of raw sentence pairs is *n*, the main space complexity of the algorithm is **O**(*n*), that is, the space occupied by the list storing the ensemble similarity scores; The time complexity of the algorithm loop is also **O**(*n*), and the time complexity of the sorting is **O**($n\log_2 n$) if the fast sort is used. In practical use, the algorithm can also use the preset ensemble similarity threshold to truncate, which can save the time and space cost of sorting.

4 Experiment

In order to verify the effectiveness of morphological semantic ensemble filtering, we implemented three filtering strategies in the experiment, filtered large-scale sentence pairs of 17 language pairs, retrained 51 neural machine translation models, and analyzed and discussed the experimental results.

4.1 Experimental Setup and Data

Using pretraining and finetuning is the most common method to make these pretrained models more suitable for downstream applications. For those practical applications lacking downstream task data, or frequently changing distribution of downstream task data, or lacking high-performance vector computing components, the cost of finetuning remains enormous. Table 1 also shows that the Base Model, Max Sequence Length, Dimensions, and Model Size of the three models are not exactly the same, especially with significant differences in Speed. Among the three models, all-MiniLM-L6-v2 has the fastest speed of 14,200, all-roberta-large-v1 has the slowest speed of 800, and all-mpnet-base-v2 has a middle speed of 2,800. In order to match the real needs of massive sentence pair filtering, we use the fastest all-MiniLM-L6-v2 pretrained model for semantic similarity scoring.

We call a low-resource MT interface with excellent performance during the experiment. We set the truncation rate of *t* to 40(%) without losing generality, which means filtering out high-quality sentence pairs with a similarity score of top 60%; and set the α in formula (5) to 0.5, which means that morphology and semantics are treated equally during ensemble filtering.

We have collected total 312,031,062 pairs of sentences parallel to Chinese from the Internet in Arabic (Ara), Czech (Ces), Filipino (Fil), Indonesian (Ind), Italian (Ita), Kazakh (Kaz), Khmer (Khm), Kyrgyz (Kir), Lao, Malay (Msa), Myanmar (Mya), Polish (Pol), Russian (Rus), Slovak (Slk), Thai (Tha), Ukrainian (Ukr) and Vietnamese (Vie). And each language has more than 15 million pairs of parallel sentences with Chinese.

4.2 Experimental Result and Discussion

We run the experimental filtering program on the above raw data, and obtain 17 sets of high-quality sentence pairs respectively. The detailed number of filtered sentence pairs is shown in Table 3. Among them, IndChi has the most pairs of sentences, with 15,244,662 pairs, while MyaChi has the least pairs of sentences, with 9,164,195 pairs. For the sake of experimental fairness, we take the same total number of sentence pairs, 60% of the number of raw sentence pairs, for each language's morphology, semantics, and ensemble filtering results, with the only difference reflected in the quality of these sentence pairs.

Table 3. Sentence Pairs and BLEU Results.

Language Pair	BLEU	Number of Filtered Sentence Pairs	$BLEU_M$	$BLEU_S$	$BLEU_E$
AraChi	33.47	13,820,536	45.29	46.13	48.15
CesChi	33.12	9,862,345	46.89	47.84	49.89
FilChi	33.73	9,936,372	46.88	47.00	49.66
IndChi	**36.35**	**15,244,662**	**47.08**	**47.98**	**50.49**
ItaChi	31.46	12,605,357	43.06	43.09	45.14
KazChi	27.82	10,068,962	39.64	40.51	42.82
KhmChi	26.55	10,008,839	39.75	40.04	43.34
KirChi	23.85	10,003,788	36.76	37.20	40.64
LaoChi	22.05	9,570,285	33.42	33.48	35.81
MsaChi	23.94	14,381,238	36.26	36.50	39.41
MyaChi	22.67	**9,164,195**	33.79	34.14	37.63
PolChi	32.81	9,880,321	46.29	46.84	48.87
RusChi	30.88	11,604,851	42.70	43.70	45.76
SlkChi	31.68	9,876,761	45.83	46.74	49.27
ThaChi	27.50	9,938,306	40.77	41.33	44.60
UkrChi	32.25	10,033,749	46.68	47.41	49.70
VieChi	26.80	11,218,070	40.04	40.95	44.02

Based on the existing 17 NMT models trained by 5 million sentence pairs respectively and their training corpus, we append filtered sentence pairs to retrain 51 NMT models without changing the main hyperparameters to further compare the three filtering effects. The BLEU values in Table 3 represent the performance of the existing models, while the $BLEU_M$, $BLEU_S$, and $BLEU_E$ values respectively represent the performance of NMT models retrained after appending 5 million original training corpus with morphological, semantic, and ensemble filtered sentence pairs. Table 3 shows that the performance of the 17 NMT models can be significantly improved under the enhancement of three kinds of filtered data. For instance, the morphological filtering results can increase by 10–14

BLEU points, with the minimum IndChi increase of 10.73 and the maximum UkrChi increase of 14.43; the semantic filtering results can increase by 11–15 BLEU points, with the minimum LaoChi increase of 11.43 and the maximum UkrChi increase of 15.16; and the ensemble filtering results can increase by 11–14 BLEU points, with the minimum ItaChi increase of 13.68 and the maximum SlkChi increase of 17.59. Table 3 also shows that the enhancement effect of morphological semantic ensemble filtering is 2–3 BLEU points higher than that of morphological filtering alone and semantic filtering alone.

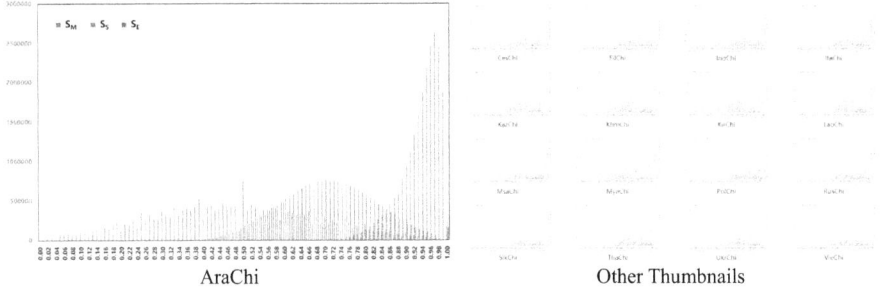

Fig. 5. Quantity Distribution of Raw Sentence Pairs.

In order to more intuitively illustrate the effects of morphological filtering, semantic filtering, and ensemble filtering, we calculate the quantity distribution of all raw sentence pairs based on the three similarity scores. The statistical results show that the 17 languages have highly similar quantity distribution. Without loss of generality, Fig. 5 selects the raw AraChi sentence pairs from thumbnails to enlarge and display the number distribution, where the abscissa represents the similarity score and the ordinate represents the number of sentence pairs. The distribution based on S_M and that based on S_S in Fig. 5 are quite different, which indicates that morphology and semantics are the computable content of two different attributes unified in one sentence pair. The distribution based on S_E indicates that the morphological semantic ensemble filtering can fully leverage the efficiency of morphological measurement derived from Levenshtein editing distance and the accuracy of semantic measurement derived from large language models, and complement each other's strengths. To sum up, our ensemble filtering method can use morphological computing to strengthen homonym data and use semantic computing to strengthen synonym data. The straightforwardly and efficiently filtered data can significantly improve the performance of NMT models.

5 Conclusion

The research in this paper only straightforwardly uses an equal-weighted linear combination of morphological similarity and semantic similarity, and its filtering results can improve the performance of the NMT model better than pure morphological filtering results and pure semantic filtering results. From this, it can be concluded that morphology and semantics are two key contents, that are related but not identical, contained in a sentence simultaneously. Combining inseparable morphology and semantics for sentence pair filtering can achieve state-of-the-art results.

The ideas and algorithms related to this paper are not only expected to be transferred to the construction of domain MT data in the future but also to upgrade and improve the classic MT evaluation of BLEU based on n-gram morphological statistics. The morphological semantic ensemble BLEU can not only handle homonyms, but also synonyms, and will transfer the existing diversity evaluation of language expression that mainly relies on multiple reference translations into an upgraded version that mainly relies on world knowledge implicitly in pretrained large language models. In addition, how to combine morphological similarity and semantic similarity more effectively is itself a machine learning problem. Further in-depth research should lead to new conclusions worth looking forward to.

Acknowledgments. The research is supported by the Humanity and Social Science Research Project of Ministry of Education of China (No. 20YJC740062, No. 20YJAZH069), the New Liberal Arts Research and Reform Practice Project of Ministry of Education of China (No. 2021060049), the Postgraduate Education and Teaching Reform Research Project of Shandong (No. SDYJG21185), and the Key Project of Undergraduate Teaching Reform Research of Shandong (No. Z2021323).

References

1. Hirschberg, J., Manning, C.D.: Advances in natural language processing. Science **349**(6245), 261–266 (2015)
2. Kaalep, H., Veskis, K.: Comparing parallel corpora and evaluating their quality. In: Proceedings of the MT Summit XI, pp. 275–279 (2007)
3. Edunov, S., Ott, M., Auli, M., Grangier, D.: Understanding back-translation at scale. In: Proceedings of the 2018 Conference on Empirical Methods in Natural Language Processing (EMNLP), pp. 489–500 (2018)
4. Liu, W., Wang, L.: Ensemble machine translation to filter low quality corpus. In: Proceedings of the 26th International Conference on Asian Language Processing (IALP), pp. 500–504 (2022)
5. Reimers, N., Gurevych, I.: Sentence-BERT: sentence embeddings using siamese BERT-networks. In: Proceedings of the 2019 Conference on Empirical Methods in Natural Language Processing and the 9th International Joint Conference on Natural Language Processing (EMNLP-IJCNLP), pp. 3982–3992 (2019)
6. Jin, K., Zhao, D., Liu, W.: Morphological and semantic evaluation of ancient Chinese machine translation. In: Proceedings of the 1st Workshop on Ancient Language Processing (ALP), pp. 96–102 (2023)
7. Peters, M.E., et al.: Deep contextualized word representations. In: Proceedings of the 2018 Conference of the North American Chapter of the Association for Computational Linguistics (NAACL), pp. 2227–2237 (2018)
8. Liu, Y., et al.: RoBERTa: A Robustly Optimized BERT Pretraining Approach. arXiv:1907.11692v1 (2019)
9. Song, K., Tan, X., Qin, T., Lu, J., Liu, T.Y.: MPNet: masked and permuted pre-training for language understanding. In: Proceedings of the 34th Conference on Neural Information Processing Systems (NeurIPS) (2020)

10. Wang, W., Wei, F., Dong, L., Bao, H., Yang, N., Zhou, M.: MiniLM: deep self-attention distillation for task-agnostic compression of pre-trained transformers. In: Proceedings of the 34th International Conference on Neural Information Processing Systems (NIPS), pp. 5776–5788 (2020)
11. Liu, W., Li, W., Wang, L.: Multiloop incremental bootstrapping for low-resource machine translation. In: Proceedings of the 19th Machine Translation Summit (MT Summit XIX), pp. 1–11 (2023)
12. Liu, W., Wang, T.: Multi-field learning for email spam filtering. In: Proceedings of the 33rd Annual International ACM SIGIR Conference on Research and Development in Information Retrieval (SIGIR), pp. 745–746 (2010)
13. Liu, W., Lin, L.: Probabilistic ensemble learning for Vietnamese word segmentation. In: Proceedings of the 37th Annual International ACM SIGIR Conference on Research and Development in Information Retrieval (SIGIR), pp. 931–934 (2014)

Improving Event-Level Financial Sentiment Analysis with Retrieval-Augmented Multipath Chain-of-Thought Prompting

Yiming Zhang[1,2,3], Xiang Ao[1,2,3(✉)], Guoxin Yu[2,3], and Qing He[1,2,3(✉)]

[1] Henan Institute of Advanced Technology, Zhengzhou University, Zhengzhou 450002, People's Republic of China
yim@gs.zzu.edu.cn
[2] Key Lab of AI Safety, Chinese Academy of Sciences (CAS), Beijing 100190, China
[3] Key Lab of Intelligent Information Processing, Institute of Computing Technology, CAS, Beijing 100190, China
{aoxiang,heqing}@ict.ac.cn

Abstract. Event-level Financial Sentiment Analysis (EFSA) aims to extract all the quintuples containing five sentiment elements from a given financial news text, which has gained prominence as an emerging domain recently. The present study utilizes a 4-hop Chain-of-Thought (CoT) prompting based on LLMs to predict sentiment elements in a fixed order, which neglects the interdependencies among the sentiment elements within a quintuple. Inspired by recent multi-view prompting (MvP) and CoT ideas, we propose a novel framework termed **R**etrieval-Augmented **M**ultipath **C**hain-of-**T**hought (RMP-CoT) that aggregates quintuples generated by LLMs through different reasoning paths, leveraging a retrieval-augmented mechanism. Specifically, RMP-CoT integrates different element orders into CoT prompting to guide LLMs in generating multiple sentiment quintuples through the utilization of retrieval-augmented mechanism, and then selects the most plausible quintuples by voting. To investigate the effectiveness of our framework, we conduct extensive experiments on four benchmark tasks of EFSA. RMP-CoT pushes the state-of-the-art by over 6% F1 on the EFSA task and also performs quite effectively on the other sub-tasks of EFSA.

Keywords: Financial Sentiment Analysis · Chain-of-Thought · Event-level Sentiment Analysis

1 Introduction

Financial Sentiment Analysis (FSA) has received much attention in both academia and industry for its great value [3,10,22]. FSA plays a crucial role in the field of sentiment analysis [3], including the study of financial textual sentiment [8] within news to predict the dynamics of financial markets. Recent

Table 1. The sub-tasks of EFSA. Company c is a text span within the sentence. Each company c is classified into a particular industry i by knowledge-based rules. Coarse-grained event e^1 and fine-grained event e^2 belong to two distinct predefined event type sets E^1 and E^2, respectively. Notably, E^2 represents a more detailed subdivision of E^1. $s \in \{\text{positive, negative, neutral}\}$ denotes the sentiment polarity.

Task	Output
Event-Level Financial Sentiment Analysis (EFSA)	(c, i, e^1, e^2, s)
Coarse-grained Event-Level Financial Sentiment Analysis (C-EFSA)	(c, i, e^1, s)
Fine-grained Event-Level Financial Sentiment Analysis (F-EFSA)	(c, i, e^2, s)
Entity-Level Financial Sentiment Analysis	(c, i, s)

studies [22] have indicated that the events described in financial news and the related sentiments are the key factors driving the impact of financial news on market volatility, thus the primary focus of FSA should sit on events extraction and related sentiment analysis. Building upon this foundation, a recent work [1] proposes a novel Event-Level Financial Sentiment Analysis (EFSA) task that can provide significant practical value in the field of financial applications, such as stock trading, stock market anomaly analysis, and enterprise risk management, etc.

EFSA aims to predict a set of quintuples for a given financial news text. Five sentiment elements constitute the main line of EFSA research: company (c), industry (i), coarse-grained event (e^1), fine-grained event (e^2), and sentiment polarity (s). In financial texts, the event often acts as the subject of sentiment, whereas the company entity is the target of the emotional impact. Extracting events from financial texts can lead to more accurate sentiment predictions. The EFSA task can be further categorized into two event-level sub-tasks, namely Coarse-grained EFSA (C-EFSA) and Fine-grained EFSA (F-EFSA), in addition to an entity-level FSA task. These tasks are outlined in Table 1.

Since the EFSA task requires predicting two events (coarse-grained event, fine-grained event) simultaneously and sentiments in the financial text are primarily implicit, even advanced Large Language Models (LLMs) like GPT-4 encounter notable obstacles in directly predicting both events and implicit emotions. A recent study devises a framework utilizing a 4-hop Chain of Thought (CoT) [1] prompting based on LLMs and achieves good performance. However, they generate sentiment elements in a fixed order, which ignores the effect of the interdependence of the elements in a sentiment tuple, as mentioned by MvP [6]. Additionally, for those uncommon types of financial news, LLMs often encounter challenges in making accurate predictions due to the limited relevant supporting information.

To address the limitations above, we propose a **R**etrieval-Augmented **M**ultipath **C**hain-of-**T**hought (RMP-CoT), which guides LLMs to generate sets of sentiment quintuples through different reasoning paths supported by a retrieval mechanism. Inspired by recent MvP and CoT [20] ideas, we consider different orders of sentiment elements and integrate them into 4-hop CoT prompting to form different reasoning paths, while alleviating the potential error accumulation via permutation of elements. In addition, RMP-CoT introduces a retrieval-augmented mechanism to guide the generation of LLMs by providing relevant financial news texts. Our method is simple yet effective. Empirical results show the superiority of RMP-CoT which outperforms the state-of-the-art by 5% and 6.66% absolute F1 scores on the sub-tasks of EFSA. Our main contributions can be summarized as follows:

1. We propose a novel RMP-CoT method that integrates element order into 4-hop CoT prompting to enhance financial quintuples prediction and then selects the most reasonable quintuples by voting.
2. To mitigate the difficulty of limited relevant supporting information, we introduce a retrieval-augmented mechanism to facilitate the RMP-CoT, in which the relevant financial news texts are fetched to guide the prediction of sentiment elements in a financial quintuple.
3. Experimental results demonstrate that RMP-CoT significantly advances the state-of-the-art performance on 4 benchmark tasks of EFSA.

2 Related Work

2.1 Financial Sentiment Analysis

Previous studies on the FSA dataset have focused either on the sentence level or the document level [2,13,17,18]. However, analyzing the entire text paragraph in FSA may lead to inaccuracies when it contains multiple financial entities with conflicting sentiments [19]. Hence, the analysis is predicated solely on the assumption that the provided text expresses a singular sentiment regarding a specific topic. Recent research on FSA shows a trend towards increasingly finer granularity [11,19], they primarily concentrate on entities and sentiments, overlooking the consideration of events. Due to the events described in financial news and the associated sentiments that dominate the impact of financial news on market volatility, a recent study [1] extends FSA to the event-level and proposes a novel Event-Level Financial Sentiment Analysis task, named EFSA.

2.2 Event-Level Sentiment Analysis

Many works [4,5,12,15,16,25] namely event-based sentiment analysis. These studies mainly focus on event detection through topic modeling and assessing the sentiment associated with the event, whether it be a category, topic, or term. Additionally, they only take into account a single event within a sentence or document. However, in reality, a text may encompass multiple events. To address these problems, some researchers [23] have proposed an end-to-end event-level sentiment analysis approach, which aims to identify events and their associated sentiments. However, they overlook the significance of events within the domain of financial sentiment analysis. A recent study [1] associates event-level sentiment analysis with the financial domain, emerging as a hot field of FSA.

3 Methodology

To enhance the comprehension of our framework, we divide the entire pipeline into three parts introduction as shown in Fig. 1, including multiple reasoning paths, retrieval-augmented mechanism, and results aggregation.

3.1 Multiple Reasoning Paths

Unlike the fixed order element prediction adopted by previous methods, inspired by MVP [6], we consider all possible orders of sentiment elements and integrate them into 4-hop CoT prompting to form different reasoning paths, as shown in Fig. 1. Under different order conditions following interactions with the LLMs, multiple quintuples can be generated from different reasoning paths. Some different reasoning paths generate the same quintuples, while others may generate less effective results and thus might be wrong, but it's unlikely to result in the same error. To put it simply, there is greater consistency among different reasoning paths regarding the correct quintuples. Leveraging this intuition, we aggregate all the predicted quintuples to obtain the final result.

3.2 Retrieval-Augmented Mechanism

As shown in Fig. 2, we introduce a retrieval-augmented mechanism to fetch relevant financial news texts to guide the generation of LLMs, which includes two parts: vector DB construction, and knowledge retrieval.

Vector DB Construction. Vector DB is a crucial component of our retrieval-augmented mechanism, employed for the efficient storage and retrieval of financial documents. We select an identical portion of each fine-grained event type from the dataset used for fine-tuning to construct the vector DB.

Knowledge Embedding. For each of the selected financial news text f_k, we obtain an embedding vector e_{fk} through a text embedding model. This vector will be stored in the database for subsequent retrieval.

$$e_{fk} = \text{TextEmbed}(f_k) \tag{1}$$

where TextEmbed refers to an embedding model, such as BGE [21] and SGPT [14]. We choose to utilize BGE as the embedding model within our framework.

Knowledge Retrieval. The input financial news text X is also fed into the same embedding model to obtain the embedding vector e_X, which is then utilized to retrieve texts from the vector DB.

$$e_X = \text{TextEmbed}(X) \tag{2}$$

We use cosine similarity to retrieve the top three financial news texts as the retrieval results. Then, We process them as the 3-shot in each step of CoT.

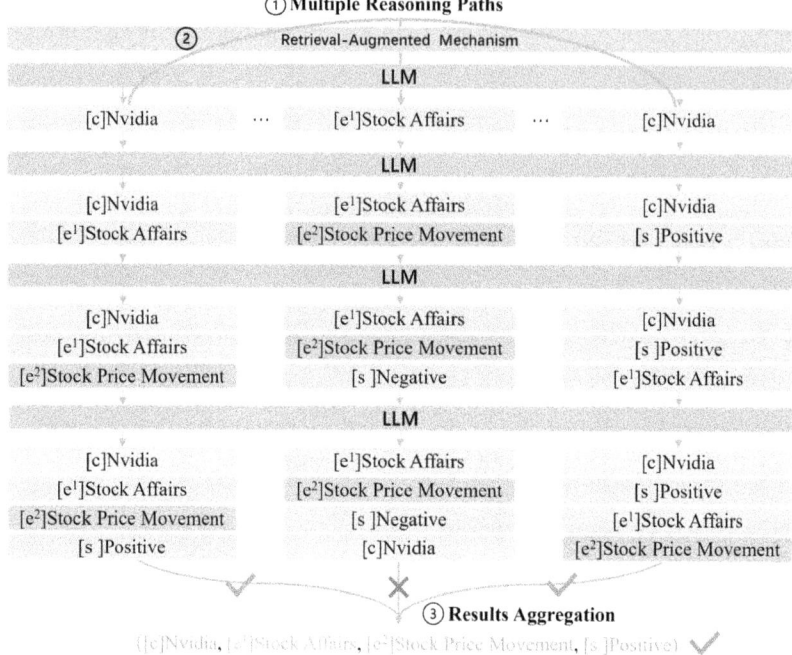

Fig. 1. The pipeline of RMP-CoT framework. RMP-CoT combines multiple reasoning paths and a retrieval-augmented mechanism to guide the LLM step by step in predicting the quadruples. RMP-CoT contains three steps: ① designs the corresponding reasoning paths based on the order of sentiment elements. ② generates 3-shot by the retrieval-augmented mechanism. See details in Fig. 2. ③ aggregates all the predictions and obtains the final output.

3.3 Results Aggregation

As financial news texts can involve events related to multiple companies, each path may predict more than one quintuple. We aggregate the results from all paths and determine the final prediction based on the quintuples that occur most frequently. In particular, for a given financial news text x, and the predicted set of quintuples for permutation p_i denoted as R'_{p_i}, which may contain one or more quintuples. We can then obtain the final aggregated result $R'_{\text{RMP-CoT}}$ using the following equation:

$$R'_{\text{RMP-CoT}} = \{r \mid r \in \bigcup_{i=1}^{n} R'_{p_i} \text{ and} (\sum_{i=1}^{n} \mathbb{1}_{R'_{p_i}}(r) \geq \frac{n}{2})\} \quad (3)$$

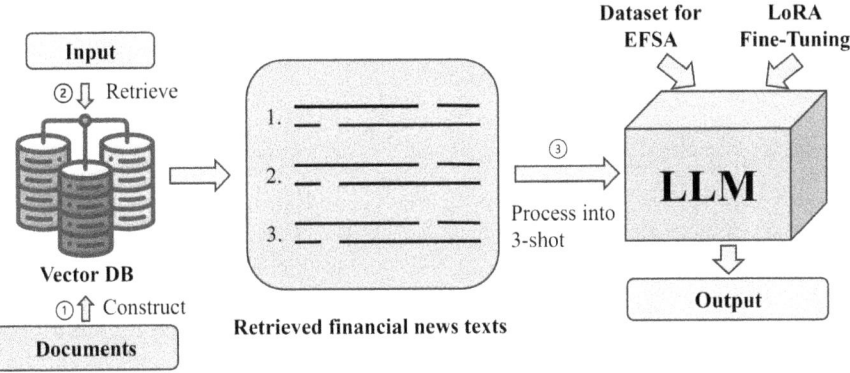

Fig. 2. An illustration of our retrieval-augmented mechanism. For an input financial news text, RMP-CoT retrieves the three most cosine-similar financial news texts from a pre-constructed VectorDB and processes them into 3-shot in each step of reasoning path to guide the generation of LLMs.

3.4 Fine-Tuning

Embedding Model Fine-Tuning. BGE-M3 [21] is the first semantic vector model that integrates the three technical features of multi-linguality, multi-granularity, and multi-functionality. We use the officially provided data format and employ unified fine-tuning to refine BGE-M3. The data used for fine-tuning BGE-M3 is the same as the dataset used for the experiments. To prevent data leakage, the data used for fine-tuning BGE-M3 is not utilized for LLM prediction.

LLMs Fine-Tuning. We design the fine-tuning process for the LLMs used in our experiment. Specifically, we create corresponding CoT datasets based on different permutations of sentiment elements to guide the LLMs to think step by step. All fine-tuning processes of the LLMs we used utilize the Low-Rank Adaptation (LoRA) [7] method.

4 Experiments

4.1 Experimental Setup

Task and Dataset. Due to the scarcity of datasets that can support EFSA tasks, we only conduct different EFSA tasks reported in Table 1 on the newly proposed Chinese financial corpus. To ensure a fair comparison, we apply the same data split as previous work [1].

Evaluation Metrics. For all sub-tasks of the EFSA task, a company prediction is deemed correct if it is found within the gold label, taking into account geographic names, stock codes, and other identifying information in the news text's company labels. Predictions of events and sentiments must align precisely with the gold label to be deemed correct.

4.2 Experiment Details

We select Chinese LLMs as pre-trained models and maintain the same hyperparameter settings as previous work across all tasks. The hyperparameters during the fine-tuning process of LLMs are as follows: train batch size 4, cosine lr scheduler, learning rate 5e−5, fp16, and 1 NVIDIA A800 80 GB PCIe.

The industry of the extracted companies is determined based on knowledge-based rules. Due to the correspondence between coarse-grained events and fine-grained events, we maintain a fixed prediction order for them and consider only three elements to design different reasoning paths: company c, events (e^1 and e^2), and sentiment polarity s. Therefore, a total of six different reasoning paths are formed. Each reasoning path is supported by a corresponding LLM that has been trained on a dataset specific to that task. In addition, taking the first reasoning path on the left side of Fig. 1 as an example, our specific prompts are provided in Appendix A.

4.3 Compared Methods

LoRA Fine-Tuning and 4-Hop CoT Prompting. LoRA fine-tuning [1] enhances performance scores by enhancing both domain-specific ability and capacity to generate structured outputs. By utilizing a 4-hop CoT prompting [1], the effectiveness of non-open-source LLMs can be significantly enhanced only through prompt-based modifications.

5 Experimental Results

5.1 Main Results and Discussions

The comparisons are shown in Table 2. We observe that our method outperforms the state-of-the-art on different EFSA sub-tasks for each Chinese LLM model listed in Table 2. Compared with 4-hop CoT, Qwen-7B-chat with RMP-CoT shows significant boosts for EFSA sub-tasks, i.e., 5% on C-EFSA and 6.66% on F-EFSA and EFSA, with an average improvement of 6.1% F1 on supervised setup. The scores for each sub-task exhibit a decreasing trend, which corresponds to the increasing complexity of the tasks and is consistent with the previously observed scoring trends. More prominently, compared with LoRA, Llama2-Chinese-7b-Chat with RMP-CoT has an impressive improvement in results, i.e., 23.73% on C-EFSA, 23.17% on E-EFSA and 23.6% on EFSA. We conjecture that the previous low scores may be attributed to the limited capability of Llama2-Chinese-7b-Chat in generating formatted outputs, even after fine-tuning. Under our framework, the outputs of LLMs are standardized by providing Q&A examples for reference through the retrieval-augmented mechanism.

Table 2. Comparison results on different EFSA tasks. The best scores are in bold and the second best ones are underlined. The 4-hop CoT framework and ours, the F-EFSA's score is identical to the EFSA's. This is because accurately predicting fine-grained events relies on correctly predicting coarse-grained events. † denotes the results are taken from Chen [1].

Settings	LLMs	Entity-Level FSA	C-EFSA	F-EFSA	EFSA
LoRA†	ChatGLM3-6B-Chat	76.83	62.26	53.11	51.18
	Baichuan2-13B-Chat	86.41	71.82	67.79	67.06
	Qwen-7B-Chat	86.14	73.22	67.32	67.28
	Llama2-Chinese-7B-Chat	61.55	43.84	36.71	36.28
4-hop CoT†	GPT-4	63.28	55.64	53.24	53.24
	ChatGLM3-6B-Chat	78.93	70.61	65.19	65.19
	Baichuan2-13B-Chat	81.74	75.66	69.52	69.52
	Qwen-7B-Chat	83.28	76.03	71.43	71.43
	Llama2-Chinese-7b-Chat	61.47	51.62	23.44	23.44
RMP-CoT	ChatGLM3-6B-Chat	82.60	75.58	71.55	71.55
	Baichuan2-13B-Chat	85.11	80.35	77.47	77.47
	Qwen-7B-Chat	**86.63**	**81.03**	**78.09**	**78.09**
	Llama2-Chinese-7b-Chat	73.96	67.57	59.88	59.88

5.2 Ablation Test

In order to verify the effectiveness of different mechanisms in our model, we conduct ablation tests on different EFSA tasks. We remove different mechanisms in turn to observe their effectiveness. As shown in Table 3, both settings are effective in RMP-CoT as expected. The scores of the model without multipath drop more than the model without retrieval-augmented. It is possibly due to the model without multipath through fine-tuning has already demonstrated impressive capabilities for the EFSA task. However, the model without multipath that only uses a retrieval-augmented mechanism has introduced errors in financial news examples. This issue is not present in the model without retrieval-augmented. Due to the high cost of GPT-4, we exclusively utilize it in the model without multipath. However, it still surpasses its performance in the 4-hop CoT.

5.3 Case Study

To better demonstrate the effectiveness of our framework, we perform a case study on Qwen-7B-Chat, as shown in Fig. 3. As shown in Example 1, Qwen with 4-hop CoT incorrectly predicts coarse-grained events, consequently leading to incorrect predictions of fine-grained events, confusing *Shareholder Affairs* with *Management Affairs*. It is possibly due to the board in the text leading to misdirection. Under our RMP-CoT framework, two types of errors have emerged:

Table 3. Results of ablation study on the model for sub-tasks of EFSA. F1 scores over one run are reported. The best scores are in bold for each setting. The reason why the F-EFSA's score is identical to the EFSA's is explained in Table 2. w/o denotes without.

Settings	LLMs	Entity-Level FSA	C-EFSA	F-EFSA	EFSA
Full Model	ChatGLM3-6B-Chat	82.60	75.58	71.55	71.55
	Baichuan2-13B-Chat	85.11	80.35	77.47	77.47
	Qwen-7B-Chat	**86.63**	**81.03**	**78.09**	**78.09**
	Llama2-Chinese-7b-Chat	73.96	67.57	59.88	59.88
w/o Multipath	GPT-4	73.42	65.66	62.44	62.44
	ChatGLM3-6B-Chat	79.01	70.10	67.80	67.80
	Baichuan2-13B-Chat	**85.21**	**79.84**	**76.42**	**76.42**
	Qwen-7B-Chat	84.02	77.53	74.03	74.03
	Llama2-Chinese-7b-Chat	65.53	54.22	42.65	42.65
w/o Retrieval-Augmented	ChatGLM3-6B-Chat	81.08	72.39	69.93	69.93
	Baichuan2-13B-Chat	85.58	79.93	**77.26**	**77.26**
	Qwen-7B-Chat	**85.74**	**80.50**	76.73	76.73
	Llama2-Chinese-7b-Chat	76.94	63.18	47.58	47.58

one is the aforementioned incorrect prediction of events, and the other is the incorrect prediction of sentiment polarity, which may be misled by *the company receives a notice of share reduction*. However, RMP-CoT only picks quintuples considered important in most paths and thus repairs the error in the single path by aggregating information from multiple paths.

Error Analysis. We observed that LLMs sometimes generate outputs that fall outside the defined label sets, as illustrated in Example 2 of our error analysis. LLMs may modify the predetermined label instead of outputting as instructed. This observation is consistent with the previous research [24], emphasizing the inherent nature of generative modeling since it does not perform "extraction" in the provided sentences. Besides, the other error is that the incorrect prediction of fine-grained events is possibly due to the presence of price-related information in the text. However, it is still the generation of labels beyond the defined set that predominantly leads to final erroneous output. This phenomenon is more evident in LLMs with smaller parameter sizes but can be alleviated through fine-tuning. Just as guiding LLMs towards structured outputs, fine-tuning markedly improves the capability of smaller-parameter LLMs to produce the desired output.

Case study

Example-1

News body: Aim Pharm(002826.SZ) announced on January 17, 2022, that the company has received a "Notice of Progress on Share Reduction" from Mr. Zhou, a member of the board. The share reduction plan has been fully implemented, and Mr. Zhou did not reduce his holdings in the company's shares in any form during the period of this reduction plan.

Gold Label: (Aim Pharm, Health Care, Shareholder Affairs, Stock Holding Adjustment, NEU)

Qwen-7B-Chat + 4-hop CoT: (Aim Pharm, Health Care, Management Affairs, Employee Dynamics, NEU) ✗

Tuples of six permutations based on Qwen-7B-Chat + RMP-CoT:

(Aim Pharm, Health Care, Shareholder Affairs, Stock Holding Adjustment, NEU)*4 — pick

(Aim Pharm, Health Care, Shareholder Affairs, Stock Holding Adjustment, NEG)

(Aim Pharm, Health Care, Management Affairs, Employee Dynamics, NEU) — drop

Final tuple : (Aim Pharm, Health Care, Shareholder Affairs, Stock Holding Adjustment, NEU) ✓

Error analysis

Example-2

News body: JPMorgan lowers the target price for Mengniu Dairy from HK$51.8 to HK$49, with an "overweight" rating.

Gold Label: (Mengniu Dairy, Food & Beverage, Compliance and Credit, Rating Adjustment, NEG)

Tuples of six permutations based on ChatGLM3-6B-Chat + RMP-CoT:

(Mengniu Dairy, Food & Beverage, Compliance and Credit, Stock Rating Adjustment, NEG)*3 — pick

(Mengniu Dairy, Food & Beverage, Compliance and Credit, Rating Adjustment, NEG)*2

(Mengniu Dairy, Food & Beverage, Stock Affairs, Stock Price Movement, NEG) — drop

Final tuple : (Mengniu Dairy, Food & Beverage, Compliance and Credit, Stock Rating Adjustment, NEG) ✗

Fig. 3. An example including input news text, gold label, quintuples predicted, and the final output by aggregating. *Pick* means that the quintuples have appeared in more than half of the predictions, while *drop* means that it has appeared less than half of the times and is discarded. The red font denotes the incorrect part of the prediction. (Color figure online)

6 Conclusion and Future Work

In this paper, we propose a novel framework RMP-CoT to enhance the performances of LLMs for the EFSA tasks. We explore the dependencies between sentiment elements in a financial quintuple and LLM's capacity to employ diverse reasoning paths to tackle the same task. Furthermore, we also introduce a retrieval-augmented mechanism that dynamically modifies the 3-shot.

Future research directions include:

1. Exploring the experimental effects of combining RMP-CoT with GPT-4.
2. Enhancing LLM's capability to tackle the EFSA task with retrieval-augmented generation (RAG) [9] techniques.
3. Supplementing datasets available for studying the EFSA task.

Acknowledgments. The research work is supported by the National Key R&D Plan No. 2022YFC3303303, the National Natural Science Foundation of China under Grant

No.61976204, U2436209. Xiang Ao is also supported by the Project of Youth Innovation Promotion Association CAS, and Beijing Nova Program 20230484430.

A RMP-CoT Prompting

Take the inference path as (company, coarse-grained event, fine-grained event, sentiment polarity) for example. Three financial news articles were returned by the retrieval mechanism: text1, text2, and text3. **Step1** As an advanced sentiment analysis model, identify company-specific events in the provided financial news article. History_step1=[{role: user, content: Acting as a fine-grained sentiment analysis model within the financial sector, I'll be presented with a segment of financial news. Identify the companies mentioned in the financial news text: text1[content]. List the company names, comma-separated, with no additional details.},

{role: assistant, content: text1[company]},

{role: user, content: Acting as a fine-grained sentiment analysis model within the financial sector, I'll be presented with a segment of financial news. Identify the companies mentioned in the financial news text: text2[content]. List the company names, comma-separated, with no additional details.},

{role: assistant, content: text2[company]},

{role: user, content: Acting as a fine-grained sentiment analysis model within the financial sector, I'll be presented with a segment of financial news. Identify the companies mentioned in the financial news text: text3[content]. List the company names, comma-separated, with no additional details.},

{role: assistant, content: text3[company]}]

Step2. Presuming you function as an advanced sentiment analysis model within the financial sector, I will supply you with a financial news article and the company name. Your objective will be to identify the key event mentioned in the news that pertains to the specified company. Financial news text as follows: [content] Identify the main event affecting [response1] in the provided financial news text. Select the primary event from the options listed: []. Respond with a tuple in the format (Company Name, Primary Event), choosing from the provided events only. Include no additional details.

History_step2=[{role: user, content: Presuming you function as an advanced sentiment analysis model within the financial sector, I will supply you with a financial news article and the company name. Identify the primary event affecting text1[company] in the financial news text: text1[content]. Choose from the provided list of primary events: []. Respond with a tuple (Company Name, Primary Event). No additional details.},

{role: assistant, content: (text1[company],text1[primary event])},

{role: user, content: Presuming you function as an advanced sentiment analysis model within the financial sector, I will supply you with a financial news article and the company name. Identify the primary event affecting text2[company] in the financial news text: text2[content]. Choose from the provided list of primary events: []. Respond with a tuple (Company Name, Primary Event). No additional details.},

{role: assistant, content: (text2[company],text2[primary event])},

{role: user, content: Presuming you function as an advanced sentiment analysis model within the financial sector, I will supply you with a financial news article and the company name. Identify the primary event affecting text3[company] in the financial news text: text3[content]. Choose from the provided list of primary events: []. Respond with a tuple (Company Name, Primary Event). No additional details.},

{role: assistant, content: (text3[company],text3[primary event])}]

Step3. Presuming you function as an advanced sentiment analysis model within the financial sector, I will supply you with a financial news article, the company name, and the primary event. Your objective will be to identify the secondary event mentioned in the news that pertains to the specified company. Financial news text as follows: [content]. The main event affecting [response1] in the provided financial news text is [response2]. Please choose the appropriate secondary event from the following list: [Appropriate Secondary Event List]. Select from the listed secondary events and provide your answer as a tuple in the format (Company Name, Primary Event, Secondary Event), including no additional information.

History_step3=[{role: user, content: Assuming you are a fine-grained sentiment analysis model in the finance domain, I will give you a piece of financial news text, the company name, and the primary event. Identify the secondary event for text1[company] in the financial news text: text1[content]. The primary event is text1[primary event]. Choose the secondary event from the given list: [Appropriate Secondary Event List]. Respond with a tuple (Company Name, Primary Event, Secondary Event). No extra information. },

{role: assistant, content:(text1[company],text1[primary event],text1[secondary event])},

{role: user, content: Assuming you are a fine-grained sentiment analysis model in the finance domain, I will give you a piece of financial news text, the company name, and the primary event. Identify the secondary event for text2[company] in the financial news text: text2[content]. The primary event is text2[primary event]. Choose the secondary event from the given list: [Appropriate Secondary Event List]. Respond with a tuple (Company Name, Primary Event, Secondary Event). No extra information.},

{role: assistant, content:(text2[company],text2[primary event],text3[secondary event])},

{role: user, content: Assuming you are a fine-grained sentiment analysis model in the finance domain, I will give you a piece of financial news text, the company name, and the primary event. Identify the secondary event for text3[company] in the financial news text: text3[content]. The primary event is text3[primary event]. Choose the secondary event from the given list: [Appropriate Secondary Event List]. Respond with a tuple (Company Name, Primary Event, Secondary Event). No extra information.},

{role: assistant, content:(text3[company],text3[primary event],text3[secondary event])}]

Step4. Presuming you function as an advanced sentiment analysis model within the financial sector, I will supply you with a financial news article, the company name, and the primary event and secondary events. Your objective will be to identify the sentiment mentioned in the news. Financial news text as follows: [content]. The primary event for [response1] in the financial news text is [response2], with the secondary event being [response]. Choose the correct sentiment from the options ['Positive', 'Negative', 'Neutral'] and present your answer as a quadruple in the format (Company Name, Primary Event, Secondary Event, Sentiment Polarity).

History_step4=[{role: user, content: Presuming you function as an advanced sentiment analysis model within the financial sector, I will supply you with a financial news article, the company name, and the primary event and secondary events. Identify the sentiment for text1[company] in the financial news text: text1[content]. The primary event is text1[primary event], and the secondary event is text1[secondary event]. Choose from ['Positive', 'Negative', 'Neutral'] and respond with a quadruple (Company Name, Primary Event, Secondary Event, Sentiment Polarity).},

{role: assistant, content:(text1[company], text1[primary event],text1[secondary event],text1[sentiment polarity])},

{role: user, content: Presuming you function as an advanced sentiment analysis model within the financial sector, I will supply you with a financial news article, the company name, and the primary event and secondary events. Identify the sentiment for text2[company] in the financial news text: text2[content]. The primary event is text2[primary event], and the secondary event is text2[secondary event]. Choose from ['Positive', 'Negative', 'Neutral'] and respond with a quadruple (Company Name, Primary Event, Secondary Event, Sentiment Polarity).},

{role: assistant, content:(text2[company],text2[primary event],text2[secondary event],text2[sentiment polarity])},

{role: user, content: Presuming you function as an advanced sentiment analysis model within the financial sector, I will supply you with a financial news article, the company name, and the primary event and secondary events. Identify the sentiment for text3[company] in the financial news text: text3[content]. The primary event is text3[primary event], and the secondary event is text3[secondary event]. Choose from ['Positive', 'Negative', 'Neutral'] and respond with a quadruple (Company Name, Primary Event, Secondary Event, Sentiment Polarity).},

{role: assistant, content: (text3[company],text3[primary event],text3[secondary event],text3[sentiment polarity])}]

References

1. Chen, T., et al.: EFSA: towards event-level financial sentiment analysis (2024). https://arxiv.org/abs/2404.08681

2. Cortis, K., et al.: Semeval-2017 task 5: fine-grained sentiment analysis on financial microblogs and news. In: Proceedings of the 11th International Workshop on Semantic Evaluation (SemEval-2017) (2017)
3. Du, K., Xing, F., Mao, R., Cambria, E.: Financial sentiment analysis: techniques and applications. ACM Comput. Surv. **56**(9) (2024)
4. Ebrahimi, M., Yazdavar, A.H., Sheth, A.: Challenges of sentiment analysis for dynamic events. IEEE Intell. Syst. **32**(5), 70–75 (2017)
5. Fukuhara, T., Nakagawa, H., Nishida, T.: Understanding sentiment of people from news articles: temporal sentiment analysis of social events. In: International Conference on Web and Social Media (2007)
6. Gou, Z., Guo, Q., Yang, Y.: MvP: multi-view prompting improves aspect sentiment tuple prediction. In: Rogers, A., Boyd-Graber, J., Okazaki, N. (eds.) Proceedings of the 61st Annual Meeting of the Association for Computational Linguistics (Volume 1: Long Papers), pp. 4380–4397. Association for Computational Linguistics, Toronto, Canada (2023)
7. Hu, E.J., et al.: Lora: low-rank adaptation of large language models (2021). https://arxiv.org/abs/2106.09685
8. Kearney, C., Liu, S.: Textual sentiment in finance: a survey of methods and models. Int. Rev. Financ. Anal. **33**, 171–185 (2014)
9. Lewis, P., et al.: Retrieval-augmented generation for knowledge-intensive NLP tasks. In: Larochelle, H., Ranzato, M., Hadsell, R., Balcan, M., Lin, H. (eds.) Advances in Neural Information Processing Systems, vol. 33, pp. 9459–9474. Curran Associates, Inc. (2020)
10. Luo, L., et al.: Beyond polarity: interpretable financial sentiment analysis with hierarchical query-driven attention. In: Proceedings of the Twenty-Seventh International Joint Conference on Artificial Intelligence (2018)
11. Maia, M., et al.: Www'18 open challenge. In: Companion of the Web Conference 2018 on Web Conference 2018 - WWW 2018 (2018)
12. Makrehchi, M., Shah, S., Liao, W.: Stock prediction using event-based sentiment analysis. In: 2013 IEEE/WIC/ACM International Joint Conferences on Web Intelligence (WI) and Intelligent Agent Technologies (IAT) (2013)
13. Malo, P., Sinha, A., Takala, P., Korhonen, P., Wallenius, J.: Good debt or bad debt: detecting semantic orientations in economic texts (2013). https://arxiv.org/abs/1307.5336
14. Muennighoff, N.: SGPT: GPT sentence embeddings for semantic search (2022). https://arxiv.org/abs/2202.08904
15. Patil, M., Chavan, H.: Event based sentiment analysis of twitter data. In: 2018 Second International Conference on Computing Methodologies and Communication (ICCMC) (2018)
16. Petrescu, A., Truica, C.O., Apostol, E.S.: Sentiment analysis of events in social media. In: 2019 IEEE 15th International Conference on Intelligent Computer Communication and Processing (ICCP) (2019)
17. Sinha, A., Kedas, S., Kumar, R., Malo, P.: Sentfin 1.0: entity-aware sentiment analysis for financial news. J. Assoc. Inf. Sci. Technol. 1314–1335 (2022)
18. Takala, P., Malo, P., Sinha, A., Ahlgren, O.: Gold-standard for topic-specific sentiment analysis of economic texts. In: Chair), N.C.C., et al. (eds.) Proceedings of the Ninth International Conference on Language Resources and Evaluation (LREC 2014). European Language Resources Association (ELRA), Reykjavik, Iceland (2014)

19. Tang, Y., Yang, Y., Huang, A., Tam, A., Tang, J.: FinEntity: entity-level sentiment classification for financial texts. In: Bouamor, H., Pino, J., Bali, K. (eds.) Proceedings of the 2023 Conference on Empirical Methods in Natural Language Processing, pp. 15465–15471. Association for Computational Linguistics, Singapore (2023)
20. Wei, J., et al.: Chain-of-thought prompting elicits reasoning in large language models. In: Proceedings of the 36th International Conference on Neural Information Processing Systems. NIPS 2022. Curran Associates Inc., Red Hook, NY, USA (2024)
21. Xiao, S., Liu, Z., Zhang, P., Muennighoff, N., Lian, D., Nie, J.Y.: C-pack: packaged resources to advance general Chinese embedding (2024). https://arxiv.org/abs/2309.07597
22. Xing, F., Malandri, L., Zhang, Y., Cambria, E.: Financial sentiment analysis: an investigation into common mistakes and silver bullets. In: Proceedings of the 28th International Conference on Computational Linguistics (2020)
23. Zhang, Q., Zhou, J., Chen, Q., Bai, Q., He, L.: Enhancing event-level sentiment analysis with structured arguments. In: Proceedings of the 45th International ACM SIGIR Conference on Research and Development in Information Retrieval (2022)
24. Zhang, W., Deng, Y., Li, X., Yuan, Y., Bing, L., Lam, W.: Aspect sentiment quad prediction as paraphrase generation. In: Moens, M.F., Huang, X., Specia, L., Yih, S.W.T. (eds.) Proceedings of the 2021 Conference on Empirical Methods in Natural Language Processing, pp. 9209–9219. Association for Computational Linguistics, Online and Punta Cana, Dominican Republic (2021)
25. Zhou, X., Tao, X., Yong, J., Yang, Z.: Sentiment analysis on tweets for social events. In: Proceedings of the 2013 IEEE 17th International Conference on Computer Supported Cooperative Work in Design (CSCWD) (2013)

POSRho: Efficient Spearman's Rho Calculation for Big Data

Xiaofei Zhao[1,2](✉) and Fanglin Guo[3]

[1] Lanzhou Petrochemical University of Vocational Technology, Gansu, China
xifizhao@gmail.com
[2] Lanzhou Jiaotong University, Gansu, China
[3] State Grid Gansu Electric Power Company, Lanzhou, China
guofanglin_xt@gs.sgcc.com.cn

Abstract. The increasing volume and complexity of data in various scientific domains necessitate robust and scalable methods for statistical analysis. Spearman's rank correlation coefficient, denoted as ρ_s, is a non-parametric measure that evaluates the monotonic relationships between variables. However, traditional methods for computing ρ_s struggle with the scalability and efficiency required for large datasets characteristic of the big data era. This paper introduces POSRho, a novel algorithm designed for the efficient and scalable computation of Spearman's rank correlation coefficient in big data settings. Leveraging parallel and distributed computing frameworks, POSRho addresses the primary challenges posed by big data, including high computational complexity, significant memory constraints, and data distribution and heterogeneity issues. We detail the algorithm's design, which utilizes data partitioning, parallel rank calculation, and efficient aggregation methods to optimize computational resources and minimize execution time while maintaining the accuracy of the correlation measure. Empirical results demonstrate that POSRho significantly reduces computation time compared to conventional methods without sacrificing accuracy, thus providing a practical solution for big data analytics in various applications such as genomics, finance, and social science research. The adaptability of POSRho across different computing environments and its integration into existing big data platforms underscore its utility and innovation in addressing the computational demands of modern data analysis.

Keywords: Spearman's rank correlation coefficient · Non-parametric statistical measures · Big data challenges · Distributed computing · Computational complexity · Data heterogeneity

1 Introduction

Spearman's rank correlation coefficient (ρ_s), a non-parametric measure introduced in 1904, has become indispensable for assessing monotonic relationships between variables across diverse scientific disciplines [21,23]. Unlike Pearson's correlation coefficient, Spearman's rho focuses on data order rather than absolute values, making it robust to outliers and suitable for non-normal distributions [4,7,9,11]. Its applications span biology (analyzing gene expression, phenotypic correlations) [10,14], social sciences (investigating socioeconomic and psychological relationships) [2,3,17], finance (analyzing correlations between market indicators) [1,16,18], and machine learning (feature selection, model evaluation) [5,12,15].

However, the traditional Spearman's rho calculation, relying on data sorting [4,7,8], becomes computationally expensive for large datasets [6,13,19,22]. This challenge is further amplified by the ever-increasing volume and velocity of big data, demanding innovative and scalable solutions [20].

To address these limitations, we propose POSRho (Parallel and Optimized Spearman's Rho). This novel algorithm leverages parallel and distributed computing to achieve efficient and scalable computation of Spearman's rho on large datasets. By overcoming the bottlenecks of traditional methods, POSRho aims to empower researchers and practitioners to unlock valuable insights from massive datasets.

The growing volume of big data presents a unique opportunity to analyze high-dimensional combinations of datasets, revealing complex relationships often missed by traditional methods. This approach holds immense potential across various fields:

Finance: Integrating transaction behaviors, consumption habits, and other data allows for accurate risk assessment and personalized financial products.
Medicine: Combining environmental data (weather, air quality) with patient information (genetics, medical records) enables better disease prediction and precision medicine.

Advancements in artificial intelligence and machine learning further enhance our ability to process and analyze high-dimensional data, making it feasible to uncover hidden relationships and extract valuable insights.

In conclusion, research on high-dimensional combination data mining is crucial for maximizing the value of big data. By developing efficient algorithms like POSRho, we can unlock the full potential of these complex datasets, leading to significant economic and social benefits across various sectors.

2 Related Work

Calculating Spearman's rank correlation coefficient for large datasets presents significant computational challenges due to the need for sorting operations, which become increasingly complex with data size. Existing methods, including traditional algorithms, recursive approaches, and approximation techniques, each have limitations in terms of scalability and efficiency, particularly when dealing with the vast volumes of data characteristic of modern applications.

To address these challenges, big data analytics frameworks such as Hadoop, Apache Flink, and various cloud-based platforms offer robust solutions for processing large-scale datasets. These platforms provide distributed file systems, parallel processing engines, and managed environments that enable the efficient implementation and deployment of scalable Spearman's rho calculation methods. For instance, Hadoop's distributed storage (HDFS) and its MapReduce programming model facilitate the handling of large datasets by distributing the workload across a cluster of machines. Apache Flink offers stream and batch processing capabilities, which are particularly useful for real-time data analysis and iterative computations.

Parallel and distributed computing technologies, such as MapReduce, Apache Spark, and the Message Passing Interface (MPI), further enhance the efficiency of Spearman's rho computation. MapReduce, with its map and reduce phases, allows for the division of tasks into smaller, independent units that can be processed concurrently across multiple nodes, significantly reducing computation time and improving scalability. Apache Spark, known for its in-memory processing capabilities, accelerates data processing tasks and supports complex analytics operations. MPI, a standard for parallel computing, enables high-performance communication between nodes in a distributed system, optimizing the execution of parallel algorithms.

These frameworks and technologies can be combined to develop efficient and scalable solutions for calculating Spearman's rho in big data environments. For example, implementing Spearman's rho calculation in a Spark-based system allows for the parallel sorting of ranks and efficient computation of the correlation coefficient across distributed datasets.

The integration of these advanced technologies and platforms provides a comprehensive foundation for overcoming the computational challenges associated with Spearman's rho calculation on large datasets. By leveraging distributed storage, parallel processing, and scalable infrastructure, researchers and practitioners can efficiently compute Spearman's rank correlation coefficient, enabling the extraction of meaningful insights from massive datasets and advancing the field of data analysis in various scientific and industrial domains.

3 POSRho Algorithm

3.1 Detailed Breakdown of the POSRho Architecture

POSRho addresses the computational challenges of calculating Spearman's rho in big data by implementing a parallel processing framework that significantly speeds up computation without sacrificing accuracy (see Fig. 1).

Initial Data Set (D): The process starts with the initial data set D, containing a large number of observations with pairs of variables for Spearman's rho calculation.

Partitioning the Data (D1, D2, ..., Dp): The data set D is divided into p partitions, breaking it down into smaller, more manageable segments. Each partition D_i can be processed independently, facilitating parallel computation.

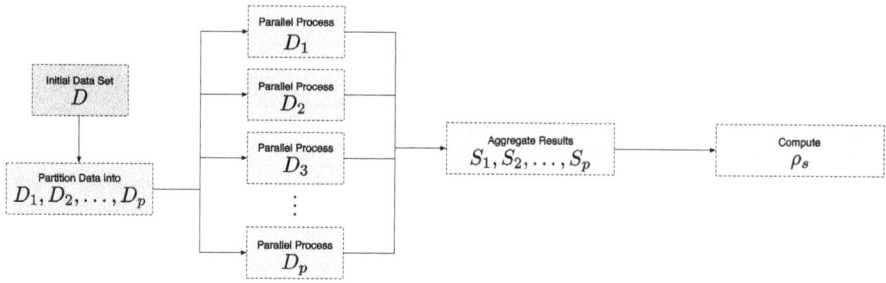

Fig. 1. Architecture of the POSRho Algorithm for Big Data

Parallel Processing: Each partition D_i undergoes parallel processing where Spearman's rho is calculated separately. This step leverages multiple processors or computational resources to handle different data segments simultaneously, reducing overall computation time.

Aggregation of Results (S1, S2, ..., Sp): After computing Spearman's rho for each partition, the results S_i are aggregated. This step synthesizes the partial results into a final Spearman's rho value for the entire data set.

Final Computation: The aggregated results are combined to produce a single Spearman's rho value that accurately reflects the rank correlation across the entire dataset.

3.2 Rank Difference Squared Calculation and Aggregation

After computing ranks for each data point, POSRho calculates the squared differences between the ranks of corresponding pairs of variables in parallel within each data partition. For each pair of data points (x_i, y_i), the squared rank difference d_i^2 is calculated as:

$$d_i^2 = (rank(x_i) - rank(y_i))^2 \tag{1}$$

Within each partition, POSRho sums these squared rank differences, resulting in a partial sum S_i:

$$S_i = \sum_{j=1}^{n_i} d_{ij}^2 \tag{2}$$

where n_i is the number of data points in partition D_i. These partial sums are then aggregated across all partitions to obtain the total sum of squared rank differences S:

$$S = \sum_{i=1}^{p} S_i \tag{3}$$

where p is the number of partitions. This aggregated value is used to compute the Spearman's rank correlation coefficient:

$$\rho_s = 1 - \frac{6 \sum_{i=1}^{p} S_i}{n(n^2 - 1)} \tag{4}$$

3.3 Overview and Design Principles

POSRho (Algorithm 1) is designed to efficiently compute Spearman's rank correlation coefficient for big data by leveraging parallel and distributed computing techniques. The algorithm follows a divide-and-conquer strategy, partitioning the data into smaller subsets, processing them in parallel, and then aggregating the results to obtain the final correlation coefficient.

Algorithm 1. Parallel Optimized Spearman Rank Correlation Coefficient Algorithm (POSRho)

Require: Data set $D = \{(x_1, y_1), (x_2, y_2), \ldots, (x_n, y_n)\}$
Ensure: Spearman rank correlation coefficient ρ_s
1: Partition the data set D into p subsets $D_1, D_2, \ldots, D_p$
2: **for all** subsets D_i execute in parallel **do**
3:     Compute the ranks rx_i and ry_i for each data point in D_i
4:     Compute the squared rank differences $d_i^2 = (rx_i - ry_i)^2$ for each data point in D_i
5:     Compute the sum of squared rank differences $S_i = \sum_{j=1}^{n_i} d_{ij}^2$ in D_i
6: **end for**
7: Aggregate the sums of squared rank differences $S = \sum_{i=1}^{p} S_i$
8: Calculate the Spearman rank correlation coefficient $\rho_s = 1 - \frac{6S}{n(n^2-1)}$
9: **return** ρ_s

The core design principles of POSRho are:

Data Parallelism: The dataset is partitioned into multiple subsets, processed concurrently on different computing nodes, leveraging modern computing systems' parallelism to reduce execution time.

Task Decomposition: The computation of Spearman's rho is broken down into smaller, independent tasks for each data subset, including calculating ranks, rank differences, and partial sums of squared rank differences.

Local Aggregation: Intermediate results are aggregated within each data subset before being combined globally, minimizing data communication overhead and improving efficiency.

Global Aggregation: Partial results from each data subset are combined to obtain the final Spearman's rho value by summing the partial sums of squared rank differences and applying the correlation coefficient formula.

These principles enable POSRho to achieve scalability, efficiency, and accuracy in computing Spearman's rho for big data.

3.4 Algorithmic Complexity Analysis

POSRho's computational complexity is determined by three key factors: data partitioning strategy, parallel rank calculation method, and aggregation steps, all contributing to the algorithm's overall efficiency.

1. **Data Partitioning**: Exhibits $O(n)$ complexity, efficiently distributing data across partitions.
2. **Parallel Rank Calculation**: Utilizes distributed sorting (e.g., quicksort or mergesort) with a complexity of $O(n \log n)$ per partition. Parallel execution significantly reduces runtime compared to traditional methods.
3. **Rank Difference Squared Calculation and Aggregation**: Both have $O(n)$ complexity within partitions.

The overall complexity of POSRho is primarily influenced by the parallel rank calculation step, dominated by $O(n \log n)$ per partition. However, parallel execution and optimizations, such as approximate ranking or rank aggregation, enhance performance compared to traditional algorithms (Table 1).

Table 1. Summary of POSRho Algorithm Complexity

Component	Complexity
Data Partitioning	$O(n)$
Parallel Rank Calculation	Distributed sorting: $O(n \log n)$ per partition. Rank aggregation might have lower complexity, with possible trade-offs in accuracy
Rank Difference Squared Calculation and Aggregation	$O(n)$
Overall Complexity	Dominated by $O(n \log n)$ per partition, effectively reduced through parallel execution

In conclusion, POSRho offers a scalable and efficient approach for computing Spearman's rho for big data. Its complexity is comparable to or lower than existing methods, leveraging parallel and distributed processing for substantial performance gains, making it valuable for large-scale data analysis.

4 Experiments

4.1 Performance Comparison

This experiment aims to compare the POSRho algorithm's performance against the traditional Spearman's rank correlation coefficient calculation on a dataset. The focus is on evaluating POSRho's efficiency and accuracy in terms of computation time and correlation results.

Dataset Description: The dataset, derived from an industrial monitoring system, contains 7,528,000 entries with multiple numerical columns representing various operational and environmental attributes of machinery in a manufacturing plant. Data preprocessing included removing incomplete records, normalizing data ranges, and encoding categorical variables.

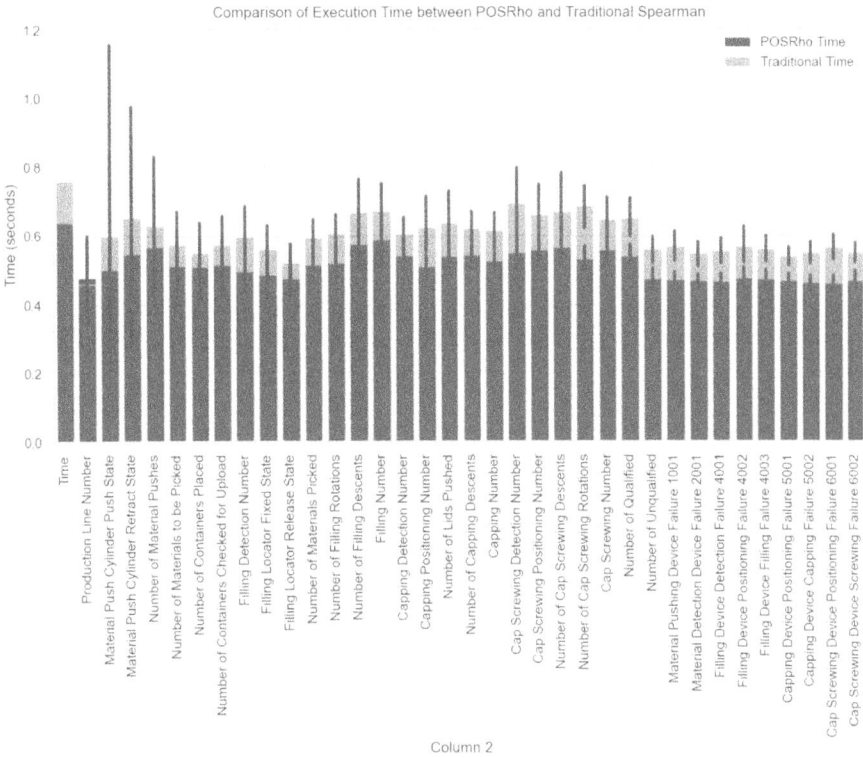

Fig. 2. Comparison of Execution Time between POSRho and Traditional Spearman.

Methodology: Two functions were implemented in Python: POSRho and conventional method. POSRho partitioned the data into four segments, calculating Spearman's rho for each and averaging the results. The traditional method processed the entire dataset in a single batch (Fig. 3).

Execution and Timing: Execution times for both methods were measured and averaged over multiple runs to ensure consistency. Results included computation times and correlation coefficients for comparison. **Results and Analysis**: The correlation results showed a high degree of agreement between POSRho and the traditional method, with data points closely clustered around the y = x line, indicating similar correlation values from both methods (Fig. 6). POSRho consistently achieved lower execution times compared to the traditional method, demonstrating its efficiency in handling large datasets and achieving approximately 13.74% time savings (Fig. 6). The absolute differences between the correlation coefficients from POSRho and the traditional method were predominantly near zero, with 95.6% of cases showing identical results, confirming POSRho's accuracy and reliability. Small variations decrease with increasing difference,

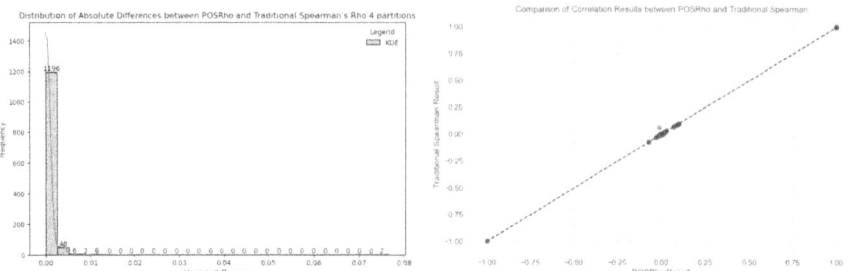

(a) Comparison of Correlation Results between POSRho and Traditional Spearman.

(b) Distribution of Absolute Differences between POSRho and Traditional Spearman's Rho (4 partitions).

Fig. 3. Comparison of Correlation Results between POSRho and Traditional Spearman.

further validating POSRho's consistency with the traditional method, even in segmented data (Fig. 2).

4.2 Application Cases

Fig. 4. Location of airports involved in the dataset.

Dataset Description: This case study examines the application of Spearman's rank correlation coefficient, enhanced by the POSRho algorithm, for analyzing Round Trip Time (RTT) relationships between airport nodes within a network (Fig. 4). This non-parametric method quantifies the strength and direction of monotonic relationships between RTTs, providing macroscopic insights into how RTT at one airport might influence another. The combination of Spearman's coefficient and POSRho proves valuable for network route optimization, anomaly detection, fault prevention, and resource allocation. This case study demonstrates its potential to enhance network performance and facilitate planning for interconnected airport nodes, ultimately improving network efficiency and stability.

Methodology: This method calculates Spearman's rank correlation coefficient, augmented by the POSRho algorithm, to assess the monotonic relationship between round-trip delay (RTT) values across different airport nodes, even in the context of large-scale network data. It involves ranking RTT observations for each node, calculating rank differences between paired observations, and finally applying the enhanced Spearman's rank correlation formula to quantify the similarity or dissimilarity in RTT rankings. The resulting coefficient provides valuable insights for network analysis and optimization, particularly in big data scenarios where traditional methods may struggle with computational demands.

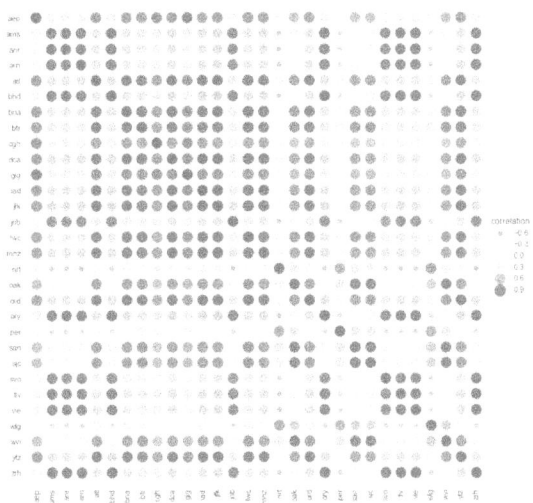

Fig. 5. Matrix of correlation coefficients for all airport networks.

Figure 5 illustrates the relationship between airport RTTs using a heatmap based on the calculated Spearman's rank correlation coefficients, enhanced by POSRho. This visualization provides a clear and intuitive representation of the correlation patterns, facilitating the identification of highly correlated, negatively correlated, and uncorrelated airport pairs.

The results, as illustrated in Figs. 7 and 8, highlight the effectiveness of the POSRho algorithm in distinguishing between highly correlated nodes and irrelevant nodes. Figure 6 shows value trends and differences for irrelevant nodes, where no significant correlation is observed. In contrast, Figs. 7 demonstrates

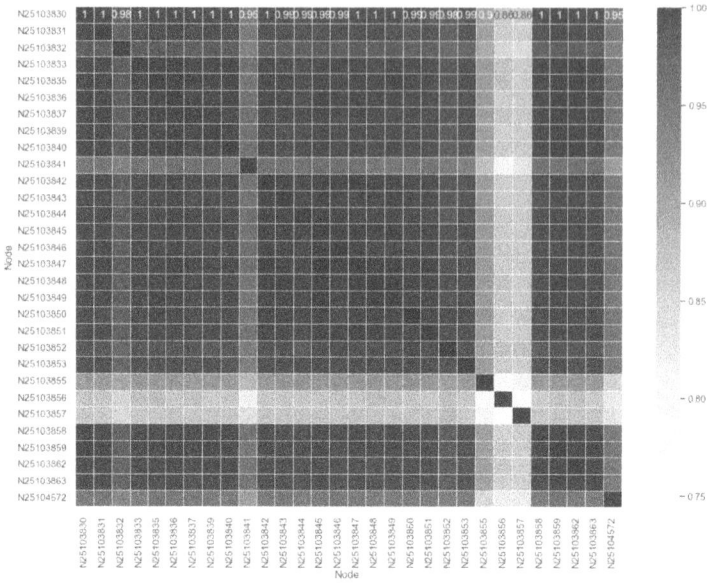

Fig. 6. Matrix of rank correlation coefficients for airport nodes.

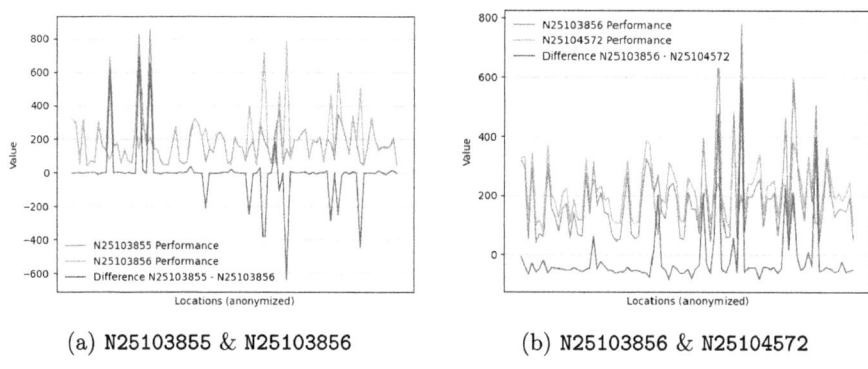

(a) N25103855 & N25103856

(b) N25103856 & N25104572

Fig. 7. Value Trends and Differences for Irrelevant Nodes.

value trends and differences for high correlation nodes, revealing strong correlation patterns that can be leveraged for optimized network performance. These findings underscore the practical implications of POSRho in real-world network scenarios, facilitating better resource management and fault prevention strategies in the aviation network.

Summary: This case study demonstrates the utility of the Spearman's rank correlation coefficient, augmented by the POSRho algorithm, in analyzing Round Trip Time (RTT) relationships between airport nodes within a network. By identifying distinct correlation patterns, the analysis reveals practical applications for

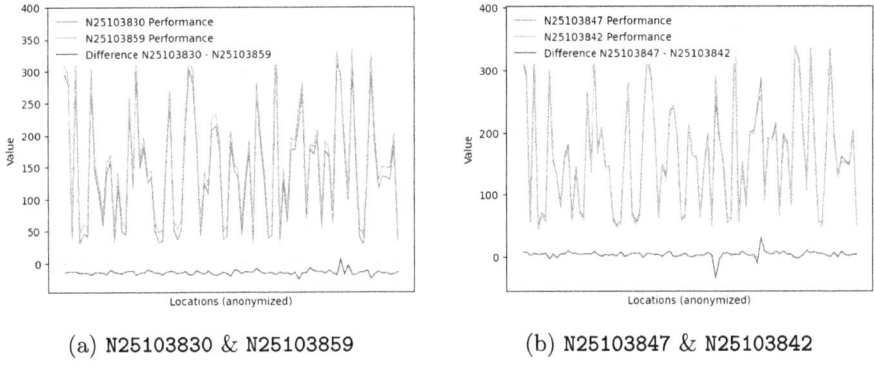

Fig. 8. Value Trends and Differences for High Correlation Nodes

network optimization, anomaly detection, and proactive resource allocation. The integration of POSRho enhances the robustness and scalability of the analysis, making it particularly effective in large-scale data environments. This highlights POSRho's significance in big data governance, providing a powerful tool for managing complex networked systems and ensuring efficient resource utilization.

5 Discussion

POSRho presents a promising solution for calculating Spearman's rank correlation coefficient on big data, offering significant advantages in scalability and efficiency. By leveraging parallel and distributed computing paradigms, POSRho effectively tackles the computational challenges posed by massive datasets, outperforming traditional methods. Its modular design and optimized algorithms contribute to reduced execution time and resource consumption, while ensuring accurate results.

Despite its strengths, POSRho faces certain limitations. Communication overhead can impact performance in bandwidth-constrained environments, and data skew may lead to load imbalance. Careful parameter tuning is crucial for optimal performance.

Future research directions for POSRho include exploring adaptive data partitioning strategies, investigating approximate ranking algorithms, and leveraging GPUs for accelerated computations. Extending POSRho to handle data streams and integrating it with machine learning pipelines hold further potential.

The applications of POSRho span diverse domains. In genomics, it can analyze large-scale genomic data to identify correlations between genes and traits. In social network analysis, it can uncover relationships and interactions within large networks. Financial modeling, recommendation systems, environmental monitoring, and healthcare analytics are other areas where POSRho can extract valuable insights from big data.

Overall, POSRho's ability to efficiently compute Spearman's rank correlation coefficient for big data opens up new possibilities for knowledge discovery and advancements across various scientific and industrial fields.

6 Conclusion

POSRho presents a novel approach for computing Spearman's rank correlation coefficient in big data settings. By leveraging parallel and distributed computing, POSRho overcomes scalability and efficiency limitations of traditional methods. Its divide-and-conquer strategy and optimization techniques ensure efficient processing of massive datasets while maintaining accuracy. The algorithm's modular design allows for adaptability to diverse big data platforms and computing environments. POSRho's ability to efficiently handle large datasets with limited resources makes it ideal for applications requiring timely and accurate correlation analysis. This empowers researchers and practitioners to uncover valuable insights from complex data relationships, advancing knowledge discovery across various domains.

Data Availibility Statement. Data supporting this study are available from the corresponding author.

Code Availability. The code supporting this study is available from the corresponding author upon request.

References

1. Ahmad, I., Tang, D., Wang, T., Wang, M., Wagan, B.: Precipitation trends over time using mann-kendall and spearman's rho tests in swat river basin, Pakistan. Adv. Meteorol. **2015**, 431860 (2015). https://doi.org/10.1155/2015/431860
2. Alibraheim, E.A.: The relationship between mothers' educational levels and their children's academic performance. Pegem J. Educ. Instr. **13**(3), 36–40 (2023). https://doi.org/10.47750/pegegog.13.03.04
3. Bergdahl, M., Bergdahl, J.: Burning mouth syndrome: prevalence and associated factors. J. Oral Pathol. Med. **28**(8), 350–354 (1999). https://doi.org/10.1111/j.1600-0714.1999.tb02052.x
4. Bishara, A.J., Hittner, J.B.: Testing the significance of a correlation with nonnormal data: comparison of pearson, spearman, transformation, and resampling approaches. Psychol. Methods **17**(3), 399–417 (2012). https://doi.org/10.1037/a0028087
5. Büyükkeçeci, M., Okur, M.C.: A comprehensive review of feature selection and feature selection stability in machine learning. Gazi Univ. J. Sci. **36**(4), 1506–1520 (2023). https://doi.org/10.35378/gujs.993763
6. Chaiken, R., et al.: Scope: easy and efficient parallel processing of massive data sets. Proc. VLDB Endow. **1**(2), 1265–1276 (2008). https://doi.org/10.14778/1454159.1454166
7. Chok, N.S.: Pearson's versus spearman's and kendall's correlation coefficients for continuous data (2010). http://d-scholarship.pitt.edu/8056. Accessed 5 May 2024

8. Critchlow, D.E.: Metric Methods for Analyzing Partially Ranked Data, vol. 34. Springer (2012)
9. Croux, C., Dehon, C.: Influence functions of the spearman and kendall correlation measures. Stat. Methods Appl. **19**(4), 497–515 (2010). https://doi.org/10.1007/s10260-010-0142-z
10. Fujita, A., Sato, J.R., Demasi, M.A.A., Sogayar, M.C., Ferreira, C.E., Miyano, S.: Comparing pearson, spearman and hoeffding's d measure for gene expression association analysis. J. Bioinform. Comput. Biol. **07**(04), 663–684 (2009). https://doi.org/10.1142/S0219720009004230
11. Hauke, J., Kossowski, T.: Comparison of values of pearson's and spearman's correlation coefficients on the same sets of data. Quaestiones Geographicae **30**(2), 87–93 (2011). https://doi.org/10.2478/v10117-011-0021-1
12. Huang, Y., et al.: Radiation pneumonitis prediction after stereotactic body radiation therapy based on 3D dose distribution: dosiomics and/or deep learning-based radiomics features. Radiat. Oncol. **17**(1), 188 (2022). https://doi.org/10.1186/s13014-022-02154-8
13. Imai, K., Lo, J., Olmsted, J.: Fast estimation of ideal points with massive data. Am. Polit. Sci. Rev. **110**(4), 631–656 (2016). https://doi.org/10.1017/S000305541600037X
14. Kraft, N.J.B., Godoy, O., Levine, J.M.: Plant functional traits and the multidimensional nature of species coexistence. Proc. Natl. Acad. Sci. **112**(3), 797–802 (2015). https://doi.org/10.1073/pnas.1413650112
15. Li, Y., et al.: Molecular subtyping of diffuse gliomas using magnetic resonance imaging: comparison and correlation between radiomics and deep learning. Eur. Radiol. 1–12 (2021). https://doi.org/10.1007/s00330-021-08237-6
16. Matkovskyy, R., Jalan, A., Dowling, M.: Effects of economic policy uncertainty shocks on the interdependence between bitcoin and traditional financial markets. Q. Rev. Econ. Finance **77**, 150–155 (2020). https://doi.org/10.1016/j.qref.2020.02.004
17. Poirier, M.J.P., Grépin, K.A., Grignon, M.: Approaches and alternatives to the wealth index to measure socioeconomic status using survey data: a critical interpretive synthesis. Soc. Indic. Res. **148**(1), 1–46 (2019). https://doi.org/10.1007/s11205-019-02187-9
18. Puth, M.T., Neuhäuser, M., Ruxton, G.D.: Effective use of spearman's and kendall's correlation coefficients for association between two measured traits. Anim. Behav. **102**, 77–84 (2015). https://doi.org/10.1016/j.anbehav.2015.01.010
19. Rajaraman, A., Ullman, J.D.: Mining of massive datasets. Autoedicion (2011)
20. Schadt, E.E., Linderman, M.D., Sorenson, J., Lee, L., Nolan, G.P.: Computational solutions to large-scale data management and analysis. Nat. Rev. Genetics **11**(9), 647–657 (2010). https://doi.org/10.1038/nrg2857
21. Spearman, C.: Correlation calculated from faulty data. Br. J. Psychol. **3**(3), 271 (1910)
22. de Winter, J.C.F., Gosling, S.D., Potter, J.: Comparing the pearson and spearman correlation coefficients across distributions and sample sizes: a tutorial using simulations and empirical data. Psychol. Methods **21**(3), 273–290 (2016). https://doi.org/10.1037/met0000079
23. Yllana-Prieto, F., González-Gómez, D., Jeong, J.S.: Influence of two educational escape room–breakout tools in PSTs' affective and cognitive domain in stem (science and mathematics) courses. Heliyon **9**(1) (2023)

A Fusion Tuning Method for Named Entity Recognition

Jitian Wang[1,2,3], Yanping Chen[1,2,3(✉)], Anqi Zou[1,2,3], Yongbin Qin[1,2,3], and Ruizhang Huang[1,2,3]

[1] Text Computing and Cognitive Intelligence Engineering Research Center of National Education Ministry, Guizhou University, Guiyang 550025, China
[2] State Key Laboratory of Public Big Data, Guizhou University, Guiyang, China
ypench@gmail.com
[3] College of Computer Science and Technology, Guizhou University, Guiyang 550025, China

Abstract. In named entity recognition, the main methods for constructing deep neural networks are fine-tuning and prompt tuning. Fine-tuning is a commonly used paradigm to optimize neural networks by using task-specific objective functions to make them more adapted to solve downstream tasks. In which pre-trained language models are used as external resources to provide contextual representations. Prompt tuning is a recent paradigm in which the downstream tasks are reformulated to look more like those solved during the original PLM training with the help of a textual prompt. It is effective to utilize potential knowledge of PLMs. In this paper, we propose the fusion approach, which sufficiently fuses the semantics obtained by extracting fine-tuning and prompt-tuning. In our approach, the model makes decisions based on the information obtained from the reconstructive fusion of the fine-tuned semantics and the semantics of the pre-trained language model. The representation obtained by fine-tuning and the representation obtained by prompt tuning are reconstructed into a multidimensional semantics using a gate fusion module (GFM). In particularly, our model has two gate components, one gate to extract the semantics of fine-tuning and the other gate to extract the semantics of prompt tuning, and finally the reconstructed semantic representation is obtained. This takes full advantage of the potential knowledge of PLM. Our model is evaluated on the CONLL2003, ACE2005 and GENIA corpora. It achieves performance close to the state-of-the-art in F1 scores.

Keywords: Prompt-tuning · Fine-tuning · Named entity recognition · Pre-trained language model

1 Introduction

Named entity recognition (NER) is a fundamental task in natural language processing (NLP) that involves recognizing named entities with predefined semantic types, e.g., people, facilities, and organizations. Due to its relevance in various NLP applications such as entity linking [1] and relation extraction [2,3], it has drawn much attention from the community. Various methods have been proposed for NER, including the sequence labeling [4,5], hypergraph-based [6,7] and span-based methods [8,9]. Among these methods, span-based methods are the most popular due to its simplicity and effectiveness. It is straightforward, usually includes all possible text spans and predicts its entity type, and is therefore suitable for a variety of NER subtasks [10].

Recently proposed models are built on pre-trained language models (PLMs), such as Bert, to provide rich semantics for NER tasks. PLMs utilize huge amounts of unlabeled corpus data to learn generalized semantic representations that contain rich knowledge contributing to NER tasks. Present paradigms using pre-trained language models are divided into two categories: one for fine-tuning and the other for prompt tuning. In the fine-tuning approach, PLMs are used to transform tokens into high-dimensional abstract representations. In deep architectures, additional parameters are usually introduced to adapt to downstream task goals, and tokens are compressed into dense vector representations according to NER task goals. Based on the dense vector representation, the output gets the probability of the specified type and the decision plane is simpler in fine-tuning. Finally, the neural network is optimized with a task-specific objective function. The deep architecture finally obtains semantics that are specific to the NER task. Prompt tuning has received a lot of attention due to its excellent performance and data efficiency and has been applied to many classification and generative tasks. In prompt tuning, the downstream task is formalized into a language modeling task, constructing completions as prompts to guide the model to stimulate PLM potential. The advantage of prompt tuning is that the downstream task goal is formalized as the goal of the PLM, and the decision-making process is the same as that of the training process of the PLM. It bridges the gap between task goals and pre-trained goals, stimulates the PLM potential from the perspective of enhancing the pre-trained original task, and helps to fully utilize the PLM for NER tasks.

The fine-tuning approach introduces additional parameters while using PLM and fine-tuning using task-specific objective functions. Prompt tuning stimulates the ability of PLM to extract shallow base features and deep abstract features. The semantic representations obtained by fine-tuning and prompt tuning are different, but also based on the knowledge obtained from the pre-trained language model, which is a multidimensional use of PLM. To exploit the semantic information of PLM in this paper in a multidimensional way, we propose a fusion tuning to utilize both fine-tuning and prompt tuning. The motivation for fusion tuning is rooted in the fact that the semantics of fine-tuning and prompt tuning are different but complementary. For better performance, we need to fuse both the fine-tuning information and the prompt tuning information using gate fusion.

In fusion tuning, sentences are followed by templates consisting of NE candidate that are recognized. They are fed into a PLM that passes through an interaction layer to obtain semantic information related to the masked token. The mask token representation and sentence representation are fed together into the gate fusion module to obtain a multidimensional semantic representation. The fusion tuning strategy combines fine-tuning and prompt tuning. The network is optimized by fine-tuning and pre-trained objectives. It has the advantage of multidimensionally stimulating the potential of PLMs to combine the semantics of different tuning to get multiple semantic representations. Using the gating fusion module to fuse the fine-tuning representation and the prompt representation, in particular our model consists of two gating mechanisms, one gating to extract the semantics of fine-tuning and the other gating to extract the semantics of prompt tuning. It is effective to distinguish overlapping features of NE candidate and take full advantage of potential knowledge of PLMs. Our model is evaluated on the CONLL2003, ACE2005 and GENIA corpora. It achieves performance close to the state-of-the-art in F1 scores.

The contributions of this paper can be summarized as follows:

(1) The paper explores the potential combination of a new traditional fine-tuning method and prompt learning, fully leveraging the prior knowledge embedded in PLM.
(2) In this paper, we use the gate fusion module to transform the different semantics of fine-tuning and prompt tuning into the same semantics to obtain a multilevel semantic.
(3) Experimental results indicate that the proposed method in this paper is applicable to both nested and flat datasets, and it significantly close to other baseline models, demonstrating the effectiveness and scalability of the proposed approach.

2 Related Work

Named Entity Recognition (NER) is a basic task of information extraction. Current named entity recognition methods can be divided into three categories, including tagging-based, span-based, hypergraph-based. Another criterion for dividing related work is based on tuning strategies, which also divides related work into two categories: fine-tuning methods and prompt tuning methods.

Named entity recognition is often considered as a sequence labeling problem, assigning a label to each word from a pre-designed labeling scheme [11]. It is difficult to cope with nested entities. [12] uses BERT for word vectors and adopts a multi-label approach for recognizing nested entities based on the inclusion of other features such as lexicality. [13] treats the label sequence of nested entities as the second best path in the range of long entities. [14] were the first to introduce hypergraph based methods to the task of nested NER. [15] proposed recursive recurrent neural networks to encode directed hypergraph representations. [16] propose LHBN model that builds multiple simpler local hypergraphs to capture named entities instead of a single complex full-size hypergraph.

The span-based methods [17] and [8] propose boundary-aware and boundary-regression strategies based on span classification respectively. A disadvantage of the span-based approach is that it usually performs poorly in determining entity boundaries. To alleviate this problem, [18] and [19] added a boundary detection component to facilitate the detection of entities. [20] proposed a pyramid structure-based model to alleviate these problems to some extent. Some other methods [10,21] perform classification on inter-word dependencies or interactions, which are essentially span classification.

Above works have also shown that better performance can be achieved by adopting a PLM, such as BERT and bioBERT. PLMs utilize massive unlabeled corpus data to learn generalized semantic representations that contain abundant knowledge contributing to NER tasks. In related works, the methods to use PLMs can be divided into two categories: fine-tuning and prompt-tuning.

In the fine-tuning method, PLM is used to convert the token into a high-dimensional abstract representation. In deep architectures, additional parameters are usually introduced to adapt to downstream task objectives, and tokens are compressed into relevant NER dense vector representations based on NER task objectives. Finally, the neural network is optimized with a task-specific objective function. The advantage of prompt tuning is that the downstream task goal is characterized by the goal of customized PLM, and the decision-making processes are the same as those of the training of the pre-trained language model. It bridges the gap between task goals and pre-trained goals, stimulates PLM potential from the perspective of enhancing the pre-trained original task, and helps to fully utilize PLM for NER tasks. [22] first applies prompt learning to NER. It proposes a straightforward way to construct separate prompts in the form of "[X] is a [MASK] entity" by enumerating all spans. [23] propose a dynamic template filling mechanism to perform bipartite graph matching between prompts and the entities.

3 Methods

The structure of our modeling framework is shown in Fig. 1, which consists of four main parts. In the first part, a boundary detection task is used to predict the NE candidate start and end positions. Then it is combined into candidate entities. For each NE candidate, markers are injected on both sides of the candidate entity for classification. In the second part, the model input consists of a template and sentences marked with NE candidate. Using the widely used PLMs as an encoder to obtain word representations. In the third part, the semantics of different dimensions are obtained using deep networks. In the fourth part, GFM is used to fuse prompt tuning with fine-tuning to get multi-dimensional representations.

3.1 NE Candidate Proposal

Let $S = [x_1, x_2, ..., x_n]$ represent a sentence. The length of is n. x_i represents the ith token in S. $E_{ij} = x_{i_j}$ is used to represent a NE candidate in S, where i and j are the start and end position of E_{ij}. All NE candidates in sentence are represented as a set $E = \{E_{ij} | 1 \leq i \leq j \leq n\}$.

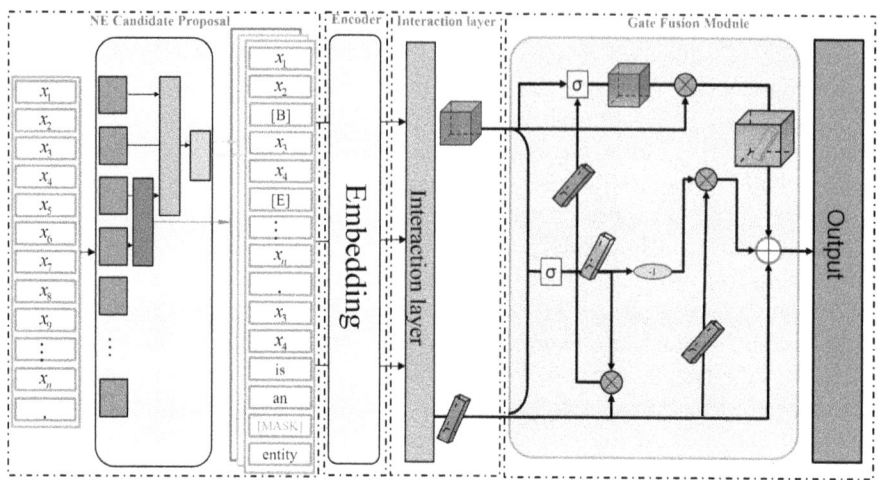

Fig. 1. Model

A simple strategy for generating E is to use enumeration, in contrase this, our approach generates E using boundary detection. This results in higher precision NE boundaries, which effectively reduces the number of ineffective NE candidates.

In our approach, we follow tht method proposed by to recognize all start and end boundaries in S, which are denoted to as $\mathbb{B}$ and $\mathbb{E}$, respectively. For all $E_{ij} \in$ E, $x_i \in \mathbb{B}$ and $x_j \in \mathbb{E}$.

For every $E_{ij} \in$ E, two tokens ($t_b =$ "[B]", $t_e =$ "[E]") are implanted as markers to indicate the position of an NE candidate in a sentence. The revised sentence is denotes as $\mathbb{S}$. $\mathbb{S}_{ij} = [x_1, ..., x_{i-1}, t_b, ..., t_e, x_{j+1}, ..., x_n]$, the subscripts i and j indicate the position of the implanted markers.

3.2 Encoder Layer

As shown in Fig. 1, the model input consists of sentence $\mathbb{S}$ and template M. The template consists of the NE candidate proposal presented in the first step and "is an [mask] entity". This paper follows the template construction method of previous related work [22].

First, the input sequence T is filled with sentences $\mathbb{S}$ and template M, and then through the word embedding layer and encoding layer of the pre-trained language model to obtain the input sequence semantics H.

$$H = BERT(T) \qquad (1)$$

$H = \{h_1, ..., h_{n+l+2}\} \in R^{(n+l+2) \times d}$, d is the dimension of h_i.

3.3 Interaction Layer

The context representation $H_{[CLS]}$ is compressed to obtain a compressed dense vector:

$$R = MLP(H_{[CLS]}) \qquad (2)$$

To enhance the interaction between the $\mathbb{S}$ and the $\mathbb{M}$, we design the H input to the interaction layer. The interaction layer consists of self-attention and BiLSTM. Subsequently, the word representation Z' is obtained:

$$\alpha = Soft\max\left((H \cdot W_q)(H \cdot W_k)^T\right) \tag{3}$$

$$Z = \alpha \cdot (H \cdot W_v) \tag{4}$$

$$Z' = BiLSTM(Z) \tag{5}$$

Its main parameters are the attention weight matrices W_q W_k and $W_v \in R^{d \times d}$.

3.4 Gate Fusion Module

According to the features of the pre-trained language model, the special word segmentation [CLS] in its input has global semantics in the output. Take the mask vector $Z'_{[MASK]}$ of Z' and go through the linear layer to get $Z''_{[MASK]}$.

$$Z''_{[MASK]} = MLP\left(Z'_{[MASK]}\right) \tag{6}$$

Secondly, inspired by the prototype representation rectification method [24] and Gated Recurrent Unit [25], the use of gated mechanisms can be used to extracted multi-dimensional semantics. Therefore, the input of the gated fusion module mainly consists of two main part, which are the output vector R of the sentence classification head [CLS] and the fusion representation $Z''_{[MASK]}$. Ours uses the gate fusion module to filter information on R and $Z''_{[MASK]}$, choosing how much local semantics to introduce and how much global semantics needs to be retained. Information filtering that preserves global semantics:

$$c = \sigma\left(W_c\left(R + Z''_{[MASK]}\right) + b_c\right) \tag{7}$$

$$C_{keep} = (1 - c) \cdot Z''_{[MASK]} \tag{8}$$

$$C_{next} = c \cdot Z''_{[MASK]} \tag{9}$$

p represents the weight proportion of global semantics, C_{keep} represents the retained global semantics, and C_{next} refers to the global semantics flowing into the next gate unit. Subsequently, our uses another gate neural unit to generate the semantics that the mask needs to preserve. Its form can be expressed as follows:

$$p = \sigma(W_m(C_{next} + R) + b_m) \tag{10}$$

$$P_{choose} = p \cdot R \tag{11}$$

In the above, P_{choose} represents the remaining mask semantics. Subsequently, the final prediction P of the entity type consists of a semantic representation of the final construction. P is mainly composed of three parts: C_{keep}, $Z''_{[MASK]}$ and P_{choose}. The process can be formalized as follows:

$$P = C_{keep} + P_{choose} + Z''_{[MASK]} \tag{12}$$

The method proposed in this paper aims to utilize prompt and fine-tuning to stimulate the potential of pre-trained language models, using gate fusion modules to fuse the semantics of different tuning. By calculating the predicted value of each NE candidate and performing a cross-entropy operation with the true value to obtain the training loss:

$$Loss = -\sum_{i=1}^{N} y_i \log(P^i) \tag{13}$$

4 Experimental Settings

In order to verify the effectiveness of our model, experiments were conducted on the data sets English ACE2005, GENIA and CoNLL03. And select named entity recognition models based on deep learning proposed in recent years for comparison. Use Precision (P), Recall (R) and F1 as the main evaluation indicators to verify the performance of this model.

4.1 Dataset

Our method conducts experiments on the nested data sets English ACE2005, GENIA and the flat data set CoNLL03. The statistical results of the number of sentences in ACE2005 English, GENIA data set and CoNLL03 data set are shown in Table 1.

(1) The ACE2005 English data set was released by the Language Data Alliance in 2006. This data set contains a total of 599 English documents, including 7 entity types, including people (PER), geographical location (LOC), transportation (VEH), political entities (GPE), weapons (WEA), organizations (ORG), Facility (FAC). The dataset is divided into training, validation and test sets using the dataset division method in [26] in the ratio of 80%:10%:10%.
(2) The CoNLL03 English data set is one of the classic named entity recognition task data sets. There are a total of 1393 English news articles, all from the Reuters corpus, which consists of Reuters news reports. In CoNLL03, entities are marked as four types: place name (LOC), organization name (ORG), person name (PER), and other (MISC).
(3) The GENIA data set is built based on the corpus GENIA3.0.2, and the data comes from biomedicine. This data set is divided into 5 types of entities according to the approach of [27], namely: DNA, RNA, protein, cellline, celltype. And divide the training set, verification set and test set according to the ratio of 81%:9%:10%.

Table 1. Information about the number of sentences in the dataset

Datasets	Train	Dev	Test	Number of Entity Types
ACE2005	7683	960	960	7
CoNLL03	17291	3453	-	4
GENIA	15023	1669	1854	5

4.2 Evaluation Metrics

The evaluation indicators used in the experiment include: precision (P), recall (R) and F1 score. The calculation formula is as follows:

$$P = \frac{TP}{(TP+FP)} \times 100\% \tag{14}$$

$$R = \frac{TP}{(TP+FN)} \times 100\% \tag{15}$$

$$F1 = \frac{2 \times P \times R}{(P+R)} \times 100\% \tag{16}$$

TP is the number of correctly recognized entities, FP is the number of incorrectly recognized entities, and FN is the number of correct entities that are not recognized.

4.3 Implementation Details

The experiments were conducted in the environment of Python 3.9 and PyTorch 1.12.0. BERT and Albert models were utilized during the training process of the boundary detection and NE candidate classification. For BERT, the models bert-large-cased and BioBERT-large-cased-v1.1 [28] with 1024 dimensions were employed on the ACE2005, CoNLL03, and GENIA datasets. For Albert, the model Albert-large-v2 [29] with 4096 dimensions was used. The optimizer employed was AdaW. The training batch size and number of training epochs were set to 32 and 20, respectively. Dropout was set to 0.33, and the learning rate was set to 3e−5.

5 Results and Analysis

5.1 Comparing with Related Works

In this section, our model is compared with other related works based on the Conll2003, GENIA and ACE05 datasets. We list their results reported with the best performance.

Our model's performance on the GENIA dataset is compared with the experimental results of related work, as shown in Table 2. Two-stage utilizes boundary information to adjust NE candidate spans through boundary regression and then improves nested named entity recognition through reclassification. The Joint model treats named entity recognition as a joint task of recognizing boundaries and entity classification. With this table, we can see that our model is better than other models. Our model is higher than other models in F1 value. This indicates the use of GFM, which can combine prompt tuning and fine-tuning to extract semantics and thus better utilize the information from the pre-trained language and ultimately make better decisions.

Table 2. Results for overlapped NER datasets

Method	GENIA			ACE2005		
	P(%)	R(%)	F(%)	P(%)	R(%)	F(%)
Boundary-aware	75.9	73.6	74.7	-	-	-
Pyramid	80.31	78.33	79.31	85.30	87.40	86.34
BiFlaG	77.4	74.6	76.0	-	-	-
MGNER	-	-	-	79.0	77.3	78.2
Two-stage	80.19	80.89	80.54	86.05	87.27	86.67
Joint model	79.74	77.70	78.70	87.87	**89.24**	88.55
Controlled model	81.92	80.49	81.19	91.61	87.94	89.73
Ours	**82.10**	**81.27**	**81.68**	**93.08**	86.76	**89.81**

Our model's performance on the ACE2005 dataset is compared with the experimental results of related work, as shown in Table 2. The two-stage approach utilizes boundary information to adjust NE candidate spans through boundary regression and then enhances nested named entity recognition through reclassification. Compared to related works, our model achieves relatively high performance in the ACE2005 corpus.

Table 3. Comparison of our model with existing models on the Conll03 dataset.

Model	P(%)	R(%)	F(%)
W2NER	92.71	**93.56**	93.05
PromptNER	92.96	93.18	93.08
DiffusionNER	92.99	92.56	92.78
Ours	**95.90**	92.05	**93.94**

Our model's performance on the CoNLL03 dataset is compared with the experimental results of related work, as shown in Table 3. DiffusionNER treats the entity recognition task as a process of denoising entity boundaries, making full use of entity boundary information. Experimental results show that our model is still valid on flat datasets.

Overall, our model shows strong and competitive performance on both nested and flat NER tasks. The reason for the improvement is that in our model, optimization is performed by both prompt tuning and pre-trained objectives. The semantics obtained from fine-tuning and the semantics obtained from prompt tuning will be extracted using GFM and fused to obtain a multi-semantic representation.

5.2 Ablations Study

In order to analyze the effects of the components in our model, we conducted ablation study experiments on GENIA and Conll03. In the "w/o template" model, it means that the inputs of template and [mask] token in the GFM are removed from the model after the sentence. In the "w/o GFM" model, the GFM is removed from the model.

Table 4. Ablations study

Method	GENIA			Conll2003		
	P(%)	R(%)	F(%)	P(%)	R(%)	F(%)
Ours	82.10	81.27	81.68	95.90	92.05	93.94
w/o template	81.80	79.95	80.35	95.98	91.66	93.77
w/o GFM	82.22	80.86	81.53	95.77	91.89	93.79

The results are displayed in Table 4. From this, we can see that by removing template, the F1 performance of the Genia dataset decreases by more than 1%, which proves that template makes the model focus more on the NE candidate part and the GFM lacks the input of template, which makes the model lack of rich semantics. Removing GFM, the GENIA dataset shows a decrease of about 0.1% in F1 and 0.4% in recall, which indicates that the model lacks the GFM part of the model, and the lack of extracting and fusing the information of prompt tuning and fine-tuning makes the model misclassify a positive NE candidate as a negative one.

5.3 Analyses

In this experiment, we analyze the influence of three issues on the performance: (1) Performance of Boundary Detection (2) Impact of the fusion method (3) case study.

Performance of Boundary Detection. In this section, we analyze the impact of boundary detection. Identifying named entities requires correct starting and ending boundaries.

The detection results of entity boundaries are shown in Table 5. Therefore, compared to entity recognition, boundary detection has higher performance. The reason is that entity boundaries are of a smaller granularity, and the semantics are less ambiguous. Experimental results indicate better performance on ACE2005 and CoNLL03. However, performance is not satisfactory on the GENIA dataset, possibly because the majority of entities in the dataset are specialized terms in the biomedical field, which include discontinuous entities and abbreviated forms. The boundary features of these specialized terms are not obvious, making it relatively challenging to achieve high-performance detection results. NE candidate are formed by pairing the predicted starting and ending boundary positions from boundary detection, as shown in Table 6. Pos represents the ratio of positive NE candidate instances to all combinations of NE candidate, while Recall represents the ratio of positive NE candidate to the number of positive entities in the test dataset. GENIA and CoNLL03 have higher Pos values, as the mentioned NE candidate filtering mechanism is used to filter out negative instances that are excessively long, and boundary detection performs well. ACE2005 and CoNLL03 exhibit good Recall performance because boundary detection correctly identifies a large number of boundaries on these two datasets.

Table 5. Entity boundary detention result

	GENIA			ACE05			CoNLL03		
	P(%)	R(%)	F(%)	P(%)	R(%)	F(%)	P(%)	R(%)	F(%)
Start	88.02	86.51	87.26	95.50	95.50	95.50	97.69	98.16	97.92
End	89.68	88.46	89.06	95.42	94.49	94.95	96.75	97.45	97.10

Table 6. Statistic information of collected NE candidates

Datasets	Pos(%)	Recall(%)
GENIA	43.23	82.96
ACE2005	33.58	93.31
CoNLL03	60.99	96.60

Impact of the Fusion Method. In this section, we analyze the effect of the fusion method. The semantic separation uses two PLMs to obtain the semantic representation of the fine-tuning model and the prompt tuning semantic representation respectively. Semantic fusion is the fusion of fine-tuning representations and prompt tuning representations in a PLM. Semantic fusion is the method we use.

Table 7. Impact of the fusion method

dataset	semantic separation			semantic fusion		
	P(%)	R(%)	F(%)	P(%)	R(%)	F(%)
GENIA	82.10	81.27	81.68	82.20	81.40	81.62
ACE05china	92.83	87.94	90.32	92.30	88.04	90.12

From Table 7 we can see that separating semantics performance does not improve. It shows that this does not fully get the fine-tuning and prompt tuning fusion semantics. Semantic fusion fuses both semantics together to further improve performance.

Case Study. In this section, we conduct a case study to demonstrate the effectiveness of our model. Our model is compared with promptNER on the GENIA dataset. We refer to the parameters of diffusion NER and run the genia dataset with the promptNER model.

Table 8. Case study

Item	Output
Gold Truth	(1) An [EBNA-2 -responsive element]$_{DNA}$ was found within the [-512 to +40 [LMP1 DNA]$_{DNA}$]$_{DNA}$ since this DNA linked to a [chloramphenicol acetyltransferase reporter gene]$_{DNA}$ was transactivated by cotransfection with an [EBNA-2 expression vector]$_{DNA}$. (2) Vitamin E therapy of acute CCl4 -induced hepatic injury in mice is associated with inhibition of nuclear factor kappa B binding.
PromptNER	(1) An [EBNA-2 -responsive element]$_{DNA}$ was found within the [-512 to +40 [LMP1 DNA]$_{DNA}$]$_{DNA}$ since this DNA linked to a [chloramphenicol acetyltransferase reporter gene]$_{DNA}$ was transactivated by cotransfection with an [EBNA-2 expression vector]$_{DNA}$. (2) Vitamin E therapy of acute CCl4 -induced hepatic injury in mice is associated with inhibition of nuclear factor kappa B] binding.
Ours	(1) An [EBNA-2 -responsive element]$_{DNA}$ was found within the [-512 to +40 [LMP1 DNA]$_{DNA}$]$_{DNA}$ since this DNA linked to a [chloramphenicol acetyltransferase reporter gene]$_{DNA}$ was transactivated by cotransfection with an [EBNA-2 expression vector]$_{DNA}$. (2) Vitamin E therapy of acute CCl4 -induced hepatic injury in mice is associated with inhibition of nuclear factor kappa B binding.

As shown in Table 8. The blue color represents DNA entities and the green color represents protein entities. Orange represents the identification of the NE candidate as a non-entity. Compared to the promptNER model, our methods are able to distinguish between different entities, as well as between entities and non-entities that are similar to entities. Although our method generates a high number of NE candidates, it is effective in recognizing most of the non-entity. For sentences with fewer entities, it is effective in identifying the entities in the sentence. This shows that our model is based on multiple semantics and is able to recognize the features of entities. It will not be easy to mistake acronyms for entities.

6 Conclusion

In this paper, we propose the fusion approach, which sufficiently fuses the semantics obtained by extracting fine-tuning and prompt-tuning. In this approach, the classification is based on information reconstructed and fused from mask semantics and sentence semantics. GFM is used to extract the representation obtained by fine-tuning and the representation obtained by prompt, in particular our model consists of two gate components, one gate to extract the semantics of fine-tuning and the other gate to extract the semantics of prompt, and finally the reconstructed semantic representation is obtained. It has the advantage of multi-dimensional stimulation of PLM, combining the semantics of different tuning to get multiple semantic representations.

Acknowledgments. This work is supported by the funds of the key Technology R&D Program of Guizhou Province No. [2024]003, National Key R&D Program of China, and the National Natural Science Foundation of China (62166007).

References

1. Hou, F., Wang, R., He, J., Zhou, Y.: Improving entity linking through semantic reinforced entity embeddings. arXiv preprint arXiv:2106.08495 (2021)
2. Tan, Z., et al.: Query-based instance discrimination network for relational triple extraction. arXiv preprint arXiv:2211.01797 (2022)
3. Yan, Z., Zhang, C., Fu, J., Zhang, Q., Wei, Z.: A partition filter network for joint entity and relation extraction. arXiv preprint arXiv:2108.12202 (2021)
4. Huang, Z., Xu, W., Yu, K.: Bidirectional LSTM-CRF models for sequence tagging. arXiv preprint arXiv:1508.01991 (2015)
5. Ju, M., Miwa, M., Ananiadou, S.: A neural layered model for nested named entity recognition. In: Proceedings of the 2018 Conference of the North American Chapter of the Association for Computational Linguistics: Human Language Technologies, Volume 1 (Long Papers), pp. 1446–1459 (2018)
6. Wang, B., Lu, W.: Neural segmental hypergraphs for overlapping mention recognition. arXiv preprint arXiv:1810.01817 (2018)
7. Luo, Y., Zhao, H.: Bipartite flat-graph network for nested named entity recognition. arXiv preprint arXiv:2005.00436 (2020)
8. Shen, Y., Ma, X., Tan, Z., Zhang, S., Wang, W., Lu, W.: Locate and label: a two-stage identifier for nested named entity recognition. arXiv preprint arXiv:2105.06804 (2021)

9. Sohrab, M.G., Miwa, M.: Deep exhaustive model for nested named entity recognition. In: Proceedings of the 2018 Conference on Empirical Methods in Natural Language Processing, pp. 2843–2849 (2018)
10. Li, J., et al.: Unified named entity recognition as word-word relation classification. In: Proceedings of the AAAI Conference on Artificial Intelligence, vol. 36, no. 10, pp. 10965–10973 (2022). https://doi.org/10.1609/aaai.v36i10.21344
11. Ma, X., Hovy, E.: End-to-end sequence labeling via bi-directional LSTM-CNNS-CRF. arXiv preprint arXiv:1603.01354 (2016)
12. Straková, J., Straka, M., Hajič, J.: Neural architectures for nested NER through linearization. arXiv preprint arXiv:1908.06926 (2019)
13. Shibuya, T., Hovy, E.: Nested named entity recognition via second-best sequence learning and decoding. Trans. Assoc. Comput. Linguist. **8**, 605–620 (2020). https://doi.org/10.1162/tacl_a_00334
14. Lu, W., Roth, D.: Joint mention extraction and classification with mention hypergraphs. In: Proceedings of the 2015 Conference on Empirical Methods in Natural Language Processing, pp. 857–867 (2015)
15. Katiyar, A., Cardie, C.: Nested named entity recognition revisited. In: 2018 Conference of the North American Chapter of the Association for Computational Linguistics: Human Language Technologies, NAACL HLT 2018, pp. 861–871. Association for Computational Linguistics (ACL) (2018)
16. Yan, Y., Cai, B., Song, S.: Nested named entity recognition as building local hypergraphs. In: Proceedings of the AAAI Conference on Artificial Intelligence, vol. 37, no. 11, pp. 13878–13886 (2023). https://doi.org/10.1609/aaai.v37i11.26625
17. Zheng, C., Cai, Y., Xu, J., Leung, H.F., Xu, G.: A boundary-aware neural model for nested named entity recognition. In: Proceedings of the 2019 Conference on Empirical Methods in Natural Language Processing and the 9th International Joint Conference on Natural Language Processing (EMNLP-IJCNLP), pp. 357–366. Association for Computational Linguistics (ACL) (2019)
18. Tan, C., Qiu, W., Chen, M., Wang, R., Huang, F.: Boundary enhanced neural span classification for nested named entity recognition. In: Proceedings of the AAAI Conference on Artificial Intelligence, vol. 34, no. 05, pp. 9016–9023 (2020). https://doi.org/10.1609/aaai.v34i05.6434
19. Xu, Y., Huang, H., Feng, C., Hu, Y.: A supervised multi-head self-attention network for nested named entity recognition. In: Proceedings of the AAAI Conference on Artificial Intelligence, vol. 35, no. 16, pp. 14185–14193 (2021). https://doi.org/10.1609/aaai.v35i16.17669
20. Wang, J., Shou, L., Chen, K., Chen, G.: Pyramid: a layered model for nested named entity recognition. In: Proceedings of the 58th Annual Meeting of the Association for Computational Linguistics, pp. 5918–5928 (2020)
21. Yu, J., Bohnet, B., Poesio, M.: Named entity recognition as dependency parsing. arXiv preprint arXiv:2005.07150 (2020)
22. Cui, L., Wu, Y., Liu, J., Yang, S., Zhang, Y.: Template-based named entity recognition using bart. arXiv preprint arXiv:2106.01760 (2021)
23. Shen, Y., et al.: Promptner: prompt locating and typing for named entity recognition. arXiv preprint arXiv:2305.17104 (2023)
24. Liu, Y., Hu, J., Wan, X., Chang, T.H.: Learn from relation information: towards prototype representation rectification for few-shot relation extraction. In: Findings of the Association for Computational Linguistics: NAACL 2022, pp. 1822–1831 (2022)
25. Cho, K., et al.: Learning phrase representations using RNN encoder-decoder for statistical machine translation. arXiv preprint arXiv:1406.1078 (2014)

26. Xia, C., et al.: Multi-grained named entity recognition. arXiv preprint arXiv:1906.08449 (2019)
27. Yan, H., Gui, T., Dai, J., Guo, Q., Zhang, Z., Qiu, X.: A unified generative framework for various NER subtasks. arXiv preprint arXiv:2106.01223 (2021)
28. Lee, J., et al.: Biobert: a pre-trained biomedical language representation model for biomedical text mining. Bioinformatics **36**(4), 1234–1240 (2019). https://doi.org/10.1093/bioinformatics/btz682
29. Lan, Z., Chen, M., Goodman, S., Gimpel, K., Sharma, P., Soricut, R.: Albert: a lite bert for self-supervised learning of language representations. arXiv preprint arXiv:1909.11942 (2019)

Design of AXI Bus-Based IP Core for Image Processing

Yutong Chen[1(✉)], Zhongchao Yi[1], Xuqiang Li[1], Xingyan Chen[1], and Yanjiang Chen[2]

[1] School of Software Engineering, University of Science and Technology, Hefei, China
{yutongchen,zhongchaoyi,xuqiangli,chenxingyan}@mail.ustc.edu.cn
[2] School of Computer Science and Technology, University of Science and Technology, Hefei, China
yjchen@mail.ustc.edu.cn

Abstract. This study addresses the significant issue of image blur through image restoration techniques that reduce or remove blur and detect blurry regions, improving image quality and analysis potential. Utilizing FPGA-based digital hardware systems, the research leverages FPGA-supported systems like ZYNQ and ARM, using the Advanced eXtensible Interface (AXI) and Advanced Microcontroller Bus Architecture (AMBA) for streamlined device communication. The design incorporates specialized Intellectual Property (IP) cores such as Square Root (Sqrt) and Shift RAM, tailored to overcome the limitations of development cycles and processing algorithms, focusing on portability across various platforms including domestic FPGAs. The practical implementation features modules for axis-stream transmission, RGB to YUV conversion, image smoothing using Gaussian, mean, and median filters, advanced image processing with Sobel edge detection, and erosion-dilation modules. A ZYNQ verification system with an OV5640 camera and HDMI output substantiates the functionality and efficiency of the designed IP core, which is further analyzed for resource and power consumption using a High-Level Synthesis (HLS) tool. The study successfully demonstrates a robust IP core for blurry image processing, significantly enhancing the field of image processing and embedded systems, laying a strong groundwork for future advancements and porting endeavors, and highlighting the efficacy of FPGA-based image processing solutions.

Keywords: AXI Bus · Image Processing · Edge Detection · IP Core

1 Introduction

In the realm of digital imaging, blur [5] is a ubiquitous problem affecting the visual quality and analytical viability of captured images. This issue arises from a myriad of objective and subjective factors, such as camera shake, focus errors,

---

Y. Chen and Z. Yi—These authors contributed equally to this work.

or motion blur, presenting a significant challenge in fields ranging from medical imaging to security surveillance. Consequently, the need for effective image blur detection and correction has been a focal point of scholarly attention, leading to extensive research on image restoration. Current methodologies predominantly aim at reducing or removing blur and identifying blurry regions within images [3, 9].

However, while advancements in parallel processing, image sensing, and storage technologies have propelled the rapid evolution of image processing, they also highlight the limitations of existing approaches, particularly in terms of computational overhead and latency. Techniques based on wavelet transforms [17], deep learning [3], and other machine learning algorithms offer improved accuracy and robustness in detecting blurred image edges but often at the expense of increased computation time and processing delays. These delays pose significant challenges in applications requiring real-time processing [6], such as video surveillance and live broadcasting.

In response to these challenges, FPGA-based digital hardware systems [12] have emerged as a powerful solution due to their strong programmability, high parallelism, and low power consumption [18]. The research and application of FPGA in image processing have gained widespread attention domestically and internationally. This approach leverages the high-efficiency communication capabilities of the AXI bus, a popular choice for managing intra-chip communications in complex system designs. This paper proposes a novel design of an AXI bus-based IP core for detecting blurred image edges using FPGA. This design not only addresses the high computational load and latency issues prevalent in current technologies but also capitalizes on the parallel computing capabilities of FPGA and the efficient communication facilitated by the AXI bus. By integrating hardware optimization strategies with the AXI bus architecture, the proposed design offers a promising direction for real-time, high-precision blurred image edge detection. The key contributions of our work in this paper are summarized as follows:

- The proposed IP core employs FPGA's parallel processing capabilities and AXI bus architecture to significantly reduce the time required for detecting blurred image edges, fulfilling real-time processing requirements.
- Utilizing Verilog HDL and a modular design approach, the IP core is adaptable and scalable, easing its integration into various FPGA platforms domestically and internationally.
- The design methodology, integrating the basic IP cores with HDL practices, shortens the development cycle and reduces effort, enhancing system compatibility and portability.

2 Related Work

The design of IP cores for blurred image edge detection has been extensively explored, focusing on two primary streams: algorithm optimization and hardware optimization [7].

2.1 Algorithm Optimization

In the realm of algorithm optimization, several methodologies have been adopted to enhance the precision and robustness of edge detection in blurred images. Notable among these are methods based on wavelet transforms, deep learning, and other machine learning algorithms. Wavelet transform techniques decompose and reconstruct the image signal to extract high and low-frequency information, effectively detecting edges within the image. Deep learning approaches, on the other hand, involve training neural network models to perform feature extraction and classification, thereby facilitating edge detection. While these algorithmic advancements significantly improve accuracy and robustness, they also introduce substantial computational overhead and processing delays. Such delays are particularly problematic in scenarios that demand real-time processing, underscoring the limitations of purely algorithmic solutions in high-stakes environments.

2.2 Hardware Optimization

Hardware optimization has predominantly focused on the use of FPGA and ASIC [8] designs to overcome the limitations noted in software-based approaches. FPGA-based solutions are celebrated for their strong programmability, high parallelism, and low power consumption, which have been widely applied in the field of image processing. ASIC designs offer even higher computational performance and lower power consumption but come at the cost of longer design cycles and higher expenses. Recently, FPGA-based methods have seen increased application due to their versatility and efficiency. The utilization of AXI bus-based designs in FPGA frameworks facilitates seamless integration with other hardware modules and processors, enhancing system scalability and versatility. The AXI bus, an efficient on-chip bus, allows for communication between multiple IP cores and with processors, enabling the integration of blurred image edge detection IP cores with other hardware components to achieve high-speed, high-precision detection [11, 14].

3 Methodology

This section discusses the methodology employed in designing IP cores for image blurring detection, focusing on the basic functionality of IP cores and the application of HDL.

3.1 Basic Functionality IP Cores

In FPGA-based image processing, the basic functionality IP cores play crucial roles. This section elaborates on the Shift RAM substitute module and the Sqrt IP substitute module, detailing their implementation using the basic IP cores and HDL.

Shift RAM Substitute Module. The Shift RAM module serves as a shift register IP core, enabling image manipulations such as translation, rotation, scaling, and convolutions for filters. Utilizing a Shift Register IP core from FPGA development platforms, the module can be configured to perform shifts in various bit lengths and directions, facilitating efficient image data processing.

To enhance the compatibility across different manufacturers' platforms, we develop a generalized Shift RAM IP core using dual-port Block RAM (BRAM), which operates based on input image data and an enable signal, storing data after a three-clock delay in first-level dual-port RAM, with output data read after a one-cycle delay. This BRAM-based design ensures versatility and high-speed data handling across various hardware platforms.

Sqrt IP Substitute Module. The Sqrt IP substitute module is essential for computing square roots, serving a key role in various image processing algorithms and feature extraction operations. This module can be instantiated using an FPGA platform's Sqrt IP core or by developing a Sqrt function in HDL, thus enabling rapid and precise square root computations.

When employing a hyperbolic coordinate system for vector mode iterative computations, the initial value of x is set to 1, and z is set to 0, with y representing the number for which the square root is to be computed. During this iteration, not only is the hyperbolic tangent function accounted for, but also the square root function. A transformation is applied where a is the square root value to be determined. Initially, x is set to $a + 1$, y to $a - 1$, and z to any arbitrary value. After iteration, the value of x becomes:

$$x = K \cdot \sqrt{(a+1)^2 - (a-1)^2} = 2K \cdot \sqrt{a} \tag{1}$$

In the implementation of the Sqrt IP substitute module, initial values are first generated using MATLAB. Initial values for x, y, and z are selected and undergo quantization and conversion to accommodate further computations and Verilog implementation as outlined in Algorithm 1. Through this process, we can determine the final square root value and apply it to practical scenarios.

3.2 Image Preprocessing Modules

Image preprocessing enhances image quality through spatial transformations and filtering techniques, including gaussian [1], mean [10], and median filtering [16]. These methods are adapted to specific requirements to improve noise reduction and image clarity.

The methodologies described leverage basic and specialized IP cores, HDL, and effective architectural designs to streamline development and enhance the performance of systems for detecting blurred images. This approach not only simplifies system design but also boosts performance and efficiency, building a robust and flexible image processing system.

Algorithm 1 Sqrt IP Substitute Module

1: **MATLAB Initial Value Generation**
2: Set initial values: $x_0 = 0.607253$, $y_0 = 0$, $z_0 = \frac{\pi}{3}$
3: Quantize and convert to a 32-bit fixed-point representation
4: **Verilog Lookup Table**
5: **for** each angle θ from $45°$ to $2.66804264453623 \times 10^{-8}$ radians **do**
6:     Assign a unique identifier to each angle and convert it to a 32-bit binary representation
7: **end for**
8: **Parameter Initialization**
9: Set the iteration count ($Pipeline$) and pre-calculate the constant value (K_n)
10: **Iterative Computation**
11: **for** each iteration i up to the count $Pipeline$ **do**
12:     Initialize registers with the initial values
13:     Update values based on the results from the previous iteration using shift and add/subtract operations
14:     **if** the angle is positive **then**
15:         Perform one set of operations
16:     **else**
17:         Perform a different set of operations
18:     **end if**
19:     Gradually approach the required square root value with multiple iterations
20:     Store the quadrant information after each iteration in the $Quadrant$ array
21:     Ensure the $Quadrant$ array contains the final quadrant information for the angle
22: **end for**
23: **Final Angle Output**
24: Output the actual angle information derived from the iterative computation

RGB to YUV Module. The RGB to YUV module is crucial for color space conversion in digital image processing. The foundational formula for this transformation is given by:

$$\begin{cases} Y = 0.299R + 0.587G + 0.114B \\ U = 0.568(B - Y) + 128 \\ V = 0.713(R - Y) + 128 \end{cases} \quad (2)$$

where R, G, and B represent the color values of a pixel, where R is the red component, G the green component, and B the blue component. Y, U, and V, on the other hand, denote the color information in the video signal, with Y indicating the luminance component and U and V the chrominance components.

Considering that FPGAs cannot directly handle floating-point numbers, an eight-bit quantization strategy is employed for discretizing the values. Therefore, Eq. 2 is transformed into the following form for practical computation:

$$\begin{cases} Y = (77R + 150G + 29B) \gg 8 \\ Cb = ((-43R - 85G + 128B) \gg 8) + 128 \\ Cr = ((128R - 107G - 21B) \gg 8) + 128 \end{cases} \quad (3)$$

The RGB to YUV color space transformation system is primarily divided into four parts: counting, window generation, RGB to YUV computation, and control with delay. Each part is detailed as Fig. 1:

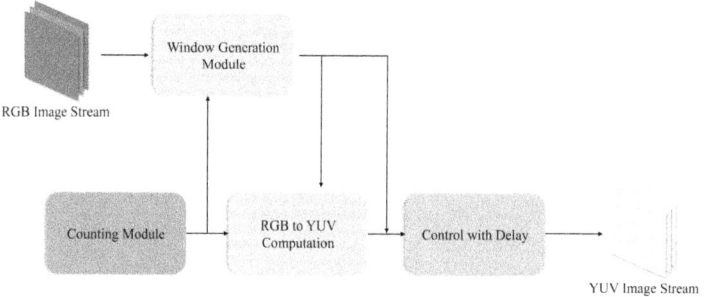

Fig. 1. Operational flowchart of the RGB to YUV Module

Filtering Modules. Filtering is a fundamental image preprocessing technique aimed at reducing noise while preserving important details.

Gaussian Filtering characterized by its utilization of a bell-shaped kernel, is a technique employed for the smoothing of images. This method is particularly adept at diminishing noise and fine details, emulating the blurring effect perceived by the human visual system. Its efficacy is most pronounced in the attenuation of Gaussian noise, which aligns with the kernel's inherent characteristics. The Gaussian filtering kernel, denoted as M_{Guass}, is defined by the matrix:

$$M_{Guass} = \begin{bmatrix} 1 & 2 & 1 \\ 2 & 4 & 2 \\ 1 & 2 & 1 \end{bmatrix} \qquad (4)$$

The computation of the filtering operation involves the multiplication of pixel data with the Gaussian kernel and the subsequent accumulation of the products. This process is orchestrated within always blocks in Verilog code, which are triggered at every rising edge of the clock signal. Variables $row1$, $row2$, and $row3$ represent the multiplication results of the kernel's first, second, and third rows with the pixel data within the window, respectively. For instance, the operation for $row1$ is given by:

$$row1 \leftarrow matrix_{p11} \times 1 + matrix_{p12} \times 2 + matrix_{p13} \times 1 \qquad (5)$$

where $matrix_{p11}$, $matrix_{p12}$, and $matrix_{p13}$ correspond to elements of the Gaussian kernel, and the numerical values represent the associated pixel data.

The final summation of the results from the multiplication steps is computed to yield the output of the filtering operation. A variable sum is defined to store the cumulative result of $row1$, $row2$, and $row3$. The summarizing operation is:

$$sum \leftarrow row1 + row2 + row3 \tag{6}$$

These operations complete the computational process of Gaussian filtering. The use of a clock signal (clk) and an asynchronous reset signal (rst_n) ensures the controlled execution and reset of computations within the Verilog code.

Mean Filtering is a fundamental image processing technique that computes the average of the pixel values within a predefined neighborhood. This method offers a straightforward yet efficacious approach for diminishing uniform noise across an image. The mean filter employs a kernel with equal weights, deviating from the Gaussian filter's weighted approach. The kernel for mean filtering, M_{Mean}, is defined as:

$$M_{Mean} = \begin{bmatrix} 1 & 1 & 1 \\ 1 & 0 & 1 \\ 1 & 1 & 1 \end{bmatrix} \tag{7}$$

Each row of the kernel is summed with the corresponding pixel data within the window, similar to Gaussian filtering but without weighted multiplication.

Median Filtering is a non-linear digital filtering technique, which serves as a robust method for the removal of so-called salt-and-pepper noise. It functions by replacing the value of a pixel with the median of intensity levels in the neighborhood of that pixel, thereby preserving vital edge information more effectively than mean filtering.

The specific implementation of median filtering is realized by invoking three *Sort* submodules to order the pixel values in each row of the processing window. Subsequently, a reapplication of the *Sort* module on the sorted rows ascertains the median value within a 3×3 window.

3.3 Edge Detection and Compensation Module

Edge detection is a crucial step in many image processing applications, including blurring detection. This module implements the Sobel edge detection method, supplemented by morphological operations for edge compensation.

Sobel Edge Detection Module. The Sobel edge detection module employs the Sobel operator to compute the gradient of the image intensity at each pixel, which effectively highlights the edges by identifying areas of significant intensity change [19]. The Sobel operator approximates the gradient of the image intensity function at each point and is implemented in digital imaging through discrete differentiation:

$$\begin{cases} xf(x,y) = f(x,y) - f(x-1,y) \\ yf(x,y) = f(x,y) - f(x,y-1) \end{cases} \tag{8}$$

The magnitude of the gradient is then calculated, providing a measure of edge strength at each pixel (Fig. 2):

$$G[f(x,y)] = |\Delta x f(x,y)| + |\Delta y f(x,y)| \qquad (9)$$

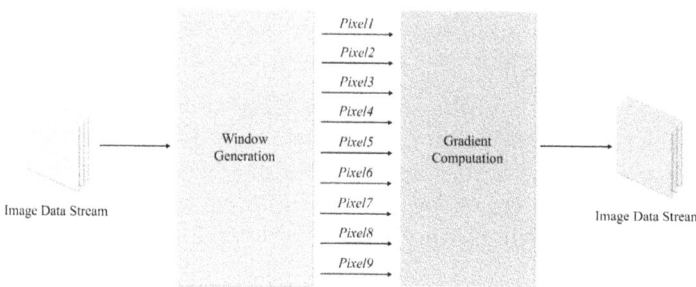

Fig. 2. Operational workflow of the Sobel Edge Detection Module

The implementation details of the Sobel edge detection IP core are primarily focused on data processing. As with previous modules, the operations necessary for data enablement are similar and will not be reiterated here. The procedure involves applying the aforementioned mathematical operations to compute the gradient vectors efficiently within each image frame, thus delineating edges with high accuracy.

Morphological Operations: Erosion and Dilation. Morphological operations such as erosion and dilation play a crucial role in refining the edge maps generated by the Sobel operator. These operations enhance the visual quality and accuracy of edges detected in binary images [2,15].

Principle of Operation. Erosion and dilation are fundamental transformations in the area of mathematical morphology. They are particularly suited for binary images, serving to either reduce or augment features within an image.

– **Dilation** ($X \oplus B$) combines two sets using vector addition, effectively increasing the size of objects within the image by expanding their boundaries. This operation is defined as the sum of all vectors where operands come from set X and structuring element B.
– **Erosion** ($X \ominus B$), on the other hand, involves vector subtraction, which serves to erode the boundaries of regions of foreground pixels, thus shrinking the objects.

Implementation Process. The operational workflow for these morphological operations is characterized by the following detailed steps:

1. **Data Acquisition:** Utilizing the Shift RAM IP core from Sect. 3.2, 3x3 binary image data blocks are fetched. This core is crucial for synchronizing data acquisition with the system clock and handling resets effectively.
2. **Row-wise Operations:** For each row within the block, bitwise AND/OR operations are executed depending on whether the operation is erosion or dilation. If the reset signal (rst_n) is low, the result variables for each row are initialized to zero to ensure a clean state.
3. **Result Compilation:** Upon completing the row-wise operations, the results are combined to derive the final output for either erosion or dilation:
   – For dilation, the row results are combined using a logical OR.
   – For erosion, the row results are combined using a logical AND.
4. **Output Synchronization:** In addition to processing, synchronization mechanisms are applied to align the module output with external interfaces or subsequent processing stages. Delays are adjusted according to system requirements and operational specifications.

The design and implementation of these operations leverage the inherent parallel processing capabilities of FPGAs, providing efficient and high-performance solutions for real-time image processing tasks.

4 Experiments

4.1 Experiment Settings

This section introduces the IP core validation system based on the ZYNQ 7010 and OV5640 cameras, discussing the implementation process in detail. The ZYNQ 7010 is a programmable logic device from Xilinx that combines the advantages of an ARM processor with an FPGA, offering robust computation and processing capabilities. The system is designed to validate the practical effectiveness of the image blurring detection IP core, with the following specific implementations:

– **Image Acquisition:** Images are captured through the OV5640 camera and transmitted to the ZYNQ 7010 for processing.
– **Image Processing:** The image is processed within the ZYNQ 7010 using the image blurring detection IP core.
– **Result Output:** The processed image results are output to an external display for observation and analysis.

Key technologies used in this system include:

– **FPGA Design:** The ZYNQ 7010 is programmed using FPGA development tools like Vivado to implement various functional modules.
– **ARM Processor:** The ARM processor in the ZYNQ 7010 manages FPGA logic and implements some software-level algorithms and processing.
– **Image Acquisition and Transmission:** The OV5640 camera captures images, and the AXI Stream protocol in the Vivado SDK is used for image data transmission.

- **Image Processing Algorithms:** The previously designed IP core is used and integrated with other modules within the ZYNQ 7010.
- **Display Output:** Processed image results are output to an external display via interfaces such as HDMI.

System Functionality. To further validate the functionality of the designed IP core, a system is set up on the ZYNQ 7010 hardware platform. The OV5640 camera captures image data, which is processed by the image processing module and output via HDMI protocol.

Hardware Setup. Figure 3 illustrates the hardware structure of the image blurring detection IP core system. The hardware platform is based on the Xilinx ZYNQ-7010 device, where the FPGA is the core component for implementing image processing and display driving modules. The system includes modules for image capture, processing, and output, integrating peripherals such as DDR3 memory, Flash storage, and a clock module.

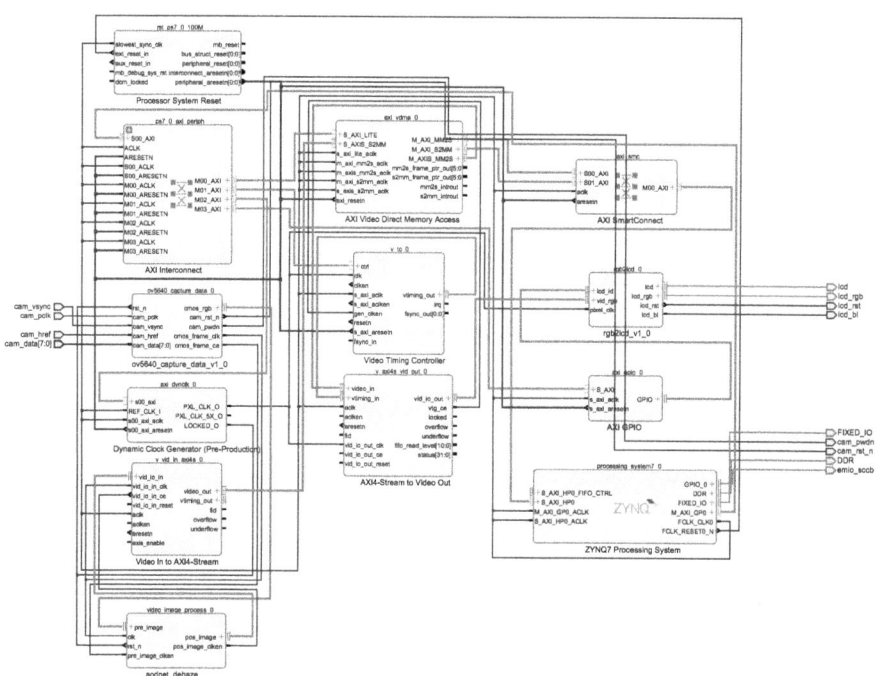

Fig. 3. Hardware System Structure of the Validation System

Embedded Software Design. Embedded software design plays a crucial role in enabling HDMI display and configuring the OV5640 camera, making the entire validation system capable of image acquisition, processing, and display. Vivado SDK is used for embedded software design, linking the designed IP core with the ZYNQ 7010 processor for system-level validation.

4.2 Results and Performance

The testing focuses on various modules like RGB to YUV conversion, filtering, and edge detection. Each module is analyzed for functionality, resource utilization, and performance:

RGB to YUV Conversion. To validate the effectiveness of the module, we assigned the converted luminance component to the RGB components of the display module to generate a grayscale image. The comparison between the original and processed images is shown in Fig. 4. Observations from the results indicate that the RGB to YUV conversion module functions as expected, aligning with our anticipated outcomes.

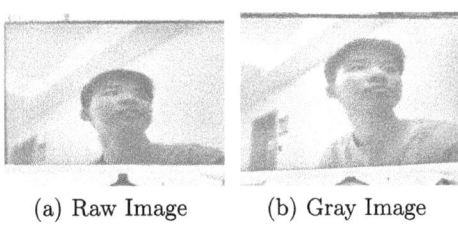

(a) Raw Image   (b) Gray Image

Fig. 4. RGB to YUV conversion result.

Additionally, we compared the usage of IP cores generated using HLS [4,13], with other related resources. The comparative analysis is presented in Table 1. Upon comparing the resource utilization, it was found that the RGB to YUV module designed in this study exhibits superior performance, utilizing fewer resources and consuming less power while maintaining similar actual effects.

Table 1. Usage of HLS-generated IP cores and other related resources.

	LUT	BUF	Register	Slice	DSP	Power
HDL	80	1	69	40	0	1.614 W
HLS	766	1	144	55	2	2.136 W

Filtering Module. In the image processing sequence, three commonly utilized filtering techniques are median filtering, mean filtering, and Gaussian filtering. To accommodate the input requirements for these filtering methods, the image data transmitted from the camera is first converted from RGB to YUV format, transforming the RGB image into a grayscale image. Subsequently, this grayscale image undergoes the respective filtering processes. The effects of these three filtering methods are illustrated in Fig. 5.

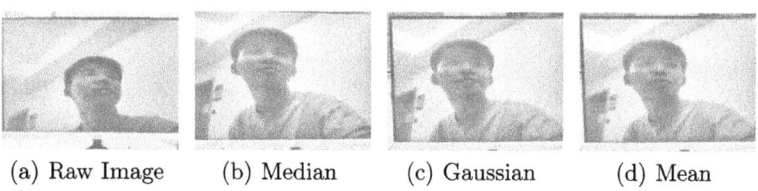

(a) Raw Image   (b) Median   (c) Gaussian   (d) Mean

Fig. 5. Filtering module verification result.

Observations from the results confirm that the filtering module functions correctly and meets the expected outcomes. However, the distinctions between the three filtering techniques are often not readily discernible to the naked eye. This is due to the different algorithms and strategies they employ in processing images, tailored to various image characteristics and application needs. Median filtering is particularly effective at removing salt-and-pepper noise, mean filtering smooths the image and reduces noise, while Gaussian filtering focuses on preserving image details while also diminishing noise. The resource usage and power consumption for the three filtering techniques are displayed in Table 2.

Table 2. Resource usage and power consumption of the three filtering methods.

		LUT	BUF	Register	Slice	DSP	Power
Mean	HDL	969	1	554	5	1	8.536 W
	HLS	7788	1	10225	12	8	12.6 W
Median	HDL	765	1	824	0	2	9.533 W
	HLS	5590	1	8995	16	6	11.611 W
Gaussian	HDL	1154	1	824	2	2	8.714 W
	HLS	6942	1	9899	1	8	12.634 W

Sobel Edge Detection. To validate the edge detection module, we initially convert the collected images into grayscale, and then this grayscale data is fed into the Sobel module. By varying the threshold values τ, differences in edge

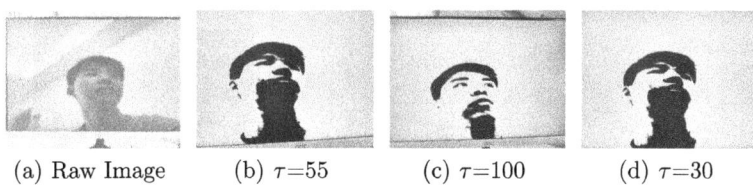

Fig. 6. Sobel edge detection verification result.

detection performance can be observed. The effects of different thresholds on edge detection are illustrated in Fig. 6.

Observations from the results indicate that the Sobel edge detection module is functioning properly, and the actual outcomes at different thresholds align with expectations. Additionally, a comparison of resource utilization for the IP cores generated using HLS is presented based on experimental outcomes, as shown in Table 3.

Table 3. Resource usage of the Sobel edge detection module.

	LUT	BUF	Register	Slice	DSP	Power
HDL	2550	1	788	2	3	10.714 W
HLS	10622	1	9899	1	8	14.634 W

From this comparison of resource usage, it can be concluded that the Sobel edge detection module designed in this study exhibits superior performance when the actual effects are comparably similar.

Image Dilation and Erosion. The image dilation module effectively enhances target regions within an image. Through the process of dilation, the pixels within these regions are expanded, and the edges are thickened, making the targets more prominent in the image. This dilation effect is particularly beneficial for applications such as object recognition and edge detection. Conversely, the image erosion module serves to reduce the size of target regions in an image. Through erosion, the pixels within these regions are contracted, and the edges are refined, leading to more detailed representations in the image. This erosion effect is crucial for detailed image processing and morphological image processing.

The erosion effects are illustrated in Fig. 7. Observations confirm that the functionality of the image compensation modules aligns with expectations. The use of resources and power consumption is displayed in Table 4.

The erosion and dilation modules designed in this study are particularly efficient in processing single-bit images. Compared to modules generated using HLS, the designs presented in this paper utilize computational resources more effectively while maintaining similar processing effects, thereby enhancing the performance of the modules.

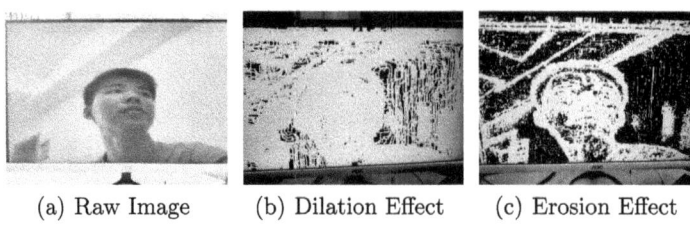

(a) Raw Image   (b) Dilation Effect   (c) Erosion Effect

Fig. 7. Image Dilation and Erosion Result

Table 4. Resource usage and power consumption of the image dilation and erosion modules.

		LUT	BUF	Register	Slice	DSP	Power
Dilation	HDL	165	0	32	3	0	1.036 W
	HLS	812	0	165	7	0	2.232 W
Erosion	HDL	141	0	21	2	0	1.016 W
	HLS	807	0	155	6	0	2.121 W

5 Conclusion

This work successfully designed and validated an IP core for blurring image processing constructed a ZYNQ verification system, and performed comparative analyses of resources and power consumption. The work significantly contributes to the fields of image processing and embedded systems and offers efficient solutions for future adaptations in blurring image edge detection. Moving forward, it is essential to address the identified shortcomings and explore advanced optimization and application strategies to enhance the system's performance and expand its practical applications.

References

1. Abuya, T.K., Rimiru, R.M., Okeyo, G.O.: An image denoising technique using wavelet-anisotropic gaussian filter-based denoising convolutional neural network for CT images. Appl. Sci. **13**(21), 12069 (2023)
2. Barkovska, O., Filippenko, I., Semenenko, I., Korniienko, V., Sedlaček, P.: Adaptation of FPGA architecture for accelerated image preprocessing. Radioelectronic Comput. Syst. **2**, 94–106 (2023)
3. Chiu, M.C., Wei, C.J.: Integrating deblurgan and CNN to improve the accuracy of motion blur X-ray image classification. J. Nucl. Sci. Technol. **61**(3), 403–416 (2024)
4. Du, C., Firmansyah, I., Yamaguchi, Y.: FPGA-based computational fluid dynamics simulation architecture via high-level synthesis design method. In: Rincón, F., Barba, J., So, H.K.H., Diniz, P., Caba, J. (eds.) ARC 2020. LNCS, vol. 12083, pp. 232–246. Springer, Cham (2020). https://doi.org/10.1007/978-3-030-44534-8_18

5. George, R.J., Charaan, S., Swathi, R., Rani, S.J.V.: Design of an IP core for motion blur detection in fundus images using an FPGA-based accelerator. In: 2023 International Conference on Bio Signals, Images, and Instrumentation (ICBSII), pp. 1–6. IEEE (2023)
6. Ghodhbani, R., Horrigue, L., Saidani, T., Atri, M.: Fast FPGA prototyping based real-time image and video processing with high-level synthesis. Int. J. Adv. Comput. Sci. Appl. **11**(2) (2020)
7. Hoshino, Y., Shimasaki, M., Rathnayake, N., Dang, T.L.: Performance verification and latency time evaluation of hardware image processing module for appearance inspection systems using FPGA. J. Real-Time Image Proc. **21**(1), 20 (2024)
8. Hu, Y., Liu, Y., Liu, Z.: A survey on convolutional neural network accelerators: GPU, FPGA and ASIC. In: 2022 14th International Conference on Computer Research and Development (ICCRD), pp. 100–107. IEEE (2022)
9. Jang, H., Tong, F.: Improved modeling of human vision by incorporating robustness to blur in convolutional neural networks. Nat. Commun. **15**(1), 1989 (2024)
10. Kumar, N., Dahiya, A.K., Kumar, K.: Modified median filter for image denoising. Int. J. Adv. Sci. Technol. **29**(4s), 1495–1502 (2020)
11. Lázaro, J., Astarloa, A., Zuloaga, A., Araujo, J.Á., Jiménez, J.: AXI lite redundant on-chip bus interconnect for high reliability systems. IEEE Trans. Reliab. **73**(1), 602–607 (2023)
12. Maxfield, C.: The design warrior's guide to FPGAs: devices, tools and flows. Elsevier (2004)
13. Molina, R.S., Gil-Costa, V., Crespo, M.L., Ramponi, G.: High-level synthesis hardware design for FPGA-based accelerators: models, methodologies, and frameworks. IEEE Access **10**, 90429–90455 (2022)
14. Noami, A., Pradeep Kumar, B., Chandrasekhar, P.: High performance AXI4 interface protocol for multi-core memory controller on SoC. In: Reddy, K.A., Devi, B.R., George, B., Raju, K.S. (eds.) Data Engineering and Communication Technology. LNDECT, vol. 63, pp. 131–140. Springer, Singapore (2021). https://doi.org/10.1007/978-981-16-0081-4_14
15. Said, K.A.M., Jambek, A.B.: Analysis of image processing using morphological erosion and dilation. In: Journal of Physics: Conference Series, vol. 2071, p. 012033. IOP Publishing (2021)
16. Tang, X., Liu, W., Ren, J., Du, Y., Chen, B.: An optimized hardware design of a two-dimensional guide filter and its application in image denoising. J. Supercomput. **78**(6), 8445–8466 (2022)
17. Wang, S., et al.: Single and simultaneous fault diagnosis of gearbox via wavelet transform and improved deep residual network under imbalanced data. Eng. Appl. Artif. Intell. **133**, 108146 (2024)
18. Wang, X., He, X., Zhu, X., Zheng, F., Zhang, J.: Lightweight and real-time infrared image processor based on FPGA. Sensors **24**(4), 1333 (2024)
19. Zhang, K., Zhang, Y., Wang, P., Tian, Y., Yang, J.: An improved sobel edge algorithm and FPGA implementation. Procedia Comput. Sci. **131**, 243–248 (2018)

Automated Clinical Summary Generation via Integrating Structured and Unstructured Data

Jiaojiao Fu[1], Bowen Yang[2,3], Yi Guo[1(✉)], Yangfan Zhou[2,3], and Xin Wang[2,3]

[1] School of Information Science and Engineering, East China University of Science and Technology, Shanghai, China
{fujj,guoyi}@ecust.edu.cn
[2] Shanghai Key Lab. of Intelligent Information Processing, Fudan University, Shanghai, China
[3] School of Computer Science, Fudan University, Shanghai, China
{20210240218,zyf,xinw}@fudan.edu.cn

Abstract. Automatically generating clinical texts can significantly reduce the time physicians spend on clinical data recording, which is particularly important for developing countries where physicians are extremely busy due to a severe shortage. This work automatically generates discharge summaries as a case to explore the methods and feasibilities of automatic clinical text summarization. Existing work typically uses either structured or unstructured data alone to generate discharge summaries. However, the content generated often has issues such as being overly verbose, lacking focus, or omitting significant information, especially key indicators and medications. This work innovatively proposes a data integration-based clinical text generation approach, using content generated from unstructured clinical data as the basis and supplementing it with text generated from structured clinical data. This study utilizes advanced natural language processing algorithms and models to create clinical texts. It addresses the challenges of lacking datasets suitable for fine-tuning pre-trained models and combining the advantages rather than the disadvantages of both types to produce discharge summaries. Experimental results show that the structured supplementation approach can effectively improve the generation of clinical texts. This work demonstrates that clinical texts generated using existing natural language processing technologies still do not meet the demands of medical practice, pointing out the need to develop further text generation technologies tailored to the characteristics of clinical data.

Keywords: Automatic text summarization · Clinical data · Data-to-text · Discharge summary

1 Introduction

During long-term chronic disease treatment, physicians have to carefully document patients' medical data to ensure continuity of care and high-quality follow-up medical services. Discharge summaries are crucial for physicians to refer to during follow-up treatments like transfer, readmission, and ongoing clinical care [5]. However, writing discharge summaries is time-consuming for physicians and can lead to errors or omissions due to heavy workloads and limited time [7,8]. Studies have shown that over 40% of medical data are omitted in handwritten discharge summaries, among which 26% are important [5]. The missed medical data are mainly structured ones like drugs (2%–40%) and test results (65%) [12], which are critical for physicians to make informed decisions.

Automatically generating clinical texts like discharge summaries can reduce the time physicians spend on documentation and reduce errors in records, thus becoming a hot research topic. However, this is a challenging task due to the complexity of medical language, terminology, and concepts [20]. Existing work tends to use either structured data or unstructured data to generate discharge summaries. For example, some research has utilized template-based methods, leveraging structured data such as examination indicators, to generate discharge summaries [2,5,6]. While these summaries are relatively comprehensive in medical content, they still suffer from data omissions, about 15%, among which 90% is important [5], and may include an overload of information. Conversely, other studies have employed pre-trained models, such as Bert, or large language models (LLMs) to create discharge summaries from unstructured data like progress notes [13,15,19]. These summaries tend to be more readable but typically overlook crucial details, including key indicators, medications, and both short- and long-term outcomes vital for clinical decision-making [16]. Additionally, LLMs can experience hallucination problems, leading to the production of excessively optimistic or ambiguous statements, particularly when summarizing lengthy texts.

To make up for the shortcomings of existing research, this study explores the methods and feasibility of integrating the strengths of generating discharge summaries from both unstructured and structured data while addressing their respective weaknesses. There are several challenges to accomplishing this integration: First, the unstructured data from inpatients is diverse in type and extensive in documentation. Processing such data to produce high-quality text summaries presents significant difficulties. Second, there is no direct correlation between patients' progress notes and their final discharge summaries. This discrepancy makes it challenging to construct datasets for fine-tuning pre-trained models. Third, the structured data, such as medications and examination reports, is voluminous. Generating text that encompasses all this data often results in overly lengthy summaries that lack focus, thus burdening physicians who need concise clinical references. Fourth, it is challenging to ensure that the integration of structured and unstructured data in discharge summaries captures the benefits of both without their disadvantages. Specifically, producing summaries that are not only comparable in length to those written by physicians but also coherent, focused, non-repetitive, and comprehensive of all essential indicators is a formidable task.

To overcome the challenges mentioned above, this work proposes a novel method for generating discharge summaries. The approach mainly relies on unstructured data to generate discharge summaries and supplements with those generated from structured data. First, we employed supervised and unsupervised extraction and abstraction algorithms and models to generate text summaries based on Progress Notes. Then, we selected a partially aligned data-to-text model to generate text summaries based on structured data. To fine-tune this model with clinical data, we need to create a <data, sentence> relational pair dataset. We used a string-matching method to get the data and the corresponding sentences. To obtain the important data for sentence generation, we built and trained a data extraction model to extract essential data from the whole set. At last, the content generated by text summary models was supplemented by the sentence generated by structured data. We utilized the MIMIC-III, an open-source clinical dataset, to conduct experiments, and used ROUGE to evaluate the effectiveness of the proposed comprehensive automatic discharge summary generation method. The experimental results indicated that when supplemented with structured data, ROUGE-L's F-value score increased by 5.6 percentage points, which attests to the effectiveness of the proposed approach. We also manually compared the similarity between automatically generated discharge summaries and those written by physicians and assessed the feasibility of clinical text generation schemes based on data integration.

The contributions of this work are mainly in the following three aspects:

- We propose generating discharge summaries based on integrated structured and unstructured data. It combines the strengths of conducting clinical text summaries with different data types. We also verify the effectiveness of this method through experiments.
- To generate clinical text based on structured data, we constructed two datasets: one containing pairs of data and sentences, and the other containing pairs of complete and important data. These two datasets make up for the lack of datasets in generating clinical text based on structured data.
- The experiments verified the feasibility of applying advanced NLP algorithms and pre-trained models to the clinical text generation task, revealed the challenges and proposed the corresponding solutions.

2 Methods

This work uses summaries generated from unstructured data as the foundational component, supplemented by sentences derived from structured data, and directly merges these elements to create the final content.

2.1 Clinical Text Summary Based on Unstructured Data

The clinical text generation framework takes physician notes as input because they contain almost all essential information. The output is a discharge summary. Text generation based on unstructured clinical data is a multi-document summary task. Each day, an inpatient has no fewer than four physician notes, including two progress notes and two respiratory care reports[1]. If a patient stays in the hospital for ten days, over forty documents need to be processed. Clinical text generation based on unstructured data is also a multi-data type summary task. For instance, progress notes keep a comprehensive record of the patient's physical condition, covering various systems such as the nervous, respiratory, endocrine, digestive, urinary, circulatory, genital, and so on.

To enhance the effectiveness of summarization, this study conducts text summaries based on categories. Different healthcare providers often use various terms to refer to the same medical category. For instance, expressions like "Resp.", "resp", "resp:", "resp care", and "respiratory care note" all refer to a patient's respiratory condition. In this work, we identify the labels commonly used to describe various body systems in progress notes through manual analysis, then use regular expressions to categorize and extract relevant content from the notes. We also manually reviewed the automatically extracted results to identify omissions. We address special cases by refining the automatic extraction methods. For instance, documents labeled as "addendum" are not treated as a separate category but are seen as extensions of an incomplete progress note. We treat the "others" as a distinct category because it may include essential overviews of patients' basic details, changes in their conditions, and details of transfers.

We employ extractive (Lead-3 and TextRank) and abstractive (BART and PEGASUS) models for text summarization to evaluate their effectiveness. TextRank utilizes three kinds of word embedding models, *i.e.*, GloVe [17], BERT [11], and ClinicalBERT [1], to assess whether different word embedding approaches impact the performance of text summarization and whether fine-tuning word embedding models based on clinical datasets can enhance text summarization quality. We also investigate whether fine-tuning pre-trained language models (BART and PEGASUS) can influence the quality of generated clinical summaries. Given the challenge of fine-tuning models on long clinical texts, we proposed an approach balancing content matching and model input requirements, truncating input text into 1000-word segments during training and concatenating the summarized segments during testing to form the final output.

2.2 Clinical Text Generation Based on Structured Data

This section describes generating supplementary content using structured clinical data through two primary components: the SDE (significant data extraction) model and the CDSG (clinical data-sentence generation) model. The process involves three stages: training the SDE model, using it to extract essential structured data, and fine-tuning the CDSG model to generate sentences from these data. The SDE model reduces the number of generated sentences by selecting significant structured data. The CDSG model generates sentences containing more information than the structured data, *e.g.*, including the causes and effects

---

[1] Many ICU patients require intubation treatment, and physicians document patients' respiratory conditions in these reports.

of medicine adjustments and test results instead of only describing the names, dosages, and values. This approach addresses the limitations of rule-based methods. The generated content is more in line with the physicians' writing pattern. The main contributions include: 1) constructing a dataset of all structured data and significant structured data and using it to train a significant data extraction model; 2) constructing a dataset of structured data and corresponding sentences and fine-tuning a data-to-text model.

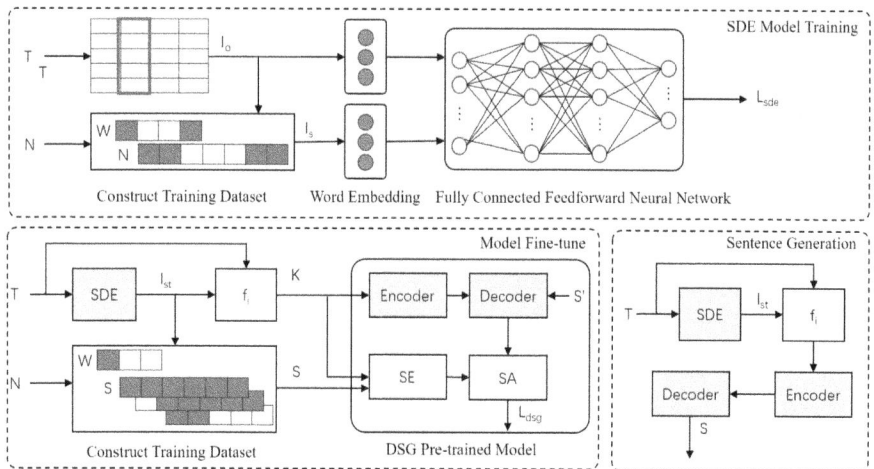

Fig. 1. Clinical text generation based on structured data

Significant Data Extraction Model. The SDE model is for extracting significant structured data. The input includes both complete structured data and partial structured data. The training objective is to extract element names, such as the medication Aspirin and the test element WBC, included in discharge summaries. This study treats the extraction of significant structured data as a binary classification problem, using one-hot encoding for vectorization before feeding into the neural network. The one-hot encoding method is selected because it is suitable for encoding discrete, unordered data, while whether structured clinical data are in discharge summaries depends solely on the data themselves, without interrelationships. The vector dimension generated by encoding is limited. For instance, discharge summaries from 33,027 hospital stays included 1,167 types of medications and 246 types of test elements. To further reduce the vector dimensions and enhance the model's computational speed, this work narrows down the inputs to include medication data present in 94.8% of discharge summaries and test data found in 95.5% of texts, aiming to filter out uncommon elements within the texts.

The process is shown in Fig. 1. I_o represents all structured data, which is obtained by extracting element names from table T, while ignoring data such as values, units, and test time. $I_o = \{i_{o1}, i_{o2}, i_{o3}, ..., i_{oj}, ..., i_{on}\}$, where i_{oj} represents the name of the j'th element, and n represents the total number. I_s represents the elements included in clinical texts, which are obtained by string matching the names in all structured data I_o with the content of discharge summaries to get the set of element names included in the text. $I_s = \{i_{s1}, i_{s2}, ..., i_{sj}, ..., i_{sm}\}$. Here, i_{sj} represents the name of the j'th element included in the text, and m represents the data of the elements included in the text, m < n. After encoding, I_o becomes a numerical matrix with dimensions of $l_o \times W$, and I_s becomes a numerical matrix with dimensions of $l_s \times W$. Here, l_o and l_s represent the number of elements contained in all structured data and significant structured data, respectively, while W denotes the length of the word vector.

This work constructs a three-layer fully connected feedforward neural network for the extraction of significant structured data. Within it, both hidden layers contain 256 neurons and utilize the ReLU activation function; the output layer employs the sigmoid activation function to constrain the probability of a certain element being present in the text between (0, 1). This work employs cross-entropy as the loss function, calculated as follows:

$$L_{sde} = -[y log y' + (1-y) log(1-y')] \qquad (1)$$

The SDE model was obtained through multiple rounds of training (see Experiment 3.4.4 for specific training process) to realize significant clinical data extraction based on patients' complete structured data.

Clinical Text Generation Model. This work fine-tunes the existing data-to-text model (DSG) to generate clinical text based on structured data. The first requirement is to construct a dataset. The inputs are Table T, including complete structured data of a patient, and Text N, the main content of a discharge summary. Based on these data, we need to construct the structured input K that meets the requirements of DSG, along with the corresponding sentence S. To this end, This work first utilizes the SDE model to extract significant elements I_{st} out of table T. Sentences containing elements from I_{st} are extracted through line-by-line string matching, ultimately obtaining the set of sentences containing elements $S = \{s_1, s_2, s_3, ..., s_j, ..., s_m\}$, where m is the number of sentences in the set. S can directly serve as input for the DSG model.

To construct the input K suitable for DSG, we analyze discharge summaries and find that they often include medication names, dosages, test indicators, and values. Extracting this information is challenging because a patient's medication and test data can change throughout hospitalization. Through content analysis, this work selects the final medication dosages and test values at discharge for dataset construction. The DSG model uses triples of (subject, attribute, attribute value) as input, so important structured data are processed into triples K, such as (patient ID, medicine, diltiazem), (patient ID, diltiazem, 60 mg), and (patient ID, white blood cells, 15.0 k/ul).

Based on this, the patient <structured data, corresponding sentence> dataset <K, S> is constructed. The i'th <K, S> pair is represented as in Eq. 2, where j indicates the number of triples for the corresponding sentence s_i:

$$< K_i, S_i > = \{\{< h_{i1}, c_{i1}, v_{i1} >, ..., < h_{ij}, c_{ij}, v_{ij} >\}, s_i\}, i < m \quad (2)$$

The choice of model is determined by the alignment of the information content between structured data and the generated text. Discharge summaries written by physicians are highly generalized, indicating a misalignment between the structured data and the generated sentences. Specifically, the information contained in clinical text sentences is more extensive than that in the data. Based on this finding, this paper opts to use the partially aligned text generation model DSG [3]. The DSG model is suited for scenarios where the generated text descriptions contain more information than the input data. Employing this model can address the issue of overgeneration, where sentences generated by well-aligned models may include content not present in the input structured data.

DSG [3] primarily utilizes a Seq-to-Seq model based on an encoder-decoder architecture to generate sentences. The DSG model comprises two key modules: the Supportive Estimator (SE) and the Supportive Adaptor (SA). SE employs a normalization layer and a feedforward network with ReLU activation function to extract features from the input triples, then computes the support matrix, and improves the balance of support scores through negative sampling. The role of SE is to optimize the supportiveness for each word. The loss function of SE is a combination of margin loss, word-consistent loss, and concentrate loss, aimed at increasing the distance between a certain word and the support scores of negative samples, enhancing its supportiveness for the same word, and reducing its support dispersion (if a certain word supports too many words, then reduce all its related supportiveness). The calculation formula is as follows:

$$L_{se} = \sum_{i=1}^{\tilde{w}} \sigma(\tilde{t}_i) - \sum_{i=1}^{w} \sigma(t_i) + a(-\sum_{i=1}^{w}\sum_{j=1}^{|K|} \mathbb{1}(S_i = K_j)(M_{i,j} - t_j)) + bmax_i \sum_{j=1}^{w} M_{i,j} \quad (3)$$

Here, $\sigma(x) = 1/(1+e^{-x})$ represents the sigmoid function, w is the number of words contained in sentence S, and $\tilde{w}$ denotes the number of negative sample words obtained after negative sampling. a and b are tunable hyperparameters. $M_{i,j}$ represents the support matrix, and t_j represents the support score vector:

$$M = F_K^T F_S, \quad F_K = NL(FW_2(ReLU(FW_1(NL(emb(K)))))) \quad (4)$$

$$t_j = \log \sum_{i=1}^{|K|} \exp(M_{i,j}) \quad (5)$$

The aforementioned NL (Normalization Layer) refers to the normalization layer, FW (Forward) denotes a linear feedforward network layer, and ReLU represents a layer with the ReLU activation function.

The role of the SA (Supportive Adaptor) module is to adjust the supportiveness scores from the SE (Supportive Estimator) module based on the output from the decoder module. The DSG model utilizes a Soft Adaptor, taking the target sentence S as input, to calculate its negative log-likelihood loss vector l, which, combined with the support scores calculated by the SE module, results in the DSG model's loss function:

$$L_{dsg} = \sum_{i=1}^{w} l_i t_i \qquad (6)$$

By fine-tuning the DSG model using the pre-trained dataset <K, S> constructed in this work, a CDSG model is obtained, tailored to better suit the characteristics of clinical data. Subsequently, this work utilizes the K obtained by extracting significant structured data from patient tables and constructing triples, employing the trained encoder-decoder framework to generate sentences that complement the content derived from text summarization.

3 Experiments

This section first introduces the dataset used for evaluating the proposed solution in this work and the experimental evaluation methods. Then, it separately discusses the results of clinical text generation based on unstructured data and the effects of generating text based on structured clinical data. Finally, it presents the results of integrated clinical text generation, which primarily involves content generated from unstructured data, supplemented by content generated from structured data.

3.1 Experimental Dataset

This work conducts research based on the open-source clinical dataset MIMIC-III (V1.4) [10], available for download on Physionet [4,9]. This work automatically generates the content of the "Hospital Course" section of discharge summaries, rather than the entire content. Through dataset analysis, we found that relying on the existing data in the MIMIC-III dataset can ensure the generated content pertains to the patient's hospital stay, including admission medication, admission physical examination, and hospital course. This study chooses to focus on automatically generating content for the hospital course section, commonly studied by many existing research [5,14]. The choice to focus on the hospital course section also because it is quite challenging to automatically generate. For example, paper [18] tried to automatically generate multiple sections of the discharge summary, such as chief complaints, history of present illness, admission medication, and hospital course, using NLP technology. The results showed that the hospital course section was the least effectively generated.

We extract the "Hospital Course" section with regular expressions. Although discharge summaries contain various pieces of information, they follow a fixed format, with each section having its corresponding chapter title. We first use regular expressions to identify the "Hospital Course" section. Then, we employ the same method to identify the subsequent section. Extracting the content between the two titles yields the "Hospital Course" section. After the content extraction, a manual review of a dozen extracted contents was performed to ensure the accuracy of this targeted content extraction method. After extracting the "Hospital Course" section, it's necessary to filter out the corresponding structured and unstructured data of the patients as input for the automatic summarization model. Since a patient's hospital admission number (HADM_ID) uniquely identifies a particular hospital stay for a patient, this paper extracts the relevant content as input by filtering through the hospital admission numbers.

3.2 Evaluation Methods

This work evaluates the similarity between the content of the "Hospital Course" section generated automatically and that written by healthcare professionals using the ROUGE metric. ROUGE encompasses a set of evaluation methods, among which ROUGE-N and ROUGE-L are the most widely used. ROUGE-N reflects the order in which words appear in the summary, with ROUGE-1 and ROUGE-2 being the most commonly used. ROUGE-1 considers the overlap of individual words, while ROUGE-2 accounts for cases where two consecutive words overlap. ROUGE-L calculates the overlap of the longest common subsequence (LCS), which does not require continuous matching and reflects the order of words in sentences. ROUGE typically calculates Precision (P), Recall (R), and F-measure (F), where Precision is the accuracy, Recall is the recall rate, and the F-measure is calculated from Precision and Recall ($F\text{-}Measure = \frac{2 \times Precision \times Recall}{Recall + Precision}$). Existing text summarization work commonly uses the F values of ROUGE-1, ROUGE-2, and ROUGE-L to evaluate the effect of automatic summarization, and this work follows the same approach.

This work is conducted on a server with 32 GB of memory, an 8-core 4.00 GHz CPU, an NVIDIA 1080 Ti graphics card, and the Ubuntu operating system. The following content first determines the summarization approach of the algorithms and models used through experimentation. Then, it presents the results of generating summaries based on structured and unstructured data. Finally, it displays the outcomes supplemented by unstructured data.

3.3 Text Summary Results Based on Unstructured Data

TextRank requires specifying the number of output sentences, while Lead-3, BART, and PEGASUS do not. The ROUGE score with TextRank varies when using different numbers of output sentences. This paper first conducts multiple experiments to calculate the relationship between the summary sentence ratio and ROUGE scores, then selects an optimal summary ratio for TextRank. The research results indicate that when the sentence extraction ratio is at 35%, the "Hospital Course" section generated by TextRank based on GloVe embedding scores the highest. Therefore, the following experiments use the TextRank with a sentence extraction ratio of 35%. The experimental results, shown in Table 1, represent the average outcome of processing data for 16 patients. The experiments demonstrate that among unsupervised methods, using the TextRank_GloVe algorithm achieves the best text summary effect for the "Hospital Course" section, with a ROUGE-L F-score of 21.5%;

Table 1. The results of clinical text summarization

Algorithm/Model	Results
Lead-3	19.0/1.7/17.2
BART	21.3/1.7/18.2
PEGASUS	19.9/1.9/17.2
TextRank_GloVe	23.2/2.7/21.5
TextRank_BERT	21.9/2.3/20.3
TextRank_ClinicalBert	21.5/2.0/20.0

Table 2. The effect of fine-tuning models on text summary results

Models	Fine-tune	No fine-tune
BART	30.4/6.9/14.5	22.4/2.1/19.6
PEGASUS	24.2/4.9/13.2	20.1/1.9/17.8

This work extracts medical notes from 1528 patients as a dataset to fine-tune the model. Among these, 93% are used as the training set, and 7% as the test set[2]. The experiment explores the effect of fine-tuning the pre-trained model on the automatically generated hospital stay results, as shown in Table 2. Exprimental results show that: 1) Fine-tuning pre-trained models can improve the F-scores of ROUGE-1 and ROUGE-2 significantly, approximately a 3%–8% increase; 2) The fine-tuned models lead to a decrease in the F-score of ROUGE-L, with a reduction of 4%–5%. Therefore, we can conclude that fine-tuning the model does not necessarily improve text summarization effects. Users can choose whether to fine-tune the model or not based on their requirements.

The aforementioned experimental results are obtained by converting all uppercase characters into lowercase. During the experiment, we also tried to remove special characters, such as "¥" and "&", from texts. However, the experimental results show that the improvement of removing special characters is not significant, ranging from −0.04% to 0.24%. Therefore, this paper adopts the same data processing method as the one used in the study [18], that is, converting all input text to lowercase.

---

[2] The same dataset is used for generating content with a non-fine-tuned model.

3.4 Text Generation Results Based on Structured Data

The MIMIC-III dataset primarily consists of structured data across 26 documents, 25 of which contain this format. It features diverse information such as patient demographics, hospital stay details, injection records, lab results, and medication data. Generating high-quality supplementary text content hinges on selecting the right structured data. This study, through manual analysis of the hospital course in discharge summaries, reveals that crucial medication details and significant lab indicators mainly originate from structured data, while other information is typically sourced from medical notes. For effective text summary generation, this paper finds that two specific tables are essential: LABEVENTS.csv, which records lab tests like white blood cell counts during hospitalization, and PRESCRIPTIONS.csv, detailing patient medication schedules, names, and dosages. However, as LABEVENTS.csv uses codes to indicate tests, consulting the D_LABITEM.csv is necessary to decipher specific lab indicators. The data from these tables is linked to corresponding discharge summaries using the hospital admission number (HADM_ID). Notably, the CHARTEVENTS.csv table, which includes vital signs like blood pressure and heart rate, was excluded as these are typically documented in medical notes by healthcare staff.

Clinical text generation based on structured data consists of two parts: the extraction of significant structured data and the generation of sentences from these data. We utilize important medication and laboratory indicators from hospital courses as outputs, with each patient's corresponding complete data as inputs. This approach has generated a dataset containing 33,027 patient admissions, each represented by pairs of complete and significant structured data. Each admission includes dozens to hundreds of medication entries and hundreds to thousands of laboratory indicator entries, culminating in approximately ten million records. We train a three-layer fully connected feedforward neural network with 80% of the dataset to extract significant structured data and the remaining 20% for validation. We conduct 100 training cycles with a batch size of 16 and a learning rate of 0.001. The medication data extraction model requires 12 h of training to achieve an accuracy of 98.50%. In contrast, the model for laboratory indicator data reaches an accuracy of 91.44% after 10 h of training. The extracted structured data is then integrated seamlessly with content generated from unstructured data.

The experimental results indicate that the extraction rates for significant medication data and laboratory data are 13.59% and 9.92%, respectively. Only a small percentage of structured data is included in the "Hospital Course" section of discharge summaries, demonstrating that extracting significant data is vital. This substantially reduces the number of sentences generated from structured data, thus decreasing the volume of information that physicians must process and ultimately enhancing their efficiency in retrieving information.

This paper employs a dataset constructed from 23,205 entries based on structured data and corresponding extracted sentences to fine-tune the DSG model. The dataset includes 8,072 entries for medication data and 15,133 for laboratory

Table 3. Automatically generated text based on structured data compared to original text

Data Type	Standard Hospitalization	Automatically generated content
Medication data	#bph: continued **finasteride**	# bph: continued home **finasteride** and terazosin
	Previously she was taking **diltiazem** for rate control	He was initially rate controlled with a **diltiazem** drip
	Amlodipine 2.5 mg daily was started [**2165-7-2**]	Patient was started on **amlodipine** 2.5 mg daily as well
	The patient was discharged on **thiamine** and **folate** supplementation	He was given **thiamine**, **folate**, and mvi
Examination data	# dm1/hyperglycemia: glucose >600 on admission, with mild **anion gap acidosis** (gap 15)	**anion gap metabolic acidosis**: likely multifactorial from lution of admission
	The patient had a history of chronic **hyponatremia** although his sodium remained between 130–140 during this admission	As per [**hospital 228**] rehab facility steroids were started to treat **low sodium**
	B12 and **folate** were within normal limits	**B12** and **folate** were normal
	Urine eosinophils negative	**Urine eosinophils negative**

data. The fine-tuning is conducted in two phases, separately based on the medication dataset and the laboratory dataset. We selected 80% of the data as the training set (6,458 entries for medication training data and 12,106 for laboratory training data). The training batch size is set to 8 (due to hardware limitations), with a learning rate of 0.0005 (the model's default rate), and the training is conducted over 300 iterations. Each fine-tuning session takes approximately 10 h. The remaining 20% of the data (1,614 entries for medication data and 2,827 for laboratory data) is used to evaluate the effectiveness of the model's fine-tuning. The results show that the average F-scores for ROUGE-L, comparing the generated sentences with the extracted sentences, are 25.14% for medication data and 24.93% for laboratory data.

Table 3 presents sentences generated using the significant data extraction model and the clinical text generation model, based on key medication and laboratory data. The sentences in the table closely resemble those written by physicians in the "Hospital Course" section, indicating that clinical text generation from structured data can be effective. For instance, it not only lists medications and laboratory trends but also provides reasons, such as diltiazem for heart rate control. However, there are notable discrepancies in some automatically generated sentences compared to the content by physicians. For example, the first and second laboratory data examples omit specific values for certain indicators, and the fourth medication example includes overly detailed drug descriptions. Thus, while the method shows potential, it needs further refinement to enhance its effectiveness.

3.5 Text Generation Results Based on Multimodal Data

This section demonstrates the results of generating the "Hospital Course" section of the discharge summary based on both unstructured data and structured data. This approach uses text generated from unstructured clinical data as a foundation and supplements it with text generated from structured clinical data. For unstructured data, this section employs a non-fine-tuned model for generating clinical text summaries by type.

Table 4. Automatic generation of clinical text results based on multimodal data

Algorithm/Model	Base Score	Lower limit of Adding Drugs and Tests	Upper Limit of Adding Drugs and Tests	Adding Extracted Drugs and Tests	Boost
Lead-3	19.2/1.7/17.6	27.5/4.6/24.9	27.3/3.7/24.6	25.6/3.2/23.2	6.4/1.5/5.6
BART	22.4/2.1/19.6	26.2/4.2/22.9	27.5/3.6/23.8	26.3/3.2/22.8	3.9/1.1/3.2
PEGASUS	20.1/1.9/17.8	26.8/4.6/23.5	27.1/3.9/23.6	25.6/3.3/22.4	5.5/1.4/4.6
TR_GloVe	23.7/2.9/21.9	24.5/4.4/22.4	27.1/4.1/24.8	**26.2/3.7/24.0**	2.5/0.8/2.1
TR_BERT	22.4/2.5/20.7	24.5/4.2/22.5	26.3/3.7/24.1	25.4/3.4/23.3	3.0/0.9/2.6
TR_ClinicalBert	22.2/2.3/20.6	24.5/4.1/22.6	26.3/3.6/24.1	25.3/3.3/23.3	3.1/1.0/2.7

Note: The pre-trained models are not fine-tuned to generate clinical text based on unstructured data. TR is short for TextRank. For example, TR_GloVe represents the TextRank algorithm that uses Glove for word embedding.

Experimental results are presented in Table 4. The first column lists the algorithms and models used. The second column shows the ROUGE scores obtained by comparing the "Hospital Course" section of the summaries generated from medical notes to the content written by physicians. The third column presents the lower limit of structured data supplementation, where all medications and laboratory indicators for a patient are used as input without data filtering. The fourth column indicates the upper limit of structured data supplementation based on the test set. The fifth column shows the results after filtering for significant content. The sixth column displays the difference between the results in the fifth column and those in the second column, illustrating the effectiveness of the structured data supplementation in the multimodal data-based clinical text auto-generation approach proposed in this paper.

The experimental results reveal: 1) Generating text with structured clinical data significantly enhances the effectiveness of the hospital course section generated from textual information (such as medical summaries and diagnostic reports). There is an improvement in the F-scores of ROUGE-1, ROUGE-2, and ROUGE-L; specifically, ROUGE-1's F-score can increase by 2.5%–6.4%, ROUGE-2's by 0.8%–1.5%, and ROUGE-L's by 2.1%–5.6%; 2) For different algorithms and models, the upper limit of utilizing structured data shows varying degrees of improvement over the lower limit. The F-score for ROUGE-L increases by −0.3% to 2.4%. Thus, extracting structured data is crucial; 3) The enhancement effect of the content generated using the significant content extraction model employed in this study does not necessarily surpass the lower limit score derived from structured data supplementation (with a variation range of −1.7% to 1.6%) and depends on the algorithms and models used. For example, with the TextRank method, the F-score for ROUGE-L using the significant content extraction model shows improvement compared to the lower limit, whereas with the Lead-3, BART, and PEGASUS methods, there is a decline, specifically by −1.7%, −0.1%, and −1.1%, respectively. However, considering the substantial reduction in text redundancy (about 90%), extracting significant content is necessary.

4 Conclusion

This work explores the feasibility and challenges of applying advanced NLP algorithms and pre-trained language models to generate clinical text summaries. The study proposes a multimodal data-based summary approach that is based on content generated from unstructured data, supplemented by sentences generated from structured data. This approach highlights the importance of generating clinical text summaries from unstructured data and validates the significant role of supplements with those generated based on structured data. The research demonstrates that effectively applying NLP technologies to clinical text summary generation requires a thorough understanding of medical practices and in-depth analysis of clinical data. The experimental results reveal that applying NLP technologies to clinical text summarization does not achieve the desired effects, indicating that future work needs to delve deeper into this area for further exploration and study.

Acknowledgments. This work was supported by the Natural Science Foundation of Shanghai (Project No. 22ZR1407900).

References

1. Alsentzer, E., et al.: Publicly available clinical BERT embeddings. In: Proceedings of the 2nd Clinical Natural Language Processing Workshop, pp. 72–78 (2019)
2. Cawsey, A.J., Webber, B.L., Jones, R.B.: Natural language generation in health care. J. Am. Med. Inform. Assoc. **4**(6), 473–482 (1997)
3. Fu, Z., Shi, B., Lam, W., Bing, L., Liu, Z.: Partially-aligned data-to-text generation with distant supervision. In: Proceedings of the 2020 Conference on Empirical Methods in Natural Language Processing. EMNLP'20, pp. 9183–9193 (2020)
4. Goldberger, A.L., et al.: PhysioBank, PhysioToolkit, and PhysioNet: components of a new research resource for complex physiologic signals. Circulation **101**(23), 215–220 (2000)
5. Goldstein, A., Shahar, Y.: An automated knowledge-based textual summarization system for longitudinal, multivariate clinical data. J. Biomed. Inform. **61**(C), 159–175 (2016)
6. Hunter, J., et al.: Summarising complex ICU data in natural language. In: AMIA Annual Symposium Proceedings 2008, pp. 323–327 (2008)
7. Hutton, C., Gunn, J.: Do longer consultations improve the management of psychological problems in general practice? A systematic literature review. BMC Health Serv. Res. **7**(1), 1–15 (2007)
8. Jacobs, M., et al.: Designing AI for trust and collaboration in time-constrained medical decisions: a sociotechnical lens. In: Proceedings of the 2021 SIGCHI Conference on Human Factors in Computing Systems. CHI'21, pp. 1–14 (2021)
9. Johnson, A., Pollard, T., Mark III, R.: MIMIC-III clinical database v1.4 - PhysioNet (2016). https://physionet.org/content/mimiciii/1.4/
10. Johnson, A.E., et al.: MIMIC-III, a freely accessible critical care database. Sci. Data **3**(1), 1–9 (2016)
11. Devlin, J., Chang, M.-W., Lee, K., Toutanova, K.: BERT: pre-training of deep bidirectional transformers for language understanding. In: Proceedings of NAACL-HLT, pp. 4171–4186 (2019)

12. Kripalani, S., LeFevre, F., Phillips, C.O., Williams, M.V., Basaviah, P., Baker, D.W.: Deficits in communication and information transfer between hospital-based and primary care physicians: implications for patient safety and continuity of care. J. Am. Med. Assoc. **297**(8), 831–841 (2007)
13. McInerney, D.J., Dabiri, B.E., Touret, A.S., Young, G., van de Meent, J.W., Wallace, B.C.: Query-focused EHR summarization to aid imaging diagnosis, pp. 632–659 (2020)
14. Moen, H., et al.: Comparison of automatic summarisation methods for clinical free text notes. Artif. Intell. Med. **67**, 25–37 (2016)
15. Patel, S.B., Lam, K.: ChatGPT: the future of discharge summaries? The Lancet Digit. Health **5**, e107–e108 (2023)
16. Peng, Y., Rousseau, J.F., Shortliffe, E.H., Weng, C.: AI-generated text may have a role in evidence-based medicine. Nat. Med. **29**, 1593–1594 (2023)
17. Pennington, J., Socher, R., Manning, C.D.: GloVe: global vectors for word representation. In: Proceedings of the 2014 Conference on Empirical Methods in Natural Language Processing. EMNLP'14, pp. 1532–1543 (2014)
18. Shing, H.C., et al.: Towards clinical encounter summarization: learning to compose discharge summaries from prior notes. arXiv preprint ArXiv:2104.13498 (2021)
19. Shing, H., et al.: Towards clinical encounter summarization: learning to compose discharge summaries from prior notes. CoRR abs/2104.13498 (2021)
20. Shubo, T., et al.: Opportunities and challenges for ChatGPT and large language models in biomedicine and health. Briefings Bioinform. **25**, 1–13 (2024)

A Review on Deep Learning for Sequential Recommender Systems: Key Technologies and Directions

Yuchen Liu, Jianpeng Qi, and Yanwei Yu(✉)

Ocean University of China, Qingdao, China
liuyuchen1092@stu.ouc.edu.cn, {qijianpeng,yuyanwei}@ouc.edu.cn

Abstract. In recent years, deep learning-based approaches have garnered significant attention in the realm of sequential recommendation. To elucidate the current trends and advancements in this field, we have systematically reviewed and classified pertinent works (especially in the past three years). This paper describes the concept of sequential recommendation, categorizes the literature according to the overall recommendation process, and evaluates key methods influencing model performance. Additionally, we analyze the role of these factors and provide a comprehensive overview of emerging challenges and future research directions. Our study offers valuable insights into the evolving landscape of deep learning-based sequential recommender systems.

Keywords: Sequential Recommendation · Deep Learning · Transformer · Contrastive Learning · Pre-training

1 Introduction

Sequential recommendation [44] is a pivotal research direction in the field of recommender systems, aiming to predict future items or content of interest to users based on their prior behaviors and preferences. In recent years, with the rapid development of deep learning technology [16] and the limitations of traditional methods in some recommendation cases [17,22], deep learning-based sequential recommendation methods have received extensive attention. The potent capability of deep (representation) learning and complex pattern capturing makes it an ideal choice for handling large-scale, high-dimensional, and dynamically changing user behavior data [53,64].

Many studies [3,60] have been previously investigated regarding recommender systems, including new technologies and improved structures, but shortcomings are still observed in existing models. For example, [20,39,83] focus on enhancing the performance of items in the model. However, the mutual relationship between users and items in the recommendation process still needs to be further considered. Moreover, despite the adoption of advanced technologies, the overall recommendation process based on deep learning remains unclear, as relying solely on complex deep learning structures does not guarantee improved performance. In this context, it is crucial to uncover the influencing factors that ensure sequential recommender systems more effective.

Based on the aforementioned issues, this review focuses on the most recent works in deep learning-based sequential recommender systems. We delve into the concept of sequential recommendation, categorize research methods, summarize crucial factors influencing model performance, and evaluate the roles of these factors. Finally, we provide a systematic outlook on the future directions and challenges. This review could offer in-depth insights and references for researchers and practitioners.

There are many surveys [49,58,61] related to sequential recommendations. Quadrana et al. [49] systematically study the current state of research in the area of sequential recommendation and provide a framework for classifying recommendation tasks and goals. Wang et al. [58] summarise the key challenges faced by sequential recommender systems, the progress made in the current work, and propose some future directions for development. The difference with the above surveys is that our survey provides a more detailed overview of the latest work in the field of sequential recommendation and focuses on a deep learning approach, suggesting some future research directions that closely follow the current hot pots. The contributions of this survey are summarized as follows:

- We propose a comprehensive classification framework for sequential recommendation, aiming to support existing classification methods and enhance understanding.
- We summarize the factors influencing typical sequential recommendation and demonstrate their impact on recommendation accuracy, guiding the research and practice of sequential recommendation.
- We summarize some existing challenges and future directions in the deep learning-based sequential recommendation.

2 Taxonomy

The process of sequential recommendation includes: 1) modeling user behavior sequences by selecting appropriate models and algorithms; 2) cleaning, converting, and pre-processing the collected user behavior data; 3) modeling the personalization, according to the user's historical behavior and interest pattern; 4) dealing with specific task, such as solving data sparsity, considering specific problems such as cold start.

Based on those four sub-processes, in this section, we categorize the phases and key considerations of constructing a sequential recommender system into four categories: basic modeling technology, data processing, personalization modeling, and specific issues. These classifications collectively offer a comprehensive understanding of the research landscape surrounding sequential recommendation, while also underscoring the importance and challenges of each phase. The methods and techniques that significantly impact sequential recommender systems, as discussed in this article, are summarized in Table 1.

Table 1. The influential learning-based methods on the sequential recommendation.

Stages	Methods	Notable Works
Basic Modeling Technology	Transformer	[71, 78]
	MLP	[31, 36]
	Contrastive Learning	[47, 70]
	Intent	[8, 34]
	Pre-training	[63, 79]
	Attention mechanism	[15, 65]
Data Processing	Denosing	[72, 76]
	Side Information	[4, 20, 39, 68, 74]
	multi-modal	[27, 36]
Personalization Modeling	Multi-behavior	[25, 66, 71, 73]
	Multi-interest	[6, 9, 46, 52, 54, 55]
Specific Issues	Cold-Start	[13, 41, 43, 57, 62, 81]
	Data Sparsity	[56]
	Bias	[11, 38, 70]

3 Basic Modeling Technology

Basic models and techniques, such as transformer, MLP, and comparative learning, can directly affect the performance and effect of recommender systems [14,21,65]. This section focuses on how these techniques can be used to build effective recommendation models that accurately predict user interests and behaviors.

3.1 Transformer-Based Models

Models based on the transformer can efficiently compute the information of all positions in the sequential in parallel, have strong global modeling capability, and can adapt to various sequence lengths. Therefore, it has been widely applied in sequential recommendation and its integration with other techniques has further enhanced the performance and adaptability of recommendation systems [66, 71, 78].

For instance, ADT [78] constructs an adaptive untangling transformer framework that utilizes mutual information estimation and auxiliary objectives to achieve adaptive untangling across different layers. This is accomplished by improving the model's learning capabilities through untangling attention heads. MBHT [71] combines multi-scale transformer and low-rank self-attention mechanisms. It introduces a hypergraph neural architecture to incorporate global multi-behavior dependencies, jointly encoding behavior-aware sequence patterns. This approach better captures hierarchical, long-term item correlations within sequences. TGT [66] leverages a behavior-aware transformer network and a temporal graph neural network to holistically consider short-term and long-term user-behavior features. The model is designed to address the challenge of multi-behavior sequential recommendation, dynamically adjusting the graph neural network to capture changes in multi-behavior interaction patterns.

These methodologies showcase the integration of transformer models with various techniques, enabling more comprehensive and dynamic modeling of user behavior sequences. The goal is to enhance the performance and adaptability of recommender systems [73,85]. However, in practice, transformer-based models are memory intensive, sensitive to sequential order, and may be slightly less capable of modeling historical information.

3.2 MLP-Based Models

The recent field of sequential recommendation has seen an influx of pure MLP-based architectures, which have become increasingly popular due to their simplicity and computational efficiency. The model based on MLP is simple and intuitive, easy to implement and train, suitable for dealing with relatively simple sequential recommendation tasks, and can achieve good performance.

AutoMLP [31] focuses on separating long-term and short-term interests using MLP blocks, MMMLP [36] utilizes MLP blocks for constructing mixers to handle multi-modal representations, and FMLP-Rec [84] employs MLP for sequence encoding in the frequency domain, aiming to achieve efficiency comparable to transformer structures in terms of receptive fields, MLP4Rec [32] employs three MLPs for interaction and sequence information discovery in separate dimensions to coherently capture sequential, cross-channel and cross-feature correlations in commodity sequences. However, the MLP-based model cannot capture the time information and long-term dependence in the sequence, the modeling ability of complex sequence data is weak, and thus the recommendation effect is limited.

3.3 Contrastive Learning-Based Models

In sequential recommendation tasks, contrastive learning [5,27] can assist the model in better understanding patterns within user behavior sequences and effectively alleviate data sparsity and noise issues. Contrastive Learning-based Models can effectively learn user preferences, improving the accuracy and robustness of personalized recommendations, especially suitable for sequential recommendation tasks with sparse data.

For example, MCLRec [47] improves model performance by introducing a learnable model enhancement module and a meta-optimization strategy on top of random data augmentation. Additionally, contrastive learning can be employed to explore relationships between different modalities of data, contributing to a more comprehensive understanding of user behavior [27]. Contrastive learning, through the clever design of sample pairs, helps alleviate the impact of overall popularity trends on recommendations. For example, DCRec [70] emphasizes the consistent relationship between the sequence view and collaboration view from the perspective of crowd awareness through contrastive learning. There are also works that apply comparative learning to negative sample sampling, multi-modal information fusion, and information optimization, providing new ideas and methods for the development of recommender systems [10,23,37,80]. Unfortunately, Contrastive Learning-based Models may require more training data and computational resources to ensure model performance.

3.4 Intent-Based Models

In recent years, some common sequential recommendation models have incorporated deep learning techniques to enable modeling and understanding of user intent [7,8,28]. Intent-based sequential recommendation models usually take into account the user's implicit intent during the recommendation process, such as the user's purchase intent, reading intent, entertainment intent, etc., to predict the user's behavior and interest more accurately [30,34,42]. Intent-based models can better understand the purposes and intentions behind user behaviors, improving recommendation accuracy and personalization, thereby enhancing user experience.

The purpose of capturing user behavior by modeling user intent is also very important. There are many models [8,42] that use clustering to model the intent. Clustering methods can group sequences of user behaviors and discover potential patterns and user groups in them to infer the user's intent and preferences. Intent-based models may require more domain expertise and manually annotated data, and have higher requirements for model design and interpretability, making it difficult to adapt to complex and changing recommendation scenarios.

3.5 Pre-training-Based Models

With the successful application of pre-trained models in the field of natural language processing, the field of sequential recommendation has also witnessed the emergence of models that utilize pre-trained models to better understand the semantic information in user behavior sequences [41,63,79]. They can leverage large-scale data for pre-training, capturing richer sequence features and semantic information, thus improving recommendation performance and generalization ability.

ASReP [41] uses pre-training to expand short sequences into long ones and then goes on to fine-tune the trained serialization model, solving the cold-start problem of sequential recommendation in this way. The disadvantage of the model based on pre-training is that it may require a lot of computational resources and time for pre-training, and the quality and diversity of pre-training data will affect the effect of the model.

3.6 Attention Mechanism-Based Models

To address long sentences or long-distance dependencies, Bahdanau et al. [1]introduced the attention mechanism. The attention mechanism plays an important role in sequential recommendation by helping the model to better understand the importance and correlation between different items in a sequence of user behavior [15,65]. In sequential recommendation, the attention mechanism is more often used in combination with other techniques [2,59,77].

Such as, By using the self-attention mechanism combined with collaborative metric learning, the AttRec [77] model can make full use of the user's historical behavior data, taking into account the user's short-term intent and long-term preference, to provide more accurate and personalized recommendation results. Models based on the attention mechanism may be limited by the length of the sequence, longer sequences increase the computational complexity and memory consumption of the model, and the calculation of attention weights may not be efficient enough.

4 Data Processing

In sequential recommendation, data processing and feature engineering are very crucial steps, in which denoising and multi-modal utilization of edge information are important research directions.

4.1 Denoising

In the data processing stage, it is crucial to deal with noises and outliers in the raw data. Data noise may negatively affect the performance of recommendation models, and several kinds of literature [72,76] work on developing various methods to recognize and deal with noise to achieve better recommendations. Such as MAERec [72] adopts an adaptive transition-path-mask strategy to dynamically extract global project transition information, which is of great help in dealing with data scarcity and noise interference in sequential recommendation scenarios.

4.2 Side Information

Side information in sequential recommendation is usually additional information related to the user or item that is used to enhance the model's understanding of the user's interests or item characteristics [18,24,75]. Such information may include the user's personal information, background information, social relationships, etc. There has been some previous work using edge information such as text and image information for recommendation [45,50]. This information can be used to enrich the user interest model so that it can better adapt to the different needs of different users. There have been some studies applying side information to recommender systems that have achieved good enhancement results [51,82].

User Information. User information refers to various data and features about a user, used to describe aspects of their personal, behavioral, and preference profiles. This encompasses but is not limited to, personal characteristics, social relationships, behavioral history, and more. For example, CoCoRec [4] comprehensively utilizes category information, user context, and collaborative filtering, considering that the user's current context influences their next action. Unlike common methods in sequential recommendation that fuse assistive information, DIF-SR [68] first uses attention mechanisms to discover sequence correlations, then combines attention weights for different attributes, and finally applies weighting to embeddings.

Item Information. Item information can include category information, tags or keywords, attribute details, and more. There are several efforts to incorporate item information into recommendations. DETAIN [39] considers an essential factor influencing user choice of items to be the extra information associated with items, defining it as the features of items. It employs a decoupling approach to separate the features in the key item sequences behind user behavior, aiming to better understand the evolution of user interests. UniSRec [20], by analyzing the relevant descriptive text of items, learns transferable sequential representations, allowing the model to adapt more effectively to various recommendation scenarios, independent of item IDs.

Dynamic Information. The characteristics of side information related to users or items may undergo changes or evolution over time. This dynamism can arise from variations in user behavior, updates to item attributes, transitions in the external environment, and other factors. DGSR [74] fully acknowledges the significance of dynamic information in modeling user sequences, employing dynamic graphs to capture interactions between users and items. It designs a Dynamic Graph Recommendation Network (DGRN) to learn the long-term and shortterm preferences of users and items.

4.3 Multi-modal

Recent research in sequential recommendation has introduced the concept of multi-modal approaches that can simultaneously leverage diverse auxiliary information. For example, MMMLP [36] utilizes a multi-modal framework to process information from images, text, and item sequences to enhance feature representation by pre-training and fine-tuning the model. By introducing an attentional mechanism, SEMI [27] allows for more flexible processing of information from different domains. It can effectively integrate multi-modal features when representing user preferences and encoding sequence behaviors.

5 Personalization Modeling

Personalization modeling in sequence recommendation aims to meet the individual needs and preferences of different users, and two aspects that have received more attention recently are multi-behavioral sequence recommendation and multi-interest sequence recommendation.

5.1 Multi-behavior

Considering various user behaviors at different time points, such as clicks, purchases, favorites, etc., to recommend items or predict the next actions that a user may be interested in the future is referred to as "multi-behavior sequential recommendation" [12,67,71]. In multi-behavior recommendation, Bayesian learning can be applied to model user behavioral patterns and preferences to provide more accurate recommendation results. For example, [48] uses it to model the distribution of users' implicit preferences over different behaviors, which can then be exploited for personalized recommendations.

MB-STR [73] addresses the limitations of the transformer by introducing multiple behavioral transformer layers, a multi-behavior sequential pattern generator, and a behavior-aware prediction module. The aim is to overcome the constraints of the transformer and comprehensively consider various types of user behaviors [26,35]. Multi-behaviors may also involve complex dynamic dependencies, meaning that the current user behavior can be influenced by a preceding series of different types of actions. TGT [66] emphasizes the explicit utilization of these dynamic dependencies to more accurately predict users' future behaviors. By considering multiple behavior types, temporal dynamics, heterogeneous relationships, long- and short-term interests, and dynamic patterns, multi-behavior recommendation aims to provide a detailed and comprehensive modeling of user behaviors, thereby improving the accuracy and personalization of recommender systems.

5.2 Multi-interest

Users' interests may relate to different topics, categories, or domains, and "multi-interest" usually refers to the diversity of users' interests in different directions, emphasizing the fact that users may have multiple interests in multiple areas [6, 9]. There has also been some recent work [46,52,54] in the area of sequential recommendation that explores multi-interest approaches.

MIMN [46] improves the Neural Turing Machine (NTM) by incorporating the Memory Sensing Unit MIU, which achieves good enhancement results in long-term user behavior modeling. MGNM [55] uses user embedding in the recommender system to incorporate the user's interest information into the graph structure and learns the user's multiple interests at different levels utilizing graph neural networks and dynamic routing to better explore the user's implicit interests and complex relationships, to improve the effect of personalized recommendation.

6 Challenges and Future Directions

6.1 Challenges

Cold-Start. The cold-start problem is classic and important in the field of recommender systems [13,43,62]. In recent years, some works [41,57,81] introduce techniques such as meta-learning, transfer learning, and integrating domain knowledge and user profiles can improve the effectiveness of recommendation in cold-start scenarios.

Data Sparsity. Data sparsity is an important problem in the field of sequence recommendation, and research has been conducted to propose solutions [33,56]. With the deepening of research and the development of technology, it is expected that there will be more effective methods to solve the data sparsity problem, so as to improve the recommendation effect and user satisfaction.

Bias. In sequential recommendation tasks, various types of biases can impact the performance and recommendations of models. Efforts to mitigate the effects of these biases are made through applying various deep learning techniques [11, 38,70], such as contrastive learning and attention mechanisms, aiming to enhance recommendation accuracy.

6.2 Future Directions

In the past years, while there have been some achievements and breakthroughs in sequential recommendation research, it has also brought to light several challenging issues. Therefore, this section summarizes the following unresolved problems, which can serve as future research directions in the field of sequential recommendation.

Variational Autoencoders. Future research can delve deeper into exploring the application of generative models such as Variational Autoencoders(VAE) [69] in sequential recommendation. These models can better capture the latent distribution behind user behavior, thereby improving the accuracy of personalized recommendations.

Loss Functions. Designing more effective loss functions [40] can also improve the effectiveness of model training. Particularly in the context of sequential recommendation tasks, loss functions that adapt to different user behavior patterns, such as multi-behavior and multi-interest, can improve the accuracy of recommendations.

Large Models. Recently, sequential recommendation methods based on big models [19,29] are enhancing personalized recommendation services through deep learning techniques and are expected to continue to progress in the future, enhancing recommendation system performance and user satisfaction.

7 Conclusions

This study systematically reviews recent advances in sequential recommendation, classifying models and examining key factors that influence performance. We propose a structured approach to analyze the task, considering state-of-the-art methods across four key areas: fundamental modeling and technology, data processing and feature engineering, user behavior and personalization modeling, and specific problems and challenges. A survey of current algorithms offers insights into the field's status quo. Finally, we discuss future directions to guide ongoing research and innovation in sequential recommendation.

Acknowledgments. This work is partially supported by the National Natural Science Foundation of China under grant No. 62176243, the Project funded by China Postdoctoral Science Foundation (No. 2023M743328), the Postdoctoral Fellowship Program of CPSF (No. GZC20232500), and the Postdoctoral Project funded by Qingdao (No. QDBSH20240102093).

References

1. Bahdanau, D., Cho, K., Bengio, Y.: Neural machine translation by jointly learning to align and translate. arXiv preprint arXiv:1409.0473 (2014)
2. Bai, T., Nie, J.Y., Zhao, W.X., Zhu, Y., Du, P., Wen, J.R.: An attribute-aware neural attentive model for next basket recommendation. In: SIGIR (2018)
3. Batmaz, Z., Yurekli, A., Bilge, A., Kaleli, C.: A review on deep learning for recommender systems: challenges and remedies. AI Rev. (2019)
4. Cai, R., Wu, J., San, A., Wang, C., Wang, H.: Category-aware collaborative sequential recommendation. In: SIGIR, pp. 388–397 (2021)
5. Chen, L., Ding, J., Yang, M., Li, C., Song, C., Yi, L.: Item-provider co-learning for sequential recommendation. In: SIGIR, pp. 1817–1822 (2022)
6. Chen, W., Ren, P., Cai, F., Sun, F., De Rijke, M.: Multi-interest diversification for end-to-end sequential recommendation. ACM TOIS (2021)
7. Chen, W., Ren, P., Cai, F., Sun, F., de Rijke, M.: Improving end-to-end sequential recommendations with intent-aware diversification. In: CIKM, pp. 175–184 (2020)
8. Chen, Y., Liu, Z., Li, J., McAuley, J., Xiong, C.: Intent contrastive learning for sequential recommendation. In: WWW, pp. 2172–2182 (2022)
9. Cheng, Y., Fan, Y., Wang, Y., Li, X.: Accurate multi-interest modeling for sequential recommendation with attention and distillation capsule network. Expert Syst. Appl. (2024)
10. Chong, L., et al.: Ct4rec: simple yet effective consistency training for sequential recommendation. In: SIGKDD, pp. 3901–3913 (2023)
11. Damak, K., Khenissi, S., Nasraoui, O.: Debiasing the cloze task in sequential recommendation with bidirectional transformers. In: SIGKDD, pp. 273–282 (2022)
12. Ding, J., et al.: Improving implicit recommender systems with view data. In: IJCAI, pp. 3343–3349 (2018)
13. Du, Z., Wang, X., Yang, H., Zhou, J., Tang, J.: Sequential scenario-specific meta learner for online recommendation. In: SIGKDD, pp. 2895–2904 (2019)
14. Fan, X., Liu, Z., Lian, J., Zhao, W.X., Xie, X., Wen, J.R.: Lighter and better: low-rank decomposed self-attention networks for next-item recommendation. In: SIGIR, pp. 1733–1737 (2021)

15. Fan, Z., et al.: Sequential recommendation via stochastic self-attention. In: WWW (2022)
16. Fang, H., Zhang, D., Shu, Y., Guo, G.: Deep learning for sequential recommendation: algorithms, influential factors, and evaluations. ACM TOIS (2020)
17. He, R., McAuley, J.: Fusing similarity models with Markov chains for sparse sequential recommendation. In: ICDM, pp. 191–200. IEEE (2016)
18. Hidasi, B., Quadrana, M., Karatzoglou, A., Tikk, D.: Parallel recurrent neural network architectures for feature-rich session-based recommendations. In: RecSys, pp. 241–248 (2016)
19. Hou, Y., He, Z., McAuley, J., Zhao, W.X.: Learning vector-quantized item representation for transferable sequential recommenders. In: WWW, pp. 1162–1171 (2023)
20. Hou, Y., Mu, S., Zhao, W.X., Li, Y., Ding, B., Wen, J.R.: Towards universal sequence representation learning for recommender systems. In: SIGKDD, pp. 585–593 (2022)
21. Hsu, C., Li, C.T.: Retagnn: relational temporal attentive graph neural networks for holistic sequential recommendation. In: WWW, pp. 2968–2979 (2021)
22. Hu, H., He, X., Gao, J., Zhang, Z.L.: Modeling personalized item frequency information for next-basket recommendation. In: SIGIR, pp. 1071–1080 (2020)
23. Huang, C., Wang, S., Wang, X., Yao, L.: Modeling temporal positive and negative excitation for sequential recommendation. In: WWW, pp. 1252–1263 (2023)
24. Huang, X., Qian, S., Fang, Q., Sang, J., Xu, C.: Csan: contextual self-attention network for user sequential recommendation. In: ACM MM, pp. 447–455 (2018)
25. Krohn-Grimberghe, A., Drumond, L., Freudenthaler, C., Schmidt-Thieme, L.: Multi-relational matrix factorization using Bayesian personalized ranking for social network data. In: WSDM, pp. 173–182 (2012)
26. Le, D.T., Lauw, H.W., Fang, Y.: Modeling contemporaneous basket sequences with twin networks for next-item recommendation. IJCAI (2018)
27. Lei, C., et al.: Semi: a sequential multi-modal information transfer network for e-commerce micro-video recommendations. In: SIGKDD, pp. 3161–3171 (2021)
28. Li, H., Wang, X., Zhang, Z., Ma, J., Cui, P., Zhu, W.: Intention-aware sequential recommendation with structured intent transition. IEEE TKDE **34**(11), 5403–5414 (2021)
29. Li, J., et al.: Text is all you need: learning language representations for sequential recommendation. arXiv preprint arXiv:2305.13731 (2023)
30. Li, J., et al.: Coarse-to-fine sparse sequential recommendation. In: SIGIR, pp. 2082–2086 (2022)
31. Li, M., et al.: Automlp: automated MLP for sequential recommendations. In: WWW, pp. 1190–1198 (2023)
32. Li, M., Zhao, X., Lyu, C., Zhao, M., Wu, R., Guo, R.: Mlp4rec: a pure MLP architecture for sequential recommendations. arXiv preprint arXiv:2204.11510 (2022)
33. Li, S., Kawale, J., Fu, Y.: Deep collaborative filtering via marginalized denoising auto-encoder. In: CIKM, pp. 811–820 (2015)
34. Li, X., et al.: Multi-intention oriented contrastive learning for sequential recommendation. In: WSDM, pp. 411–419 (2023)
35. Li, Z., Zhao, H., Liu, Q., Huang, Z., Mei, T., Chen, E.: Learning from history and present: next-item recommendation via discriminatively exploiting user behaviors. In: SIGKDD, pp. 1734–1743 (2018)
36. Liang, J., et al.: Mmmlp: multi-modal multilayer perceptron for sequential recommendations. In: WWW, pp. 1109–1117 (2023)

37. Lin, G., et al.: Dual contrastive network for sequential recommendation. In: SIGIR, pp. 2686–2691 (2022)
38. Lin, G., et al.: Dual-interest factorization-heads attention for sequential recommendation. In: WWW, pp. 917–927 (2023)
39. Lin, K., Wang, Z., Shen, S., Wang, Z., Chen, B., Chen, X.: Sequential recommendation with decomposed item feature routing. In: WWW, pp. 2288–2297 (2022)
40. Liu, Y., Walder, C., Xie, L.: Determinantal point process likelihoods for sequential recommendation. In: SIGIR, pp. 1653–1663 (2022)
41. Liu, Z., Fan, Z., Wang, Y., Yu, P.S.: Augmenting sequential recommendation with pseudo-prior items via reversely pre-training transformer. In: SIGIR (2021)
42. Ma, J., Zhou, C., Yang, H., Cui, P., Wang, X., Zhu, W.: Disentangled self-supervision in sequential recommenders. In: SIGKDD, pp. 483–491 (2020)
43. Meng, W., Yang, D., Xiao, Y.: Incorporating user micro-behaviors and item knowledge into multi-task learning for session-based recommendation. In: SIGIR (2020)
44. Nasir, M., Ezeife, C.: A survey and taxonomy of sequential recommender systems for e-commerce product recommendation. SN Comput. Sci. (2023)
45. Nguyen, H.T.H., Wistuba, M., Grabocka, J., Drumond, L.R., Schmidt-Thieme, L.: Personalized deep learning for tag recommendation. In: Kim, J., Shim, K., Cao, L., Lee, J.-G., Lin, X., Moon, Y.-S. (eds.) PAKDD 2017. LNCS (LNAI), vol. 10234, pp. 186–197. Springer, Cham (2017). https://doi.org/10.1007/978-3-319-57454-7_15
46. Pi, Q., Bian, W., Zhou, G., Zhu, X., Gai, K.: Practice on long sequential user behavior modeling for click-through rate prediction. In: SIGKDD (2019)
47. Qin, X., et al.: Meta-optimized contrastive learning for sequential recommendation. arXiv preprint arXiv:2304.07763 (2023)
48. Qiu, H., Liu, Y., Guo, G., Sun, Z., Zhang, J., Nguyen, H.T.: BPRH: Bayesian personalized ranking for heterogeneous implicit feedback. Inf. Sci. (2018)
49. Quadrana, M., Cremonesi, P., Jannach, D.: Sequence-aware recommender systems. ACM Comput. Surv. 1–36 (2018)
50. Rawat, Y.S., Kankanhalli, M.S.: Contagnet: exploiting user context for image tag recommendation. In: ACM MM, pp. 1102–1106 (2016)
51. Sun, Z., et al.: Research commentary on recommendations with side information: a survey and research directions. Electron. Commer. Res. Appl. (2019)
52. Tan, Q., et al.: Sparse-interest network for sequential recommendation. In: WSDM, pp. 598–606 (2021)
53. Tang, J., Wang, K.: Personalized top-n sequential recommendation via convolutional sequence embedding. In: WSDM, pp. 565–573 (2018)
54. Tang, Z., Wang, L., Zou, L., Zhang, X., Zhou, J., Li, C.: Towards multi-interest pre-training with sparse capsule network. In: SIGIR, pp. 311–320 (2023)
55. Tian, Y., Chang, J., Niu, Y., Song, Y., Li, C.: When multi-level meets multi-interest: a multi-grained neural model for sequential recommendation. In: SIGIR, pp. 1632–1641 (2022)
56. Wang, H., Shi, X., Yeung, D.Y.: Collaborative recurrent autoencoder: recommend while learning to fill in the blanks. Adv. Neural Inf. Process. Syst (2016)
57. Wang, H., Wang, N., Yeung, D.Y.: Collaborative deep learning for recommender systems. In: SIGKDD, pp. 1235–1244 (2015)
58. Wang, S., Cao, L., Wang, Y., Sheng, Q.Z., Orgun, M.A., Lian, D.: A survey on session-based recommender systems. ACM Comput. Surv. (2021)
59. Wang, S., Hu, L., Cao, L., Huang, X., Lian, D., Liu, W.: Attention-based transactional context embedding for next-item recommendation. In: AAAI (2018)

60. Wang, S., Hu, L., Wang, Y., Cao, L., Sheng, Q.Z., Orgun, M.: Sequential recommender systems: challenges, progress and prospects. arXiv preprint arXiv:2001.04830 (2019)
61. Wang, S., Zhang, Q., Hu, L., Zhang, X., Wang, Y., Aggarwal, C.: Sequential/session-based recommendations: challenges, approaches, applications and opportunities. In: SIGIR, pp. 3425–3428 (2022)
62. Wang, Y., Yao, Q., Kwok, J.T., Ni, L.M.: Generalizing from a few examples: a survey on few-shot learning. ACM Comput. Surv (2020)
63. Wang, Z., Wu, Q., Zheng, B., Wang, J., Huang, K., Shi, Y.: Sequence as genes: an user behavior modeling framework for fraud transaction detection in e-commerce. In: SIGKDD, pp. 5194–5203 (2023)
64. Wu, S., Tang, Y., Zhu, Y., Wang, L., Xie, X., Tan, T.: Session-based recommendation with graph neural networks. In: AAAI, vol. 33, pp. 346–353 (2019)
65. Wu, Y., et al.: Linear-time self attention with codeword histogram for efficient recommendation. In: WWW, pp. 1262–1273 (2021)
66. Xia, L., Huang, C., Xu, Y., Pei, J.: Multi-behavior sequential recommendation with temporal graph transformer. IEEE TKDE (2022)
67. Xia, Q., Jiang, P., Sun, F., Zhang, Y., Wang, X., Sui, Z.: Modeling consumer buying decision for recommendation based on multi-task deep learning. In: CIKM (2018)
68. Xie, Y., Zhou, P., Kim, S.: Decoupled side information fusion for sequential recommendation. In: SIGIR, pp. 1611–1621 (2022)
69. Xie, Z., Liu, C., Zhang, Y., Lu, H., Wang, D., Ding, Y.: Adversarial and contrastive variational autoencoder for sequential recommendation. In: WWW (2021)
70. Yang, Y., Huang, C., Xia, L., Huang, C., Luo, D., Lin, K.: Debiased contrastive learning for sequential recommendation. In: WWW, pp. 1063–1073 (2023)
71. Yang, Y., Huang, C., Xia, L., Liang, Y., Yu, Y., Li, C.: Multi-behavior hypergraph-enhanced transformer for sequential recommendation. In: SIGKDD (2022)
72. Ye, Y., Xia, L., Huang, C.: Graph masked autoencoder for sequential recommendation. In: SIGIR. pp. 321–330 (2023)
73. Yuan, E., Guo, W., He, Z., Guo, H., Liu, C., Tang, R.: Multi-behavior sequential transformer recommender. In: SIGIR, pp. 1642–1652 (2022)
74. Zhang, M., Wu, S., Yu, X., Liu, Q., Wang, L.: Dynamic graph neural networks for sequential recommendation. IEEE TKDE **35**(5), 4741–4753 (2022)
75. Zhang, Q., Wang, J., Huang, H., Huang, X., Gong, Y.: Hashtag recommendation for multimodal microblog using co-attention network. In: IJCAI (2017)
76. Zhang, S., Yao, D., Zhao, Z., Chua, T.S., Wu, F.: Causerec: counterfactual user sequence synthesis for sequential recommendation. In: SIGIR, pp. 367–377 (2021)
77. Zhang, S., Tay, Y., Yao, L., Sun, A., An, J.: Next item recommendation with self-attentive metric learning. In: AAAI, vol. 9 (2019)
78. Zhang, Y., Wang, X., Chen, H., Zhu, W.: Adaptive disentangled transformer for sequential recommendation. In: SIGKDD (2023)
79. Zhao, Q.: Resetbert4rec: a pre-training model integrating time and user historical behavior for sequential recommendation. In: SIGIR, pp. 1812–1816 (2022)
80. Zheng, Y., et al.: Disentangling long and short-term interests for recommendation. In: WWW, pp. 2256–2267 (2022)
81. Zheng, Y., Liu, S., Li, Z., Wu, S.: Cold-start sequential recommendation via meta learner. In: AAAI, pp. 4706–4713 (2021)
82. Zhou, C., et al.: Atrank: an attention-based user behavior modeling framework for recommendation. In: AAAI (2018)

83. Zhou, J., et al.: Graph neural networks: a review of methods and applications. AI open (2020)
84. Zhou, K., Yu, H., Zhao, W.X., Wen, J.R.: Filter-enhanced MLP is all you need for sequential recommendation. In: WWW, pp. 2388–2399 (2022)
85. Zou, J., Kanoulas, E., Ren, P., Ren, Z., Sun, A., Long, C.: Improving conversational recommender systems via transformer-based sequential modelling. In: SIGIR, pp. 2319–2324 (2022)

TrajUT: Intruder Trajectory Recovery on Utility Tunnel via Video Surveillance Systems

Wenbin Song, Baijian Yin, Xinwei Li, Shuai Wang, Shuai Wang[✉], and Zhao-Dong Xu

Southeast University, Nanjing 211189, Jiangsu, China
{wenbinsong,220234766,230238546,shuaiwang_iot,shuaiwang,
xuzhdgyq}@seu.edu.cn

Abstract. The security management of underground utility tunnels is crucial, especially in tracking and recovering intruder trajectories. Traditional trajectory recovery methods rely on GPS data, which are not suitable for underground environments with inferior signals. Due to the city's widely deployed distributed video surveillance systems, we focus on leveraging them to recover intruders' trajectories, but there are still challenges. Firstly, the records data generated from videos lacks clear identity labels, leading to ambiguous correspondence between intruders and video records. Secondly, the uneven and incomplete distribution of surveillance cameras results in data sparsity issues. In this paper, we propose a novel two-stage framework, TrajUT, to address these challenges. We introduce the multi-criteria-based trajectory segmentation to segment the raw record sequence into sub-sequences belonging to different groups of intruders, based on the continuity of sequences. In addition, designing a speed-constrained (Hidden Markov Model) HMM-based method to recover the movement trajectory of intruders by leveraging the distance and dynamic speed relationship among records. We conduct experiments using real-world data from Suzhou's underground utility tunnels, which show our framework's effectiveness, achieving a 12.9% precision improvement over the best baseline method.

Keywords: Trajectory recovery · Trajectory segmentation · Utility tunnel

1 Introduction

The underground utility tunnel plays an increasingly vital role in urban cities and about 69 Chinese cities have invested 12.9 billion U.S. dollars in building utility tunnels [14]. Ensuring the security of utility tunnels emerges as a crucial topic, closely tied to urban safety and the sustainable development of cities. One of the significant threats is human intrusion [15], which leads to equipment damage and economic losses, thereby disrupting city functions. Recovering the movement trajectories of intruders allows for a review of the entire destruction event, facilitating early detection of damaged equipment and identifying areas of management oversight.

Traditional trajectory recovery methods [4,5,10] are mostly based on GPS data from costly dedicated devices, and the quality of the data is greatly affected by signal strength, making accurate trajectory recovery a challenge. Additionally, GPS data collection typically requires the active cooperation of the moving object, rendering it unsuitable for passive collection scenarios such as intruder tracking. Existing works have not been able to address these challenges simultaneously. Therefore, it drives us to design an alternative method to recover trajectory in utility tunnel scenarios.

It is worth noting that video surveillance systems are widely deployed in modern utility tunnels, continuously recording all passers with timestamps and camera information, as shown in Fig. 1. This enables us to develop a framework for recovering intruder trajectories using distributed video surveillance systems.

(a) Video frames captured in color mode

(b) Video frames captured in night vision mode

Fig. 1. The utility tunnel includes some pipelines and a narrow path, with installed cameras generally equipped both color mode and night vision mode for low-light conditions.

However, the environmental impact of utility tunnels and the inherent factors of video surveillance systems lead to two challenges: **1) Video records without identification labels.** The video records data generated from the raw video in a time-ordered sequence with location information but lacks unique identity markers (such as GPS device IDs). This results in the original video sequence containing sub-sequences from multiple intrusion events, making it challenging to accurately segment sub-sequences belonging to different events. **2) Uneven distribution of surveillance cameras.** Due to cost considerations, the distribution of surveillance cameras is incomplete and uneven, with denser deployments at critical positions. The sparse video records extracted from the uneven distribution of cameras present multiple possible paths, posing a significant challenge for high-precision trajectory recovery.

To address the above challenges, we propose a video-based **Tra**jectory recovery framework for **U**tility **T**unnels (TrajUT), which includes two modules focused on trajectory segmentation and trajectory recovery tasks. For the trajectory segmentation task, we employ a multi-criteria-based method to detect discontinuity positions in the raw video record sequence, segmenting it into subsequences corresponding to different groups of intruders, and addressing the issue of unclear record relationships caused by the lack of identity labels. In the speed-constrained HMM-based trajectory recovery module, we recover the optimal trajectories from each video record sequence by using spatial information and dynamic speed constraint to calculate each trajectory's underlying cost, which tackles the problem caused by sparse video record points.

Overall, we summarize our contributions as follows:

- To the best of our knowledge, we are the first to propose an intruder trajectory recovery framework for utility tunnels, TrajUT, through a distributed video surveillance system.
- We design a trajectory segmentation module that segments the origin record sequences into independent sequences belonging to different intrusions based on the continuity of the record sequence. We introduce a speed-constrained HMM-based trajectory recovery module that considers the spatial relationship and dynamic speed variation between video records, enabling higher precision trajectory recovery within sparse trajectory data.
- We conduct experiments on real-world data from the utility tunnels in Suzhou City. The result demonstrates the effectiveness of our method, showing a 12.9% increase in precision compared to the best baseline method.

2 Problem Formulation

In this section, we describe the key definitions: Road Network, Video Record, Candidate Road Point, and Recovered Trajectory. Then, we introduce the problem statement.

Definition 1 (Road Network). We represent the road network as $G = (V, E)$, where V and E denote the set of nodes and edges, respectively. Each node $v \in V$ corresponds to specific road points, while each edge $e \in E$ means the connection between two adjacent road points.

Definition 2 (Video Record). Each video record, extracted from surveillance video R, is represented as $p = (d, t, num, lat, lon)$, meaning that num intruders are captured at time t by camera d at coordinate (lat, lon). A video record sequence can be denoted as $S_i = (p_1, p_2, \ldots, p_n)$ where $1 \leq i \leq N$ and N is the number of record sequences, n is the number of records in one sequence.

Definition 3 (Candidate Road Point). Candidate road point represents as c_i^l which denotes the l-th road point near the video record p_i, $1 \leq i \leq n$, $1 \leq l \leq M$ and M is the number of candidate points that the record p_i has. The candidate road segment r_i^l is a segment where the c_i^l is located.

Definition 4 (Recovered Trajectory). A recovered trajectory is denoted as $T_i = (r_1^l, r_2^j, \ldots, r_n^k)$, representing the actual trajectory of intruders from the video record sequences S_i where $1 \leq i \leq N$, $1 \leq l, j, k \leq M$.

Problem Statement. Given the raw surveillance videos R collected by cameras and a utility tunnel map which are converted into road network G. We aim to obtain $S_i = (p_1, p_2, \ldots, p_n)$ by segmenting from R and recover the multiple intruder trajectory $T_i = (r_1^l, r_2^j, \ldots, r_n^k)$ from $S_i = (p_1, p_2, \ldots, p_n)$ based on the constraint of road network G.

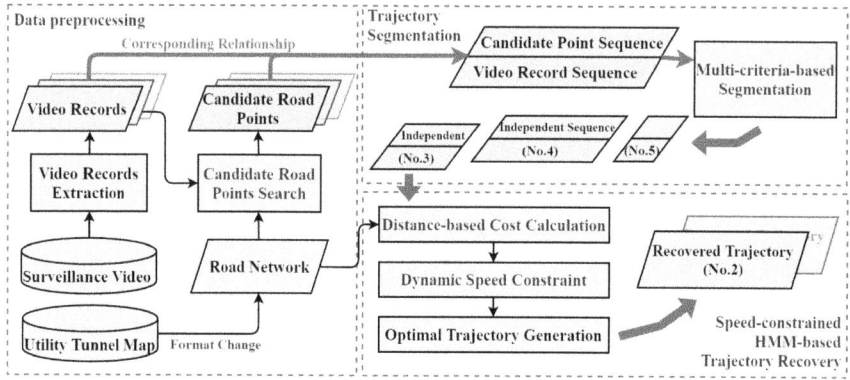

Fig. 2. The framework of TrajUT consists of three modules: data preprocessing, trajectory segmentation, and trajectory recovery. Firstly, video records and corresponding candidate road point sequences are extracted from the surveillance video and map. Then, based on the continuity evaluation results, divide the initial sequence into small segments aligning with the intrusion event. Finally, trajectory recovery process is executed on each sequence to obtain the reconstructed trajectory of the intruder's movement.

3 Methodology

3.1 Framework Overview

The framework of TrajUT, as illustrated in Fig. 2 comprises three modules: **1) Data Preprocessing**, convert the original surveillance video and utility tunnel map to a format suitable for subsequent processing. For candidate road points search, the road points within a circular area with p_i as the center and r_z meters as the radius are candidate points. **2) Trajectory Segmentation**, employing a multi-criteria-based method to evaluate the continuity between two adjacent records based on movement speed, time interval, and the number of intruders, to segment record sequences in low-continuity positions into several sub-sequences belonging to different intrusion events. **3) Speed-constrained HMM-based Trajectory Recovery**, treats the video record sequence as an observation sequence and corresponding candidate road segments as hidden states in the Markov chain. We calculate two distance-based costs between video records and candidate road segments, adjusting for dynamic speed constraints. Using dynamic programming, we find the optimal trajectory with the lowest cost and backtrack to generate the recovered trajectory.

3.2 Trajectory Segmentation

We assume that video records of intruders from different groups do not overlap. Within a single group, the video record sequence is homogeneous in continuity level and conforms to certain criteria, but distinct in the junction position of two different groups due to different movement behavior. Inspired by sequence continuity, we design our approach for segmenting video records.

Multi-criteria-Based Segmentation. We establish three criteria to initially assess the continuity of sequences independently. When any attribute exceeds the established criteria, assign the same value for marking accordingly.

Movement speed continuity evaluates video record continuity based on intruder speeds, hypothesizing that human movement speeds typically fall within a specific range. Specifically, for two adjacent video records p_i and p_{i+1}, we calculate the movement speed $\tilde{v}_{i,i+1}^{j,l}$ through corresponding candidate road points c_i^j and c_{i+1}^l by

$$\tilde{v}_{i,i+1}^{j,l} = \frac{\left\|c_i^j - c_{i+1}^l\right\|_{road}}{t_i - t_{i-1}} \quad (1)$$

Here $\|\cdot\|_{road}$ represents the distance between two candidate road points and t_i denotes the timestamp of record p_i. For the movement speed criterion, $\tilde{v}_{i,i+1}^{j,l}$ is expected to be lower than the threshold γ, namely $0 < \tilde{v}_{i,i+1}^{j,l} < \gamma$. If a value exceeds the threshold γ, we set a marker for variable CS_{i+1}.

Time interval continuity reflects the continuity of intruder movement in temporal level. Briefly, a shorter sampling interval between two adjacent video records strongly suggests movement by the same group of intruders. For this criterion, the two adjacent points' timestamp t_i and t_{i+1} is expected to fall in the range from 0 to the threshold τ, namely $0 < t_{i+1} - t_i < \tau$. We apply markers for variable CT_{i+1} if the criterion are not met.

Intruder number consistency evaluates video record continuity based on intruder composition. In intrusion events, intruders typically move in groups of two to four individuals together. For any adjacent video records p_i and p_{i+1}, the number num_i and num_{i+1} is expected to be consistent. If two records with different numbers, we assign a marker to CN_{i+1}.

The weighted average continuity score δ_{i+1} is calculated among three marker variables, due to their varying impacts, which is calculated as follows:

$$\delta_{i+1} = w_t \cdot CT_{i+1} + w_s \cdot CS_{i+1} + w_n \cdot CN_{i+1} \quad (2)$$

Here $w_t = 0.474$, $w_s = 0.368$ and $w_n = 0.158$ are weights, calculated by ground truth data, representing the different importance of three attributes.

The value of δ_{i+1} indicates the degree of continuity between two points p_i and p_{i+1}. If δ_{i+1} exceeds a predefined continuity evaluation threshold θ, it suggests the junction position between sequences of different groups of intruders. Consequently, p_{i+1} is classified as the start of a new sequence, where p_i is the end of the previous sequence.

3.3 Speed-Constrained HMM-Based Trajectory Recovery

We design a speed-constrained HMM-based recovery module, referencing Koller's method [10]. Our approach simplify the HMM methods by reformulating the problem to find the hidden state sequence with the minimum joint cost rather than the highest probability to reduce the adjustment of tunable parameters. The cost represents distance, and the minimum joint cost trajectory, based on multiple distance relationships, is the most likely movement path for intruders.

Distance-Based Cost Calculation. The joint cost is composed of emission and transition costs. To be specific, "Emission" refers to the selection of a candidate point and its road segment, while "Transition" denotes the movement from one candidate point and its road segment to the next. The record position and distance relationship are shown in Fig. 3.

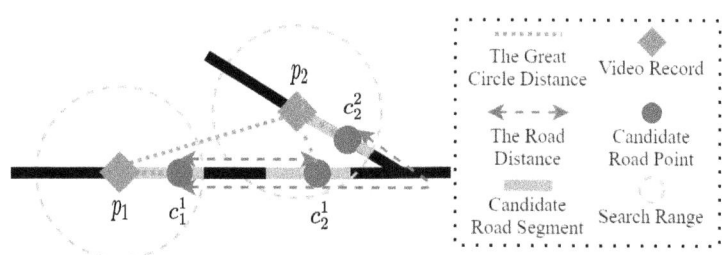

Fig. 3. The distance relationship between video records and candidate road points. The brown dashed line between the red squares and blue dots represents the emission cost. Transition cost is the difference in distance between two video records (a brown dashed line between two red squares) and their respective candidate road points (arrow green dashed line between two blue dots). (Color figure online)

The calculation method for emission cost is as follows:

$$f_e(p_i, c_i^j) = \left\| p_i - c_i^j \right\|_{great\ circle} \tag{3}$$

Here $\|\cdot\|_{great\ circle}$ represents the great circle distance. The emission cost usually appears in pairs, which can be simplified as $f_{ee}(c_i^j, c_{i+1}^l) = f_e(p_i, c_i^j) + f_e(p_{i+1}, c_{i+1}^l)$.

Transition cost reflects the likelihood of transitioning from c_i^j to c_{i+1}^l and also indicates the degree of alignment with the observation points. The transition cost is defined as

$$f_t(c_i^j, c_{i+1}^l) = \ln\left|\left\|c_{i+1}^l - c_i^j\right\|_{road} - \|p_{i+1} - p_i\|_{great\ circle}\right| \qquad (4)$$

Since the distance differences have a wide range of values, we apply a logarithmic transformation [4] to achieve a more uniform distribution.

Dynamic Speed Constraint. The speed of movement is a critical characteristic of intruders, reflected in video records and the reconstructed trajectories also preserve this feature. To dynamic speed constraint between p_i and p_{i+1}, we begin by calculating the average speed $\bar{v}_{i+1,i}$ and $\bar{v}_{i,i-1}$ from p_{i-1} to p_{i+1} using Eq. 1, and then their weighted average speed $v'_{i,i+1}$ is calculated by

$$v'_{i,i+1} = w_a \cdot \bar{v}_{i,i-1} + (1 - w_a) \cdot \bar{v}_{i-1,i-2} \qquad (5)$$

Here $w_a = 0.75$ obtained through experiments. Using a weighted average speed more accurately reflects the intruders' consistent speed over long distances, reducing errors from camera sampling discrepancies. Furthermore, we estimate the average speed $\tilde{v}_{i,i+1}^{j,l}$ of the feasible path between their candidate points c_i^j and c_{i+1}^l by Eq. 1.

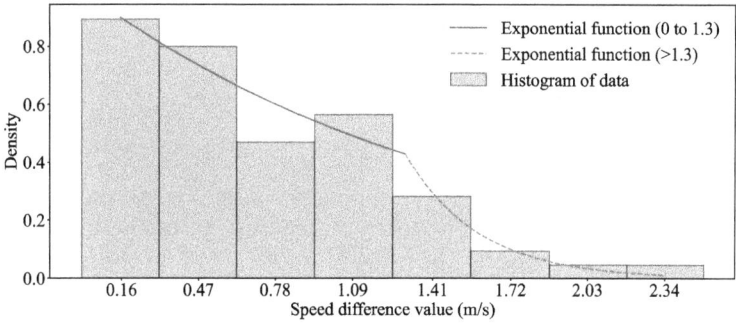

Fig. 4. The histogram of the speed difference and piecewise exponential fitting curve on historical data.

After obtaining the two speed values, we focus on their difference $\chi_{i,i+1}^{j,l} = \left|v'_{i,i+1} - \tilde{v}_{i,i+1}^{j,l}\right|$. A smaller difference indicates higher similarity between the candidate road and the actual trajectory. Given that the scale of speed differences is typically small, we convert them into a probabilistic form to better represent transitioning likelihood from speed perspective. Since the distance difference following an exponential probability distribution [4], we infer the same for the speed difference. As shown in Fig. 4, the historical speed difference data histogram fits well with the piecewise exponential probability distribution, with an average R-squared value of 0.864 and RMSE of 0.051, effectively capturing the data distribution. The speed difference of 1.3 is the function's segmentation point, determined through data distribution analysis. Then, we convert new speed difference $\chi_{i,i+1}^{j,l}$ into corresponding probability values $\rho_{i,i+1}^{j,l}$ via this equation, calculated as

$$\rho_{i,i+1}^{j,l} = \begin{cases} \frac{1}{\alpha_1} \cdot e^{\frac{\chi_{i,i+1}^{j,l}}{\beta_1}}, & 0 < x \leq 1.3 \\ \frac{1}{\alpha_2} \cdot e^{\frac{(\chi_{i,i+1}^{j,l}-1.3)}{\beta_2}}, & x > 1.3 \end{cases} \qquad (6)$$

Here $\alpha_1 = 1$, $\alpha_2 = 2.33$, $\beta_1 = -1.55$ and $\beta_2 = -0.29$ are obtained from fitting. Finally, The joint cost $f(c_i^j, c_{i+1}^l)$ is calculated as follows

$$f(c_i^j, c_{i+1}^l) = \begin{cases} f_{ee}(c_i^j, c_{i+1}^l) + f_t(c_i^j, c_{i+1}^l), & 1 \leq M < \varepsilon \\ (f_{ee}(c_i^j, c_{i+1}^l) + f_t(c_i^j, c_{i+1}^l)) \cdot (1 - \rho_{i,i+1}^{j,l}), & M \geq \varepsilon \end{cases} \qquad (7)$$

Here M is the number of candidate road points of p_i and ε is a tunable parameter that determines when to use dynamic speed constraints to adjust cost. When M is small, indicating clear matching and cost relationships, using dynamic speed constraint functions may be counterproductive, leading to significant cost changes and potential misjudgments.

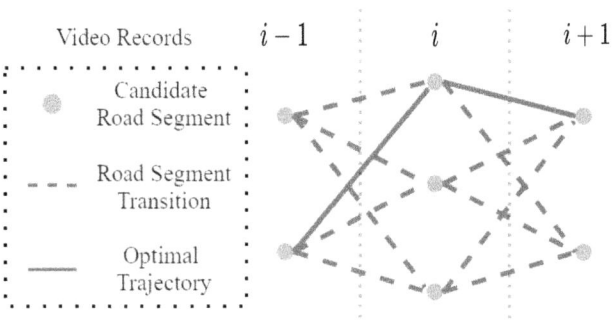

Fig. 5. The transition relationship between candidate road segments for three video records and the lowest cost transition, representing the optimal trajectory.

Optimal Trajectory Generation. As Newson [4] indicates, it is feasible to identify an optimal trajectory among the numerous transitions from the first layer to the last, as illustrated in Fig. 5, representing the most probable and also the lowest cost trajectory.

We employ the Viterbi algorithm [13] to solve the HMM-based trajectory recovery problem. The idea is to iteratively calculate the joint cost of states using distance-based cost and dynamic speed constraints, recording the previous cost of each current hidden state to ultimately determine the endpoint of the optimal path, the lowest total cost path. By backtracking from this endpoint using the recorded information, we obtain the recovered trajectory.

4 Experiments

4.1 Experimental Setting

Datasets. We collect 245 videos in a day (8 a.m. to 8 p.m.) from 69 surveillance cameras within Suzhou's utility tunnels, with a total of 108 video record points. The road network is constructed from the map of the utility tunnel, around 10 km, which encompasses 357 road nodes and 238 road segments.

Metrics. We utilize precision, recall, and F1-score to assess the performance of both trajectory segmentation and trajectory recovery, by comparing the recovered trajectory E_R with the ground truth E_G. Precision is defined as $precision = \frac{|E_R \cap E_G|}{E_R}$, and Recall is denoted as $recall = \frac{|E_R \cap E_G|}{E_G}$. The F1-score is calculated as $F1 = 2 \cdot \frac{Precision \cdot Recall}{Precison + Recall}$.

4.2 Baseline

Due to existing video-based trajectory recovery methods focus on traffic scenario, are unsuitable for utility tunnels. Thus, we design a two-stage comparison framework: trajectory segmentation stage and trajectory recovery stage.

For trajectory segmentation baseline: **1) K-means++** [12]: It achieves faster convergence speed based on the K-means algorithm and is commonly used for segmentation tasks. **2) HDBSCAN** [11]: It is a memory-optimized parallel DBSCAN-based clustering algorithm. **2) ST-DBSCAN** [1]: This is an improved version of DBSCAN that incorporates the time dimension.

For the baseline of trajectory recovery: **1) Nearest Matching** [9]: This represents a simple method but widely use method. **2) HMM-based** [8]: This is a typical HMM-based method that represents mainstream.

For the variants of TrajUT: **1) TrajUT-S:** We replace the multi-criteria-based module with a sole movement speed criterion to reveal the significance of multiple criteria. **2) TrajUT-T:** Using a sole time interval criterion. **3) TrajUT-T:** Using a sole number of intruder criterion. **4) TrajUT-w/oDSC:** We remove the dynamic speed constraint from TrajUT to detect its importance.

4.3 Results

Table 1 shows the overall performance of the proposed method compared to baselines in terms of Precision, Recall, and F1-score. Table 2 illustrates the importance of our multi-criteria segmentation and dynamic speed constraint. We have the following observations:

1. Our proposed method achieved a precision of 0.9661 and a recall of 0.9501, surpassing all baseline combinations, as shown in Table 1. It improved the F1-score by 9.63% and precision by 12.9% compared to the best baseline, ST-DBSCAN+Nearest Matching. This demonstrates that our multi-criteria segmentation and speed-constrained trajectory recovery effectively address the challenges. ST-DBSCAN performs better than other trajectory segmentation methods due to its spatiotemporal clustering capability. The Nearest Matching method slightly outperforms the HMM-based approach as it is better suited for environments with more straight-line paths.
2. From Table 2, TrajUT performs better than TrajUT-S, TrajUT-T, and TrajUT-N, indicating that its multi-criteria-based module is better at measuring sequence continuity and accurately segmenting trajectories than single criteria. TrajUT outperforms TrajUT-w/oDSC in all three metrics, showing that dynamic speed constraints effectively adapt to complex situations.

Table 1. Performance comparison of our method and baselines

Methods	Precision	Recall	F1-score
K-means++ [12]+Nearest Matching [9]	0.1882	0.2082	0.1973
HDBSCAN [11]+Nearest Matching [9]	0.5863	0.4444	0.5048
ST-DBSCAN [1]+Nearest Matching [9]	0.8554	0.8946	0.8737
K-means++ [12]+HMM-based [8]	0.1891	0.1654	0.1761
HDBSCAN [11]+HMM-based [8]	0.5821	0.3973	0.4712
ST-DBSCAN [1]+HMM-based [8]	0.8333	0.6951	0.7576
TrajUT	**0.9661**	**0.9501**	**0.9579**

4.4 Parameter Sensitivity Analysis

Threshold of Time Interval Continuity τ. Results from Fig. 6(a) indicate that setting the threshold to 900 s effectively distinguishes the video records of different intruders from the perspective of temporal continuity. When the threshold for the time interval continuity is set at 1200 s, performance deteriorates sharply and remains at a similar level thereafter, reflecting that the shortest intervals between intrusions by different groups of intruders range from 900 s to 1200 s.

Table 2. Performance comparison of our method and variants

Variants	Precision	Recall	F1-score
TrajUT-S	0.8120	0.9289	0.8614
TrajUT-T	0.6253	0.6329	0.6288
TrajUT-N	0.3231	0.4903	0.3414
TrajUT-w/oDSC	0.9615	0.9471	0.9542
TrajUT	**0.9661**	**0.9501**	**0.9579**

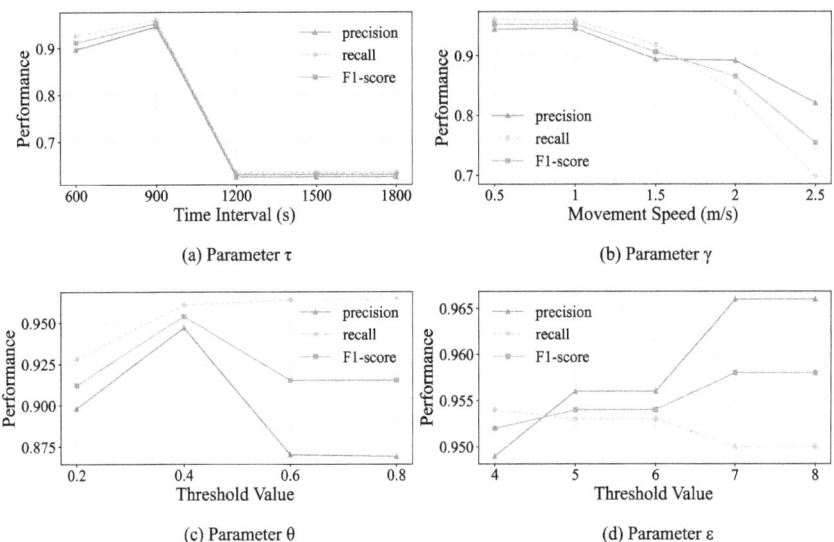

Fig. 6. The sensitivity analysis of four parameters

Threshold of Movement Speed Continuity γ. The results from Fig. 6(b) reveal that when the threshold is set at 1 m/s or lower, the final trajectory recovery yields better performance. A larger threshold allows for more speed variation, so intruders who should be divided into two groups may be assigned to the same group. It leads to a decrease in trajectory segmentation accuracy and subsequently lowers trajectory recovery recall, as shown by the significant decline in the recall curve in Fig. 6(b).

Threshold of Continuity Evaluation θ. For the threshold of Continuity Evaluation θ, the optimal balance of precision and recall is achieved at 0.4, as shown in Fig. 6(c). A threshold of 0.2 tends to cause the method to divide the record sequence more extensively, resulting in a higher number of sub-sequences than the ground truth. When the threshold exceeds 0.4, it reduces the number of sub-sequences, slightly improving overall recall but significantly reducing precision by failing to separate sequences belonging to different intruders.

Threshold of Using Dynamic Speed Constraint ε. For the threshold of Using Dynamic Speed Constraint ε, F1-score is the highest, achieving the best balance between precision and recall when $\varepsilon = 7$, as illustrated in Fig. 6(d). This threshold determines the use of dynamic speed constraints to adjust joint cost, impacting optimal trajectory selection. When the threshold is larger than 6 means that video records have a greater number of candidate road point combinations, using dynamic speed constraints can adjust the inaccuracy of distance-based cost in complex transition situation from a speed perspective, improving overall precision.

5 Related Work

5.1 Trajectory Recovery

The trajectory recovery task is restoring low sampling rate trajectory data to high sampling rate, reconstructing the actual movement trajectory of the object. The existing work can be roughly divided into three categories based on the type of data used. The first category is mainstream [4,5,7,10], based on GPS data, which is collected from specialized GPS devices. Newson and Krumm [4] utilize the Hidden Markov Model to solve such problems from GPS data. Sun et al. [7] utilize historical data to improve the recovery effect. The second category of work utilizes cellular data [17], which is usually sparse and has low accuracy. The third category is video-based work [18–20], mostly focusing on traffic scenario. Yu et al. [18] design an iterative framework to jointly optimize vehicle re-identification and trajectory recovery.

However, the first two categories assume high confidence in the trajectory data identity, which is unsuitable for low-confidence data such as video. The third category is also video-based but significantly differs from our scenario.

5.2 Trajectory Segmentation

In trajectory segmentation problems, the objective is to subdivide an original trajectory into the fewest possible sub-trajectories, each exhibiting homogeneity. For criteria-based trajectory segmentation methods [2,3,6], Buchin et al. [3] focus on various geographical and geometric features. Mainstream work regards it as a clustering problem and proposes clustering-based methods. Density-based clustering methods [1,11,16] are commonly used in segmentation.

However, due to the uneven distribution of records, missing points, and complex movement patterns of intruders, existing methods are not suitable.

6 Conclusion

In this paper, we propose the TrajUT, an intruder trajectory recovery framework for utility tunnels via distributed video surveillance systems. We design a multi-criteria-based trajectory segmentation method to solve the problem of ambiguous relationships between records and intruders, a speed-constrained HMM-based trajectory recovery method for the challenge of uneven data distribution. Experiments on real-world surveillance videos from utility tunnels in Suzhou show that TrajUT outperforms baseline methods, with a 12.9% increase in precision.

Acknowledgments. This work is supported by the Major Project of Fundamental Research on Frontier Leading Technology of Jiangsu Province under Grant BK20222006.

References

1. Cakmak, E., Plank, M., Calovi, D. S., Jordan, A., Keim, D.: Spatio-temporal clustering benchmark for collective animal behavior. In: The 1st ACM SIGSPATIAL International Workshop on Animal Movement Ecology and Human Mobility, pp. 5–8 (2021). https://doi.org/10.1145/3486637.3489487
2. Aronov, B., Driemel, A., Kreveld, M.V., Löffler, M., Staals, F.: Segmentation of trajectories on nonmonotone criteria. ACM Trans. Algorithms (TALG) **12**(2), 1–28 (2015)
3. Buchin, M., Driemel, A., van Kreveld, M.J., Sacristán, V.: Segmenting trajectories: a framework and algorithms using spatiotemporal criteria. J. Spatial Inf. Sci. **3**, 33–63 (2011)
4. Newson, P., Krumm, J.: Hidden Markov map matching through noise and sparseness. In: Proceedings of the 17th ACM SIGSPATIAL International Conference on Advances in Geographic Information Systems, pp. 336–343 (2009). https://doi.org/10.1145/1653771.165381
5. Lou, Y., Zhang, C., Zheng, Y., Xie, X., Wang, W., Huang, Y.: Map-matching for low-sampling-rate GPS trajectories. In: Proceedings of the 17th ACM SIGSPATIAL International Conference on Advances in Geographic Information Systems, pp. 352–361 (2009). https://doi.org/10.1145/1653771.1653820
6. Buchin, M., Driemel, A., Van Kreveld, M., Sacristán, V.: An algorithmic framework for segmenting trajectories based on spatio-temporal criteria. In: Proceedings of the 18th SIGSPATIAL International Conference on Advances in Geographic Information Systems, pp. 202–211 (2010). https://doi.org/10.1145/1869790.1869821
7. Wenbin, S.U.N., Ting, X.I.O.N.G.: A low-sampling-rate trajectory matching algorithm in combination of history trajectory and reinforcement learning. In: Acta Geodaetica et Cartographica Sinica, vol. 45, pp. 1328–1334 (2016)
8. Meert, W., Verbeke, M.: HMM with non-emitting states for map matching. In: European Conference on Data Analysis (ECDA), Germany (2018)
9. White, C.E., Bernstein, D., Kornhauser, A.L.: Some map matching algorithms for personal navigation assistants. Transp. Res. Part C: Emerg. Technol. **8**(1–6), 91–108 (2000)
10. Koller, H., Widhalm, P., Dragaschnig, M., Graser, A.: Fast hidden Markov model map-matching for sparse and noisy trajectories. In: 2015 IEEE 18th International Conference on Intelligent Transportation Systems, pp. 2557–2561 (2015). https://doi.org/10.1109/ITSC.2015.411
11. Wang, Y., Yu, S., Gu, Y., Shun, J.: Fast parallel algorithms for Euclidean minimum spanning tree and hierarchical spatial clustering. In: The 2021 International Conference on Management of Data, pp. 1982–1995 (2021). https://doi.org/10.1145/3448016.3457296

12. Hämäläinen, J., Kärkkäinen, T., Rossi, T.: Improving scalable K-means++. Algorithms **14**(1), 6 (2020)
13. Johnson, L.S., Eddy, S.R., Portugaly, E.: Hidden Markov model speed heuristic and iterative HMM search procedure. BMC Bioinform. **11**, 1–8 (2010)
14. Wang, T., Tan, L., Xie, S., Ma, B.: Development and applications of common utility tunnels in China. Tunn. Undergr. Space Technol. **76**, 92–106 (2018)
15. Shahrour, I., Bian, H., Xie, X., Zhang, Z.: Use of smart technology to improve management of utility tunnels. Appl. Sci. **10**(2), 711 (2020)
16. Chen, W., Ji, M.H., Wang, J. M.: T-DBSCAN: a spatiotemporal density clustering for GPS trajectory segmentation. Int. J. Online Eng. **10**(6) (2014)
17. Hoteit, S., Secci, S., Sobolevsky, S., Ratti, C., Pujolle, G.: Estimating human trajectories and hotspots through mobile phone data. Comput. Netw. **64**, 296–307 (2014)
18. Yu, F., et al.: City-scale vehicle trajectory data from traffic camera videos. Sci. Data **10**(1), 711 (2023)
19. Yu, F., Ao, W., Yan, H., Zhang, G., Wu, W., Li, Y.: Spatio-temporal vehicle trajectory recovery on road network based on traffic camera video data. In: the 28th ACM SIGKDD Conference on Knowledge Discovery and Data Mining, pp. 4413–4421 (2022). https://doi.org/10.1145/3534678.3539186
20. Tong, P., Li, M., Li, M., Huang, J., Hua, X.: Large-scale vehicle trajectory reconstruction with camera sensing network. In: the 27th Annual International Conference on Mobile Computing and Networking, pp. 188–200 (2021). https://doi.org/10.1145/3447993.344861

Spatio-Temporal Graph Fusion Network-Based Multivariate Time Series Forecasting of Environmental Factors in Utility Tunnels

Wenbin Song, Peiyi Zhao, Xinwei Li, Shuai Wang, Shuai Wang[✉],
and Zhao-Dong Xu

Southeast University, Nanjing 211189, Jiangsu, China
shuaiwang@seu.edu.cn

Abstract. With the rapid expansion of urban utility tunnels, the safety of their environments has become an important research area. We collect diverse environmental sensor data for multivariate time series forecasting, ensuring tunnel safety. However, there is an inherent spatial dependency among variables from different sensor locations. Static sensor networks usually utilize fixed distances to reflect the spatial correlation, and it is difficult to capture the dynamic spatial correlation among them. In addition, different environmental factors in the utility tunnels exhibit temporal features of different periods, and traditional methods are incapable of capturing the temporal features of different periods that exist among these different environmental factors. To overcome these challenges, this paper proposes a spatio-temporal model that combines a graph convolution module and a temporal convolution module. The model adaptively learns graph structures to capture dynamic spatial correlations and utilizes the temporal convolution module to capture temporal features of different periods among different environmental factors in utility tunnels. To better capture the trend and seasonal variations among environmental factors in utility tunnels, we also integrate the linear capture module with a nonlinear neural network in parallel, resulting in our proposed graph and temporal convolution-linear capture networks (GTCLNs). We conduct experiments on the real-world datasets collected in Suzhou utility tunnels. The experimental results surpass the performance of existing baseline methods, proving the effectiveness of our model.

Keywords: Utility Tunnels · Multivariate Time Series Forecasting · Graph Neural Networks

1 Introduction

Environmental factors in utility tunnels include humidity, temperature, harmful gases, and equipment like water pumps and fans, crucial for safety assessments. Given the enclosed nature of these tunnels and limited emergency exits, access for maintenance is only allowed once the safety of internal conditions is verified. Thus, accurate prediction of these factors is essential for ensuring tunnel safety.

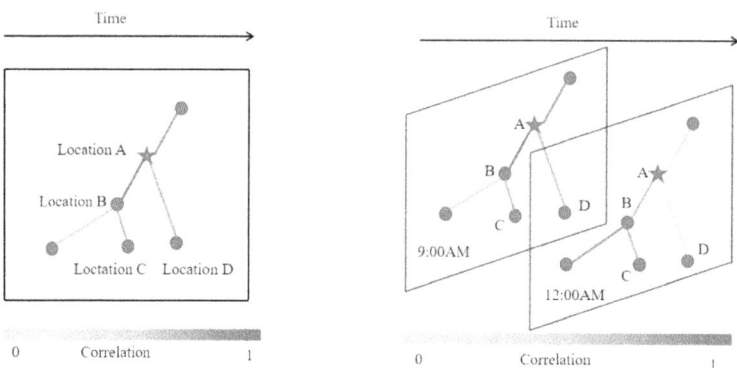

(a) Distance-based Static Sensor Networks   (b) Real-situation Dynamic Sensor Networks

Fig. 1. The strength of correlation between sensors in utility tunnels.

Figure 1(a) shows that static sensor networks usually reflect the spatial location correlation between different sensors in terms of a fixed distance, and this relationship does not change over time. But as shown in Fig. 1(b), the spatial correlation between the four sensors changed during the period from 9 a.m. to 12 p.m., indicating that spatial correlation between different sensors in the real world changes over time. Traditional distance-based static sensor networks cannot represent this complex spatial correlation. In addition, different factors in utility tunnels have temporal features of different periods, and it is difficult for traditional approaches to capture the temporal features of different periods to meet the demand for accurate prediction. To summarise, environmental data from utility tunnels exhibits significant dynamic correlations in both spatial and temporal dimensions. Mining these nonlinear and complex spatio-temporal data, revealing their intrinsic patterns, and making accurate predictions accordingly constitutes a challenging problem.

Extensive efforts have been devoted by many researchers to do time series forecasting by utilizing the data. Earlier, statistical methods [8–10] were utilized for time series forecasting problems. However, it is difficult for them to capture the nonlinear relationship among environmental variables in utility tunnels. Subsequently, more sophisticated machine learning techniques were utilized to model complex data, yet these methods [11,15,16] struggle with grasping the intricate spatio-temporal correlations among sensor networks in utility tunnels. Recently, deep learning methods [3,12,13] have been increasingly utilised by many researchers to manage complex spatio-temporal data with high dimensions, but it is difficult to capture the temporal features of different periods that exist among different environmental factors and fail to reflect the complex dynamic spatial correlations among different environmental factors on static sensor networks.

To address these challenges, this paper presents a novel framework called Graph and Time Convolution-Linear Capture Networks (GTCLNs) to predict environmental factors in utility tunnels. We design the graph generation module and stacked graph convolutional module that do not require a predefined graph structure and can adaptively learn the dynamic spatial correlations among sensors in utility tunnels. In addition, we develop the temporal convolutional module to capture the temporal features of different periods among environmental factors in utility tunnels. Finally, we introduce a linear capture module to capture the trend and seasonal variations among environmental factors in utility tunnels. In summary, our contributions are as follows:

- To the best of our knowledge, we are the first to propose a novel spatio-temporal graph fusion network-based model for multivariate time series prediction of environmental factors in utility tunnels.
- The design GTCLNs is capable of capturing (i) the dynamic spatial correlations and (ii) the temporal features of different periods among environmental factors in utility tunnels. Through the integration of the temporal convolutional module and graph convolutional module, GTCLNs fully exploits the complex spatio-temporal correlations among environmental factors in utility tunnels.
- Experiments utilize the real-world data which contains two different datasets collected from different road sections in Suzhou utility tunnels. The results validate the effectiveness of the model which outperforms existing baseline methods.

2 Problem Formulation

Definition 1. - Graph. In utility tunnels, sensors at different locations form a graph. A graph is formalized as $G = (V, E)$, where V represents the sensors in utility tunnels, and E denotes the set of edges among different sensors.

Definition 2. - Adjacency matrix. The adjacency matrix reflects the strength of the correlation among different sensors in utility tunnels. An adjacency matrix A of size $N \times N$ is employed to describe the relationship among sensors in utility tunnels.

Problem Formulation. In the context of underground utility tunnel environments, different sensors consistently monitors time series data of various gases, temperature, humidity, and water levels at a uniform sampling frequency.

In the context of underground utility tunnel environments, different sensors consistently monitors time series data of various gases, temperature, humidity, and water levels at a uniform sampling frequency. The prediction problem for utility tunnel environments with H variables can be defined as follows: given a series of historical observations over P time steps of environmental data variables in the tunnel, denoted by $X = \{X_{t-P}, X_{t-2}, \ldots, X_{t-1}\}$ with $X \in \mathbb{R}^{H*P}$. A represents the adjacency matrix with $A \in \mathbb{R}^{H*H}$. Our goal is to predict the node values in a multivariate temporal graph $\mathcal{G} = (X, A)$ for the Q-step ahead timestamp, denoted by $\hat{X}_Q$. The result can be inferred by the forecasting model F with parameter Φ and a temporal graph structure $\mathcal{G}$. Equation (1) illustrates our problem definition.

$$\hat{X}_Q = F(X_{t-K}, \ldots, X_{t-1}; \mathcal{G}; \Phi) \tag{1}$$

3 Methodology

3.1 Model Architecture

Firstly, we provide a detailed description of the overarching framework. Our GTCLNs design comprises a Graph Generation Module (GGM), stacked Graph Convolutional Module (GCM) and Temporal Convolutional (TCM) Module, and a Linear Capture Module (LCM), as illustrated in Fig. 2. To uncover the hidden dynamic spatial correlation among different environmental factors in utility tunnels, the GGM computes the graph adjacency matrix, which then serves as the input for the GCM. The GCM and TCM are alternately connected to capture spatio-temporal dependencies among different environmental factors in utility tunnels. Furthermore, to better capture the trend and seasonal variations among environmental factors in utility tunnels, the LCM is introduced. Finally, our input passes through several convolution modules and the linear capture module in parallel, resulting in the final output. More detailed descriptions of the core components of our model will be provided in the following sections.

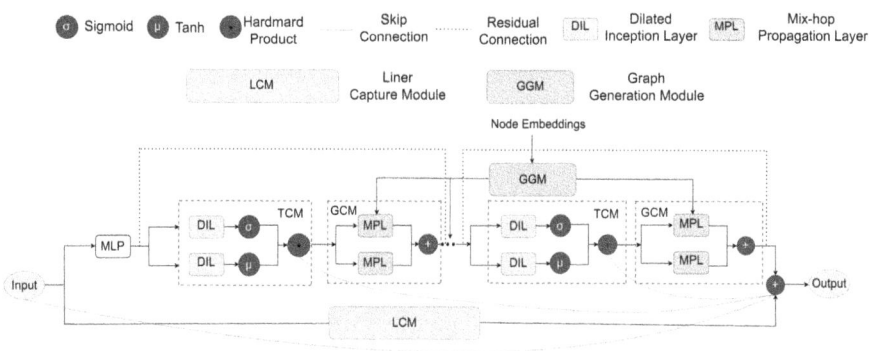

Fig. 2. The overall architecture of graph and temporal convolution-linear capture networks (GTCLNs).

3.2 Graph Generation Module

The graph generation module is capable of adaptively learning the graph adjacency matrix to identify the hidden relationships among environmental factors in utility tunnels. We utilize the node embedding method to construct the graph adjacency matrix, as illustrated in the following equation:

$$C_1 = \tanh(\alpha E_1 \theta_1) \tag{2}$$

$$C_2 = \tanh(\alpha E_2 \theta_2) \tag{3}$$

$$A = \text{ReLU}(\tanh(\alpha(C_1 C_2^T - C_2 C_1^T))) \tag{4}$$

Specifically, E_1 and E_2 denote randomly initialized node embeddings, while θ_1 and θ_2 represent the parameters of the model. The hyperparameter α is employed to modulate the saturation level of the activation function. The function argtopk($-$) is designed to return the indices of the top k environmental factors that have the strongest spatial correlation with the current utility tunnel environmental factors. Equation (4) denotes the adjacency matrix we obtain through node embeddings.

3.3 Temporal Convolution Module

The temporal convolution module consists of two dilated inception layers [15], which are utilized to extract temporal characteristics of different periods among different environmental factors in utility tunnels. For environment variable i in utility tunnels, the dilated convolution is defined as follows:

$$Z_i * f_{1 \times m}(t) = \sum_{\tau=0}^{m-1} f_{1 \times m}(\tau) Z_i(t - s \times \tau) \tag{5}$$

Here the term s denotes the expansion factor and τ represents the offset relative to the current time point t, while Z_i represents the input sequence at a specific layer of environment variable i. The term $f_{1 \times m}$ refers to a one-dimensional convolutional filter kernel of size m.

Additionally, we employ multiple filters of varying sizes to extract temporal characteristics of different periods among different environmental factors in utility tunnels. Consequently, the dilated inception layer is defined as follows:

$$\hat{\xi}_i = \text{concat}(\mathbf{Z}_i * f_{1 \times m_1}, \mathbf{Z}_i * f_{1 \times m_2}, \ldots, \mathbf{Z}_i * f_{1 \times m_\omega}) \tag{6}$$

Here $[m_1, m_2, \ldots, m_\omega]$ represents the sizes of ω distinct filters, with the outputs of these various filters concatenated along the channel dimension. $\hat{\xi}_i$ represents the output of the dilates inception layer. Subsequently, the resulting output is processed through two different activation functions, namely the sigmoid and tanh functions, culminating in an element-wise multiplication to produce the output of the temporal convolution module.

3.4 Graph Convolution Module

The graph convolutional module receives an adjacency matrix generated by the graph generation module and extracts the strength of spatial correlation among different environmental factors from this matrix to achieve better prediction results. Each graph convolution is realised through mixed-hop propagation layer [15], comprising two steps: information propagation and information selection. This propagation is defined as:

$$H^{(k)} = \beta H_{in} + (1-\beta)\hat{A}H^{(k-1)} \quad (7)$$

Here β functions as a hyperparameter, governing the proportion of the current environmental factor's original state that is retained and $1-\beta$ controls the proportion of information the current environment factor receives from its neighboring environment factors when updating its state. H_{in} is the input hidden state from the previous layer's output. The term $\tilde{A}$ is defined as $\tilde{D}^{-1}(A+I)$, where $\tilde{D}_{ii} = 1+\sum_j A_{ij}$. Such a propagation step can help integrate environmental information from sensors in nearby utility tunnels. The information selection step is defined as follows:

$$H_{out} = \sum_{k=0}^{K} H^{(k)} W^{(k)} \quad (8)$$

Here K indicates the depth of propagation, , and H_{out} is the output hidden state for the current layer, initializing with $H^{(0)} = H_{in}$.

3.5 Liner Capture Module

To further capture the trend and seasonal variations among environmental factors in utility tunnels, we have incorporated a linear capture module, conceptualising the final prediction of GTCLNs as a parallel amalgamation of linear and non-linear components. We use the classical autoregressive (AR) component to capture the linear component. For environment variables in utility tunnels, t time steps of environmental data variables in utility tunnels are as follows:

$$X_t = \sum_{j=1}^{p} \phi_i X_{t-j} + \varepsilon_t \quad (9)$$

Here X_t is the value of the time series of environmental data variables in utility tunnels at time t and $\phi_1, \phi_2, \ldots, \phi_p$ are the autoregressive coefficients. ε_t is the white noise error term with mean zero and constant variance. The order p represents the number of lags or past values included in the model.

4 Experiments

4.1 Datasets and Setup

Datasets: The Pip1 dataset, which encompasses environmental data collected by sensors situated at various locations in utility tunnels on Hujin Road, includes temperature, humidity, hydrogen sulphide gas and carbon monoxide. And the Pip2 dataset, which includes similar data in utility tunnels on Chengbei Road. Both datasets were compiled within the timeframe from March 1, 2022, to August 2, 2022. Table 1 provides a concise statistical overview of these datasets. Our goal is to predict the future values of the four variables in utility tunnels. For our evaluation metrics, we employed the Root Relative Squared Error (RSE) and the Empirical Correlation Coefficient (CORR). Lower RSE values indicate superior performance, while higher CORR values suggest enhanced performance.

Table 1. The overall information for two datasets.

Datasets	Nodes	Timesteps	Granularity	Partition	Input Length	Output Length
Pip1	4	9417	20min	6/2/2	168	1
Pip2	4	14869	15min	6/2/2	168	1

Setup: The experiment is conducted 10 times, with the mean of the evaluation metrics reported. The initial learning rate is set at 0.001, with an L2 regularization factor of 0.0001. Following each temporal convolutional layer, dropout is set at 0.3 to prevent overfitting. Layernorm is incorporated subsequent to each graph convolution layer to standardize layer outputs. For the hybrid hop propagation layer, a depth of 2 and a retention rate of 0.05 are set. In the graph learning layer, the saturation rate of the activation function is specified as 3. We employ 5 modules each of graph and temporal convolutions, doubling the dilation index with each iteration. The training runs for 30 epochs. The batch size is set to 4.

4.2 Baseline

- **AR-GARCH** [1]: A time series model that combines autoregressive (AR) and generalized autoregressive conditional heteroskedasticity (GARCH) techniques.
- **VAR-MLP** [11]: A hybrid model that blends vector autoregression with multilayer perceptrons for enhanced time-series forecasting.
- **LSTNet** [2]: A deep learning model that combines CNNs and RNNs to capture complex patterns in time-series data.
- **MTNet** [12]: A deep learning model for forecasting multivariate time-series data, using attention mechanisms and RNNs.

- **TPA-LSTM** [3]: A deep learning model that combines LSTM with attention mechanisms to focus on key temporal patterns in time-series data.
- **STGCN** [14]: A spatio-temporal model, combining graph convolutions and one-dimensional convolutions.
- **MTGNN** [15]: A spatio-temporal graph model that does not require a predefined graph structure.

4.3 Main Results

This research includes a comparative analysis between GTCLNs and other multivariate time series models, where our model is evaluated against seven baseline methods on Pip1 and Pip2. Table 2 presents the average forecasting results across various time horizons.

As can be seen from Table 2, our GTCLNs achieve excellent performance in both datasets for all evaluated metrics. We observe that traditional time series analysis methods often yield unsatisfactory prediction results, indicating their limitations in modelling complex and nonlinear spatio-temporal data. In comparison, deep learning-based methods achieve superior prediction results compared to traditional time series analysis techniques. Among them, models that take into

Table 2. Comparison with baselines on our two datasets.

Dataset		Pip1				Pip2			
		Horizon				Horizon			
Methods	Metrics	3	6	12	24	3	6	12	24
AR-GARCH [1]	RSE	0.0588	0.0681	0.0705	0.0783	0.0398	0.0434	0.0467	0.0412
	CORR	0.7235	0.7023	0.6435	0.5827	0.7823	0.7526	0.7422	0.7023
VARMLP [11]	RSE	0.0632	0.0689	0.0771	0.1022	0.0369	0.0386	0.0478	0.0418
	CORR	0.7416	0.7516	0.6554	0.6102	0.7896	0.7587	0.7432	0.7226
LSTNet [2]	RSE	0.0582	0.0698	0.0740	0.0757	0.0264	0.0342	0.0322	0.0353
	CORR	0.7503	0.7272	0.7370	**0.7236**	0.8093	0.7762	0.7644	0.7441
MTNet [12]	RSE	0.0528	0.0557	**0.0670**	0.0730	0.0223	0.0234	0.0276	0.0302
	CORR	0.7856	0.7505	0.6945	0.6042	0.7978	0.7759	0.7221	0.7034
TPA-LSTM [3]	RSE	0.0532	0.0665	0.0734	0.0742	0.0275	0.0335	0.0325	0.0342
	CORR	0.7630	0.6862	0.5358	0.4413	0.7935	0.7880	0.7532	0.7476
STGCN [14]	RSE	0.0528	0.0608	0.0684	0.0753	0.0224	0.0245	0.0258	0.0302
	CORR	0.7762	0.7462	0.7244	0.6648	0.8065	0.7892	0.7692	0.7464
MTGNN [15]	RSE	0.0518	0.0564	0.0696	0.0736	0.0215	0.0226	0.0245	0.0284
	CORR	0.7830	0.7635	0.7368	0.6913	0.7935	0.7880	0.7765	0.7483
GTCLNs	RSE	**0.0511**	**0.0545**	0.0683	**0.0724**	**0.0198**	**0.0205**	**0.0231**	**0.0269**
	CORR	**0.8028**	**0.7832**	**0.7415**	0.7170	**0.8147**	**0.7928**	**0.7774**	**0.7519**

account spatial correlation, including STGCN, MTGNN, and our model, outperform deep learning models such as LSTNet and TPA-LSTM. And our model GTCLNs also outperforms models that take spatial correlation into account, demonstrating that our model is able to capture dynamic spatial correlation and incorporate temporal characteristics of different periods among different environmental factors, which further subtracts the forecasting error.

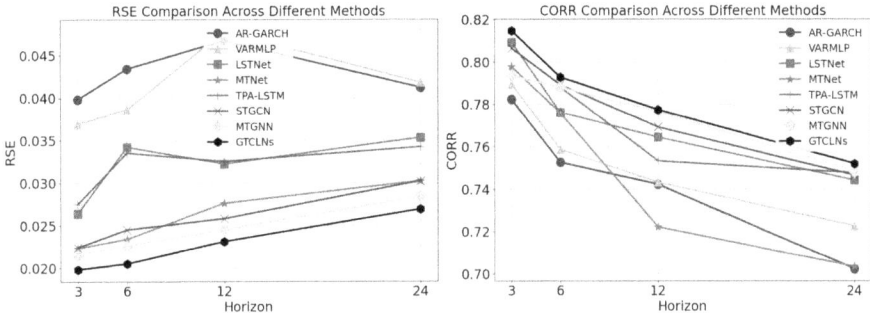

Fig. 3. The prediction performance of different horizons on Pip2 dataset.

Figure 3 more intuitively show the changes in the prediction performance of various methods on the Pip2 datasets as the prediction horizon increases. For the data set, traditional machine learning methods only consider time correlation, and their prediction errors decrease sharply as the time step increases. In contrast, the prediction error of deep learning methods increases slowly with increasing time steps. Our model GTCLNs always achieves the best prediction performance on the Pip2 dataset, which shows that the strategy can better mine the dynamic spatio-temporal patterns of the utility tunnel data.

4.4 Ablation Study

To validate the key components, we performed ablation studies on the Pip2 dataset. We named the variants of GTCLNs as follows:

- **w/o Liner Caputure Moudule:** Removing the linear capture module, the ultimate prediction results are solely constituted by the nonlinear component.
- **w/o Inception:** This variant of GTCLNs lacks the inception component in the dilated inception layer, instead utilizing a sole 1×7 filter while keeping the output channel count unchanged.
- **w/o Dynamic Graph:** Node embedding of dynamic graphs is not performed in the graph generation module; instead, pre-defined static graphs are utilized.

Each experiment is repeated 10 times, with the Root Square Error (RSE) and Correlation (CORR) averages and standard deviations report in Table 3. From this data, it is discernible that the linear capture component plays a significant role in the final prediction outcomes. This is attributed to its parallel integration with nonlinear components, effectively studying the trend and seasonal variations among environmental factors in utility tunnels.

From the experimental data, we can conclude that using multiple filters of different sizes to learn temporal features of different periods among different environmental factors in utility tunnels is better than a single temporal feature learned by a single filter.

Table 3. GTCLNs ablation study on Pip2 dataset.

Method	RSE	CORR
w/o LCM	0.0208 ± 0.0012	0.8075 ± 0.0120
w/o Inception	0.0216 ± 0.0021	0.8045 ± 0.0140
w/o Dynamic Graph	0.0210 ± 0.0023	0.8026 ± 0.0156
GTCLNs	**0.0198 ± 0.0017**	**0.8147 ± 0.0132**

The effectiveness of utilizing predefined static graphs is inferior to that of our model for learning dynamic spatial correlations. This is because there are dynamic spatial correlations among environmental factors in utility tunnels that static graph methods fail to capture.

5 Related Work

5.1 Operation and Maintenance Safety Strategies in Utility Tunnels

In the field of risk assessment for integrated utility tunnels, a novel expert system has been developed that merges color coding, the Delphi method, and the Analytic Hierarchy Process (AHP) to analyze the criticality and threats associated with these tunnels [4]. Furthermore, a fuzzy comprehensive evaluation model based on Bayesian networks [5] has been constructed for disaster risk assessment in integrated utility tunnels, enabling the categorization of disaster risk levels. Additionally, several studies [6,7] have also conducted comprehensive risk evaluations of underground utility tunnels from diverse perspectives, covering comparative analyses of operation and maintenance across different regions and examining various aspects of tunnel operation and maintenance.

5.2 Multivariate Time Series Forecasting

In the field of multivariate time series forecasting, traditional machine learning methods such as Vector Autoregression (VAR) [10], VAR-MLP [11], KNN [16] and SVM [17] have been deployed for multivariate forecasting tasks. However, these conventional approaches struggle to capture the complex non-linear spatio-temporal correlations among multiple variables. Consequently, approaches utilizing deep learning have surfaced as a promising alternative.

LSTNet [2], combining Recurrent Neural Networks (RNNs) and Convolutional Neural Networks (CNNs), further enhanced prediction accuracy, especially in handling long-term dependencies. MTNet [12] incorporates memory networks and attention mechanisms to provide interpretability and address the complexity of multivariate data. Huang et al. [13], employing a dual self-attention mechanism, improved the model's comprehension of time series data, excelling in high-dimensional data processing. Shabani et al. [18] propose a Transformer-based model that iteratively refines time series at multiple scales to better capture temporal dependencies. The work STGCN [14], designed as a multi-convolutional block architecture, models spatial and temporal dependencies with graph convolution and convolutional sequence layers, respectively. MTGNN [15] allows multivariate time series forecasting without predefined graph structures.

6 Conclusion

To enhance the precision and efficiency of environmental safety forecasts in utility tunnels, this study presents GTCLNs, a novel multivariate time series prediction model. GTCLNs effectively learns spatial correlations among environmental factors without requiring a predefined sensor graph. It combines temporal and graph convolution modules to capture complex nonlinear and temporal dynamics. Additionally, a linear capture module assesses trends and seasonal changes. Empirical tests on two datasets confirm that GTCLNs outperform existing methods, suggesting its potential for broader real-world applications in diverse environments.

Acknowledgments. This work was supported in part by the Major Project of Fundamental Research on Frontier Leading Technology of Jiangsu Province under Grant BK20222006.

References

1. Huang, G., Liu, R., Liu, M., et al.: Modeling and simulating nonstationary thunderstorm winds based on multivariate AR-GARCH. J. Wind Eng. Ind. Aerodyn. **211**, 104565 (2021)
2. Lai, G., Chang, W.-C., Yang, Y., Liu, H.: Modeling long-and short-term temporal patterns with deep neural networks. In: Proceedings of SIGIR, pp. 95–104. ACM (2018). https://doi.org/10.1145/3209978.3210006
3. Shih, S.-Y., Sun, F.-K., Lee, H.-y.: Temporal pattern attention for multivariate time series forecasting. Mach. Learn. **108**(8-9), 1421–1441 (2019)
4. Julian, C., Jorge, C., Vicente, C.: Criticality and threat analysis on utility tunnels for planning security policies of utilities in urban underground space. Expert Syst. Appl. **40**(11) (2013)
5. Chen, Y., Li, H., Wang, W., Xue, B., Li, G.: Bayesian network-based disaster risk analysis of integrated pipe corridor operation and maintenance. Saf. Environ. Sci. **18**(6), 2109–2114 (2018)

6. Yuan, X., Xu, S., Lei, L., Xu, R., Jiya, S.: Comparative analysis of integrated pipe corridor operation and maintenance management in Taiwan and mainland China. Municipal Technol. **37**(1), 172–174 (2019)
7. Guo, J., Qian, Y., Wang, Z., Dong, Z., Liu, X.: Research on common operation and maintenance disasters and countermeasures of urban underground integrated pipe corridors. Disaster Sci. **34**(1), 27–33 (2019)
8. Wei, W.W.: Time series analysis. The Oxford Handbook of Quantitative Methods in Psychology **2**, (2006)
9. Williams, B.M., Hoel, L.A.: Modeling and forecasting vehicular traffic flow as a seasonal ARIMA process: theoretical basis and empirical results. J. Transp. Eng. **129**(6), 664–672 (2003)
10. Zivot, E., Wang, J.: Vector autoregressive models for multivariate time series. In: Modeling Financial Time Series With S-Plus, pp. 385–429 (2006)
11. Zhang, G.: Peter: time series forecasting using a hybrid ARIMA and neural network model. Neurocomputing **50**, 159–175 (2003)
12. Chang, Y. Y., Sun, F. Y., Wu, Y. H., et al.: A memory-network based solution for multivariate time-series forecasting. arXiv preprint arXiv:1809.02105 (2018)
13. Huang, S., Wang, D., Wu, X., et al.: DSANet: dual self-attention network for multivariate time series forecasting. In: Proceedings of the 28th ACM International Conference on Information and Knowledge Management, pp. 2129–2132 (2019). https://doi.org/10.1145/3357384.3358132
14. Yu, B., Yin, H., Zhu, Z.: Spatio-temporal graph convolutional networks: a deep learning framework for traffic forecasting. arXiv preprint arXiv:1709.04875 (2017)
15. Wu, Z., Pan, S., Long, G., et al.: Connecting the dots: multivariate time series forecasting with graph neural networks. In: Proceedings of the 26th ACM SIGKDD International Conference on Knowledge Discovery & Data Mining, pp. 753–763 (2020). https://doi.org/10.1145/3394486.3403118
16. Van Lint, J., Van Hinsbergen, C.: Short-term traffic and travel time prediction models. Artif. Intell. Appl. Critical Transp. Issues **22**(1), 22–41 (2012)
17. Jeong, Y.-S., Byon, Y.-J., Castro-Neto, M.M., Easa, S.M.: Supervised weighting-online learning algorithm for short-term traffic flow prediction. IEEE Trans. Intell. Transp. Syst. **14**(4), 1700–1707 (2013)
18. Shabani, A., Abdi, A., Meng, L., Sylvain, T.: Scaleformer: iterative multi-scale refining transformers for time series forecasting. arXiv preprint arXiv:2206.04038 (2022)

Improving Spatial Co-location Pattern Mining with Enhanced Neighbor Relationship Measures

Liang Xu[1], Lizhen Wang[2(✉)], Vanha Tran[3], and Hongmei Chen[1]

[1] School of Information Science and Engineering, Yunnan University, Kunming 650091, China
[2] School of Science and Technology, Dianchi College, Kunming 650228, China
`lzhwang@ynu.edu.cn`
[3] FPT University, Hanoi 155514, Vietnam
`hatv14@fe.edu.vn`

Abstract. The primary objective of mining spatial prevalent co-location patterns (SPCP) is to extract subsets of spatial features from large datasets, which are often found in close geographical neighbor. Traditional SPCP mining techniques have overlooked the intrinsic attributes of spatial instances. This paper investigates the impact of elevation attributes on SPCPs and proposes the ISCPM-ENM (Improved Spatial Co-location Pattern Mining with Enhanced Neighbor Relationship Measures) method, designed to improve the assessment of neighboring relationships among spatial instances. More importantly, this paper proposes a pioneering post-mining methodology that employs ISCPM-ENM to mitigate efficiency impacts when incorporating additional attributes. It also seamlessly integrates elevation attributes into the mining process for experimental purposes. Extensive experiments demonstrate that ISCPM-ENM offers higher accuracy than traditional methods, and the new post-mining method not only preserves time efficiency on par with conventional techniques but also significantly cuts memory usage, reducing it to approximately half that of traditional methods across datasets of any size.

Keywords: Spatial Data Mining · Spatial Co-location Patterns · Neighbor Relations · Elevation · Post-mining

1 Introduction

Spatial data contains richer and more complex semantic information than general data, making it of significant importance in real life. With the development of global satellite positioning systems and remote sensing technology, humans are generating a vast amount of spatial data every day. Discovering interesting knowledge from this massive spatial dataset has become a crucial task. Compared to transactional data mining, mining spatial data is more complex. Spatial data mining aims to extract a large amount of potentially useful information from spatial databases and has been widely applied in fields such as environmental science [1,2], earth science [3], disease prevention and control [4], and e-commerce [5].

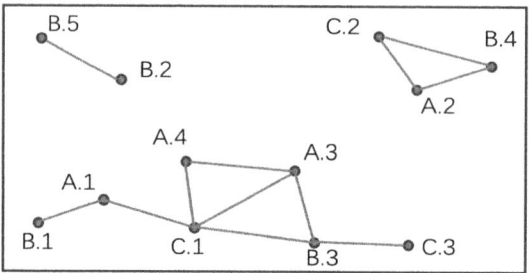

Fig. 1. The spatial instances distribute example

Spatial prevalent co-location patterns (SPCPs) are an important direction in spatial data mining. A spatial co-location pattern is a subset of spatial feature sets whose instances prevalently appear together in each other's neighborhood in space. For example, Fig. 1 shows a spatial instance distribution example, where each instance is represented by its belonging to feature type f, instance number id, and the spatial coordinate information of the instance, e.g., such as instance A.1 representing the first instance of feature A. These instances represent the specific occurrences of certain feature objects in space. For instance, if the places where Tricholoma matsutake grows always have Termites, then a SPCP {Tricholoma matsutake, Termites} can be identified. The recognition of this pattern reveals the potential connections in species distribution within ecosystems. However, to precisely understand and quantify these connections, we need to further explore the proximities among instances. The neighbor relationship between two instances can be determined by different measurement methods according to different application needs, such as Euclidean distance [6] and Manhattan distance. In Fig. 1, two instances connected by a solid line indicate that these two instances satisfy the neighbor relationship.

The prevalence of spatial co-location patterns can be measured by specific numerical indicators or scoring methods. This measurement reflects the prevalence of a particular spatial combination pattern occurring in the dataset. When this indicator exceeds a minimum prevalence threshold given by users, we consider this combination of spatial features to be prevalent, i.e., a SPCP. For example, Fig. 1 shows a dataset containing three features, A, B, and C, with a total of 12 observation instances. We examine the combination of feature set {A, B, C}, and the co-location instances of this feature set are {A.2, B.4, C.2} and {A.3, B.3, C.1} since the instances in each of them form a complete graph with mutual associations between each pair. The strength of this association (measured by a numerical indicator named participation index, PI, which will be introduced later) reaches 0.4. If the minimum prevalence threshold set by a user to determine whether a spatial feature set is prevalent is 0.3, then according to this standard, the specific feature set {A, B, C} can be considered a SPCP.

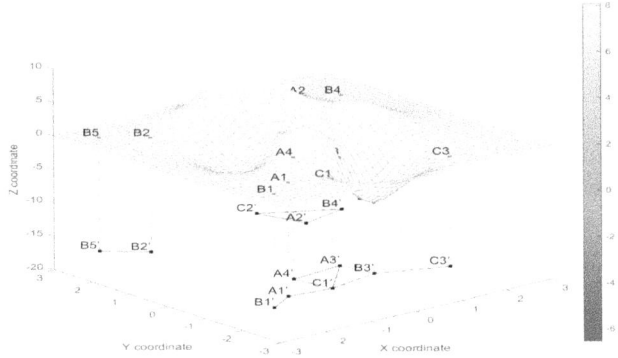

Fig. 2. Example of spatial instance distribution considering elevation

However, traditional spatial co-location pattern SPCP mining algorithms have primarily focused on the spatial feature instances within the same plane, neglecting the distribution of data in three-dimensional space, especially the important dimension of elevation. Such bias may limit our ability to fully understand spatial features, for example, the growth condition of vegetation is closely related to its elevation. As elevation increases, environmental factors such as temperature, light intensity, and rainfall exhibit different trends, which in turn affect vegetation growth, resulting in the so-called "vertical zonation" phenomenon [7]. This phenomenon is characterized by vegetation distribution bands with clear boundaries along the elevation gradient. For instance, Fig. 2 shows an example of spatial instance distribution with elevation attributes, where spatial instances are distributed on a three-dimensional surface. A.1 is the projection of spatial instance A.1 on the XY plane. All spatial data on the three-dimensional surface have the same relative position in their projection on the XY plane as the spatial instances in Fig. 1. Measuring the neighbor relationship between two instances requires using the Euclidean distance in three-dimensional space. When considering the elevation attribute, instances that were previously neighbors may no longer have a neighbor relationship (as shown by the dashed lines connecting instances in Fig. 2). In Fig. 2, the instances of the spatial co-location pattern {A, B, C} become {A.2, B.4, C.2}, and compared to the spatial co-location pattern {A, B, C} shown in Fig. 1, under the same distance threshold conditions, the instance {A.3, B.3, C.1} is missing in Fig. 2, resulting in a reduced association strength of 0.2. When the user-defined threshold for prevalence is 0.3, the pattern {A, B, C} is not considered a SPCP. Thus, considering the elevation attribute of spatial instances can also change the instances of spatial co-location patterns, thereby affecting the results of spatial co-location pattern mining. Therefore, when mining spatial data that is significantly influenced by elevation, such as vegetation distribution, using traditional co-location pattern mining algorithms may encounter the following issue: vegetation that is adjacent in horizontal space but grows at different elevations might be incorrectly classified as being in the same vegetation zone. This could lead to non-existent neighbor relationships being considered as present, causing the prevalence of co-location patterns to be overestimated, thereby misjudging those spatial co-location patterns that should not be considered prevalent.

Given the background and considerations above, this study introduces a novel method, ISCPM-ENM, to enhance spatial co-location pattern mining by incorporating elevation attributes, thereby improving neighbor relationship measures between spatial instances. By integrating elevation into neighbor relationship calculations, our experiments on both synthetic and real datasets show that our method better meets practical needs and offers greater application value. Additionally, this work presents a post-mining method that re-mines traditional spatial patterns with added elevation attributes, significantly boosting time and space efficiency compared to direct mining methods.

The paper is structured as follows: Sect. 2 reviews spatial co-location pattern mining basics and related research. Section 3 provides a detailed exposition of our proposed algorithm and its pseudocode, along with a discussion on the workflow and implementation of the post-mining approach. Section 4 provides experimental analysis, and Sect. 5 concludes the paper and suggests directions for future research.

2 Related Work and Basic Concepts

2.1 Related Work

There are many way to determine the neighbor relationship between spatial instances. Euclidean distance is one of the most commonly used measures of distance, used to measure the straight-line distance between two points in multidimensional space. Methods based on fuzzy theory [8,9] transform neighbor relationships into degrees of nearness, i.e., fuzzy membership degrees, which can alleviate the irrationality of fixed global adjacent distance thresholds to some extent. A method for calculating neighbor membership degrees based on fuzzy theory and d-grid, which can avoid calculating Euclidean distance, was proposed [10]. Adaptive neighbor generation methods represented by Delaunay triangulation and Voronoi diagrams [11] use geometric theory and consider spatial heterogeneity to adaptively generate spatial neighbor relationships. Based on verification statistical methods, it was proposed to use multi-distance mining of co-location patterns throughout the entire research space [12]. A method based on neighbor-based approaches and spatial autocorrelation theory methods using the K function to estimate suitable neighbor distance thresholds was proposed [13]; Qian et al. [14] proposed regional co-location pattern mining based on k-nearest neighbors, using k-nearest neighbors instead of a single distance threshold, but setting an appropriate k value is also very difficult.

2.2 Basic Concepts

Spatial features refer to different types of things in space (for example, supermarkets, restaurants, etc.), and a spatial feature set is a collection of different types of things in space, denoted as $F = \{f_1, f_2, \ldots, f_n\}$. A spatial instance is the occurrence of a spatial feature at a specific location in space, denoted as $S = s_1 \cup s_2 \cup \ldots s_n$, where s_i is the collection of spatial instances for feature f_i. R is the spatial neighbor relationship on S. If the Euclidean distance is used to measure the spatial neighbor relationship, then $R = \{(i_j, i_k)|d(i_j, i_k) \leq \{min\_dist, i_j \in S, i_k \in S\}$, where d represents the Euclidean distance between instances, and min_dist represents the distance threshold given by the user. $cl = (i_1, \ldots, i_k)$ is a group of instances in space. When any two spatial instances in cl satisfy the spatial neighbor relationship R, is called a clique. A spatial co-location pattern c ($c \subseteq F$) is a subset of the spatial feature set F, and the order of c is the number of features in c. A clique cl contains all the spatial features in c and any subset of cl does not contain all the spatial features in c, then cl is called a row instance of c. All row instances of c constitute the table instance $\mathbf{T(c)}$. The participation ratio $PR(c, f_i)$ measures the participation of feature f_i in the k-size spatial co-location pattern $c = f_1, f_2, \ldots, f_k$, $PR(c, f_i) = \frac{Number\ of\ distince\ objects\ of\ f_i\ in\ instances\ of\ C}{Number\ of\ objects\ of\ f_i}$. The minimum value of the participation ratio of all spatial features in spatial co-location pattern c is defined as the participation index PI, i.e., $PI(c) = \min_{f_i \subseteq c} PR(c, f_i)$. When $PI(c)$ is not less than the given minimum prevalence threshold min_prev, the spatial co-location pattern c is considered a SPCP.

Figure 1 shows the distribution of a spatial dataset with three features A, B, and C. A has 4 instances A.1, A.2, A.3, and A.4; B has 5 instances B.1, B.2, B.3, B.4 and B.5; C has 3 instances C.1, C.2 and C.3. Two instances with a spatial neighbor relationship are connected by a line. In Fig. 1, The table instance of the pattern {A,B,C} is {A.2,B.4,C.2},{A.3,B.3,C.1}, $PR\{\{A,B,C\},A\} = \frac{|\{A.2,A.3\}|}{|\{A.1,A.2,A.3,A.4\}|} = \frac{2}{4}$, $PI(\{A,B,C\}) = \min\{PR\{\{A,B,C\},A\}, PR\{\{A,B,C\},B\}, PR\{\{A,B,C\},C\}\} = 0.4$, when min_prev is less than 0.4, $\{A,B,C\}$ is a SPCP.

3 Proposed Algorithm

In this section, we developed two algorithms. First, we introduce the ISCPM-ENM method, which is based on the classic Joinless algorithm. This method incorporates elevation attributes when assessing neighbor relationships and sets them as a baseline. Building on this baseline algorithm, and anticipating the future introduction of more spatial instance attributes with vast data, we implemented post-mining method to reduce the computational burden of adding new attributes, with special consideration given to the elevation attribute.

3.1 Neighbor Relationship Definitions

Definition 1. *Two-dimensional Euclidean distance.*

$$d(i_j, i_k) = \sqrt{(x_{i_j} - x_{i_k})^2 + (y_{i_j} - y_{i_k})^2} \tag{1}$$

Based on the conditions defined by Eq. 1, it is possible to accurately determine whether there is a neighbor relationship between spatial instances. The definition of this neighbor relationship is based on the relative position of instances in two-dimensional space.

Definition 2. *The neighbor relationship R_h for instances in three-dimensional space.*

This paper extends the concept of spatial neighbor relationships in three-dimensional space and defines the neighbor relationship R_h. Within this method, the spatial neighbor relationship R_h in three-dimensional space satisfies Eq. 2, Where $d_h(i_j, i_k)$ satisfies Eq. 3.

$$R_h(i_j, i_k) \Leftrightarrow (d_h(i_j, i_k) \leq min\_dist) \tag{2}$$

$$d_h(i_j, i_k) = \sqrt{(x_{i_j} - x_{i_k})^2 + (y_{i_j} - y_{i_k})^2 + (z_{i_j} - z_{i_k})^2} \tag{3}$$

3.2 ISCPM-ENM Implementation

The ISCPM-ENM employs a star partition model to materialize spatial neighbor relationships, replacing the computationally intensive instance join operations with a lookup mechanism when calculating instances of co-location patterns.

Algorithm 1. ISCPM-ENM

Input:
$F = f_1, f_2, \ldots, f_n$: a set of spatial feature types
S: a set of spatial objects
R_h: a spatial neighbor relationship
min_prev: prevalence threshold
Output:
A set of all co-location rules with participation index
Variables:
$SN = SN_{f_1}, \ldots, SN_{f_n}$: a set of feature f_i star neighborhoods
k: co-location size
C_k: a set of size k candidate colocations
SI_k: a set of star instances of size k candidates
CI_k: a set of clique instances of size k candidates
P_k: a set of size k prevalent colocations
1: $SN = gen\_star\_neighborhoods(F, S, R_h)$;
2: $P_1 = F; k = 2$;
3: **while** not empty P_{k-1} **do**
4:    $C_k = gen\_candidate\_colocations(P_{k-1})$;
5:    **for** i in 1 to n **do**
6:      **for** $t \in SN_{f_i}$ where $f_i = cf_1$, cf_1 is the first feature of $C_k(cf_1, \ldots, cf_k)$ **do**
7:        $SI_k = filter\_star\_instances(C_k, t)$;
8:    **if** $k = 2$ **then**
9:      $CI_k = SI_k$
10:   **else**
11:      $C_k = select\_coarse\_prevalent\_colocations(C_k, SI_k, min\_prev))$;
12:      $CI_k = filter\_clique\_instances(C_k, SI_k))$;
13:   $P_k = select\_prevalent\_colocations(C_k, CI_k, min\_prev)$;
14:   $k = k + 1$;
15: **return** $\cup(P_2, \ldots, P_k)$

The ISCPM-ENM is divided into three stages. The first stage materializes a star neighbor set from the input spatial dataset based on the star neighbor partition model (line1), adopting the Euclidean distance that considers elevation when calculating the neighbor relationships between instances. The second stage collects candidate co-location pattern's star instances from the materialized star neighbor set and implements a rough filtering based on the participation index of the star instances (line4–line12). The third stage filters co-location instances from the star instances, identifies prevalent co-location patterns, and generates co-location rules while updating the flag field of each spatial instance in the spatial instance set (line 13). The second and third stages are repeated as the order of co-location increases. After the mining process concludes, we obtain SPCPs that consider the elevation attribute.

In terms of the correctness and time complexity of the algorithm, this study builds upon Joinless, with a primary improvement in the calculation of neighbor relationships between spatial instances. Therefore, for the time complexity and correctness of this method, one can refer to the literature related to Joinless.

3.3 Post-Mining Method

Lemma 1. *If $d_h(i_j, i_k) \geq d(i_j, i_k)$ and $d(i_j, i_k) > min\_dist$, then $d_h(i_j, i_k) > min\_dist$*

Proof. Based on Definition 1 and Definition 2, we can come to the conclusion: $d(i_j, i_k)^2 = (x_{i_j} - x_{i_j})^2 + (y_{i_j} - y_{i_k})^2$, $d_h(i_j, i_k)^2 = (x_{i_j} - x_{i_j})^2 + (y_{i_j} - y_{i_k})^2 + (z_{i_j} - z_{i_k})^2$. Because of $(z_{i_j} - z_{i_k})^2 \geq 0$, and if and only if $z_{i_j} \neq z_{i_k}, (z_{i_j} - z_{i_k})^2 > 0$. Therefore, $d_h(i_j, i_k)^2 = d(i_j, i_k)^2 + (z_{i_j} - z_{i_k})^2 > d(i_j, i_k)^2$. Take the square root on both sides, we can get the conclusion that $d_h(i_j, i_k) \geq d(i_j, i_k)$, and if $d(i_j, i_k) > min\_dist$, then $d_h(i_j, i_k) > min\_dist$.

Definition 3. *Spatial Instances with Flag.*

$$Spatial\ Instance\ = <Spatial\ Feature, InstanceID, Position, Flag>$$

Based on Lemma 1, we designed a labeled spatial feature instance storage structure as Definition 3. After the first mining, all instances that do not satisfy the neighbor relationship threshold min_dist with other instances are marked as False. This is done in order to exclude them from the second mining process, ensuring they no longer participate in the computation.

Figure 3 presents the post-mining method proposed in this paper. Within the post-mining method, the process initially enters the first mining stage (as shown in Fig. 3, part 1), where, if it is found that a spatial instance does not form any instance of a co-location pattern, the label of that instance is updated to False (as shown in Fig. 3, part 2). After the first mining is completed, we obtain a spatial dataset with labels. Then, if an elevation attribute is added to the dataset (as shown in Fig. 3, part 3), in the subsequent mining process, the

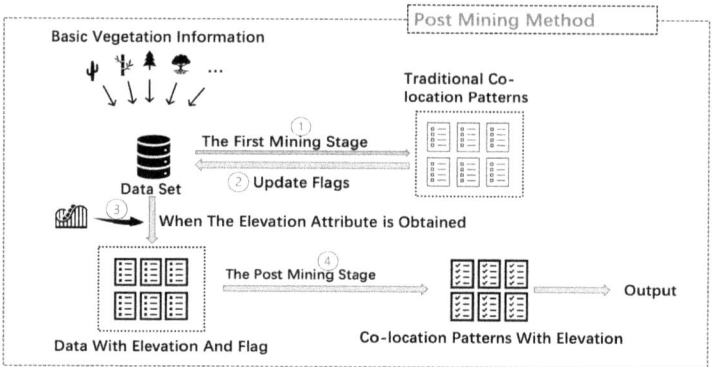

Fig. 3. Post-mining method

Algorithm 2. select_prevalent_colodations(C_k, CI_k, min_prev)

Input: C_k: a set of size k candidate colocations
CI_k: a set of clique instances of size k candidates
min_prev: prevalence threshold
Output: $Instance$: the instances after update the Flag
P_k: a set of size k prevalent colocations
1: $pariticipationIndex$ = calculateParticipation(C_k, CI_k)
2: **if** $pariticipationIndex > min\_prev$ **then** $P_k = C_k$
3:    **for** $Instance\ in C_k$) **do**
4:       Flag in Instance = true;
5: **return** $P_k, Instance$

---

post-mining method will first check whether each instance is marked with False. This labeling mechanism allows the algorithm to first filter all spatial instances, and those marked as False will no longer be included in the computation of co-location patterns (as shown in Fig. 3, part 4). In ISCPM-ENM's line 13, we update every instances's Flag.

For each candidate pattern, we calculate its PI value (line 1). Subsequently, for each candidate pattern that meets the minimum participation threshold, we update its Flag to true (lines 2, 3 and 4). Finally, we return the current k-size prevalent patterns and the updated spatial instance information (line 5). In this code segment, we focus on analyzing and processing the time complexity of the search and marking process for instances of k-size prevalent patterns. Considering that instances of n k-size prevalent patterns need to be processed, we employ binary search to mark each prevalent pattern instance. The time complexity of binary search is $O(\log n)$, and the search for an individual instance can be completed in logarithmic time. When facing m prevalent patterns that need this operation, the overall time complexity is $O(m \log n)$, where m represents the number of prevalent patterns, and n represents the number of k-size prevalent pattern instances.

4 Experiments

We conducted a comprehensive and detailed comparative analysis of ISCPM-ENM and post-mining method against the traditional Joinless. The concept proposed in this paper is theoretically applicable to all subsequent algorithms that require the computation of neighbor relationships. Given that this paper is based on Joinless, we primarily compared it with Joinless. All experiments were conducted on an Apple Silicon M1 processor with 64 GB of memory, and were compiled and run using C++17.

4.1 Experimental Dataset

Synthetic Dataset. Based on the two-dimensional synthetic dataset generation method described in [15], this paper extends it to three-dimensional space.

1. First, n_{parent} parent nodes are randomly generated within a three-dimensional space with a volume of 100*100*100, as shown in Fig. 4(a).
2. Subsequently, centered on each parent node, n_{child}^i child nodes are generated around it within a radius of r_{child}, as demonstrated in Fig. 4(b).
3. Next, centered on each child node, $n_{grandchild}$ grandchild nodes are generated within a radius of $r_{grandchild}$, and all child nodes are removed, as shown in Fig. 4(c).
4. Finally, centered on each parent node, $n_{interference}$ interference nodes are generated outside a radius of $r_{interference}$ (which is the sum of r_{child} and $r_{grandchild}$), as shown in Fig. 4(d).

Throughout the generation process, n_{parent} determines the size of the simulation range, r_{child} and n_{child}^i together dictate the density of instances within each area. Additionally, $r_{grandchild}$ and $n_{grandchild}$ control the number of features and the distance between instances.

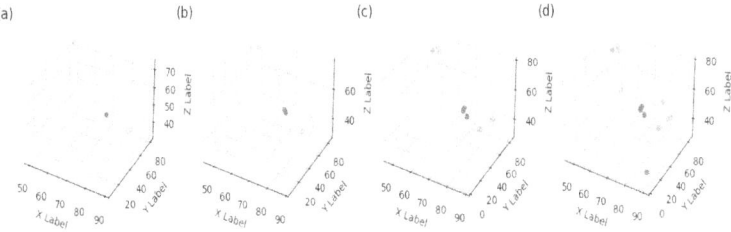

Fig. 4. Synthetic dataset generate

In this study, we defined Interference Patterns(IP) as specific patterns that, after considering the height attribute of spatial objects, should have been excluded from the mining results.

Real Dataset. This paper employs the Gaoligong Mountain vegetation dataset as a real-world ecological data source [3]. The dataset encompasses 25 different features and contains a total of 13,349 data instances. The experimental design randomly selects 2000–10000 samples from the real dataset, generating 9 subsets of the real dataset, with each subset having a difference of 1000 samples between them. Such an experimental design aims to determine the proposed algorithms' ability to process data of varying sizes, thereby comprehensively evaluating the algorithms' performance in handling datasets of different sizes.

4.2 Comparision of Accuracy

First, we compare the accuracy of mining results on a synthetic dataset. The accuracy Precision = CP/(CP + IP), where CP (Correct Patterns) represents the number of true co-location patterns predefined. By comparing the accuracy of ISCPM-ENM and Joinless under the influence of different numbers of interfering features (IP), we noticed that the accuracy of Joinless shows a downward trend as the number of interfering features increases. This result can be interpreted as: Joinless's method of calculating neighbor relationships ignores the attribute of instances at different elevations, is limited to analyzing instance distributions in two-dimensional space. In contrast, ISCPM-ENM incorporates elevation as a parameter when assessing the neighbor relationship between instances, which significantly mitigates the adverse effects brought by interfering features. Figure 5 shows the curve of changes in accuracy between ISCPM-ENM and Joinless as the number of interfering characteristics increases. It is observed that as the number of interference patterns rises, the accuracy of Joinless steadily declines. In contrast, our method maintains its accuracy consistently at a level of 1.

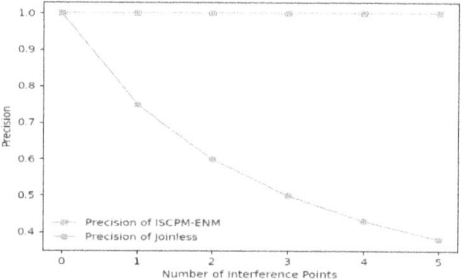

Fig. 5. The accuracy of mining results on a synthetic dataset

Secondly, we analyze the performance of ISCPM-ENM and Joinless on a real dataset. Table 1 shows the comparison data of participation rates in the mining results of Joinless versus ISCPM-ENM, using 3000 randomly selected instances from the Gongshan vegetation dataset, with a distance threshold of 2000 and a minimum participation rate threshold of 0.22. The second and third columns of the experimental results display the comparison of participation rates when the same co-location patterns are discovered using Joinless and ISCPM-ENM, respectively. Rows marked with "-" indicate that no co-location patterns meeting the minimum participation rate threshold were found. The results reveal differences in participation rates between Joinless and ISCPM-ENM when mining the same co-location patterns. Generally, ISCPM-ENM results in lower participation rates than Joinless, primarily because Joinless considers only the horizontal Euclidean distance between spatial instances, which must be within a set threshold. However, with the elevation attribute introduced, the same instances might exceed this threshold in ISCPM-ENM, leading to fewer instances in co-location patterns and thus impacting participation rates. Larger elevation differences more significantly impact neighbor relationship calculations. It is noted that ISCPM-ENM did not mine the Oak Species, Alder and Fir, Mixed Shrubbery patterns at the set participation rate threshold of 0.22, because all row instances of these two patterns, after considering the elevation attribute, no longer meet the minimum participation rate threshold. Figure 6 illustrates the difference in the number of row instances when the same co-location patterns are mined by Joinless and ISCPM-ENM.

Table 1. Comparison of Mining Results between Joinless and The ISCPM-ENM on Real Datasets.

Co-location	ID	Joinless's PI	ISCPM-ENM's PI
{Yunnan Pine, Other Broadleaf Forests}	01	0.667411	0.651786
{Fir, Hemlock}	02	0.557099	0.527778
{Other Broadleaf Forests, Alder}	03	0.517857	0.513393
{Yunnan Pine, Alder}	04	0.490868	0.474886
{Other Broadleaf Forests, Hemlock}	05	0.440546	0.423002
{Yunnan Pine, Hemlock}	06	0.358674	0.342593
{Fir, Arrow Bamboo}	07	0.353395	0.341131
{Sichuan Pepper, Iron Walnut}	08	0.333333	0.333333
{Mixed Shrubbery, Arrow Bamboo}	09	0.333333	0.32197
{Yunnan Pine, Other Broadleaf Forests, Alder}	10	0.332589	0.314732
{Yunnan Pine, Mixed Shrubbery}	11	0.285714	0.268707
{Other Broadleaf Forests, Fir}	12	0.259259	0.241497
{Other Broadleaf Forests, Mixed Shrubbery}	13	0.255102	0.239766
{Mixed Shrubbery, Hemlock}	14	0.251462	0.238839
{Oak Species, Alder}	15	0.235294	–
{Fir, Mixed Shrubbery}	16	0.228395	–

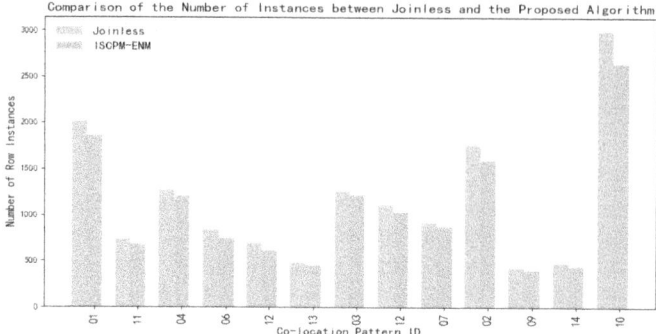

Fig. 6. Comparison of the row instances between Joinless and ISCPM-ENM

As can be observed from Fig. 6, the number of row instances obtained when mining co-location patterns (represented by ID in Fig. 6)using ISCPM-ENM is always less than or equal to the number of row instances obtained under the same co-location patterns by Joinless. This indicates that some row instances, which should not appear, are filtered out after considering the elevation attribute. Furthermore, as shown in Fig. 7(b), based on the results of Joinless, the spatial neighbor relationship between Yunnan pine and Other Broadleaf Forests is significantly identified. However, in our algorithm evaluation, when using ISCPM-ENM, this neighbor relationship is deemed non-existent. The different interpretations of spatial feature associations by these two algorithms are represented in the illustrations by dashed lines, highlighting the difference in algorithm analysis results when considering additional attribute factors. The impact of incorporating the elevation attribute of spatial objects into the neighbor relationship calculation is significant.

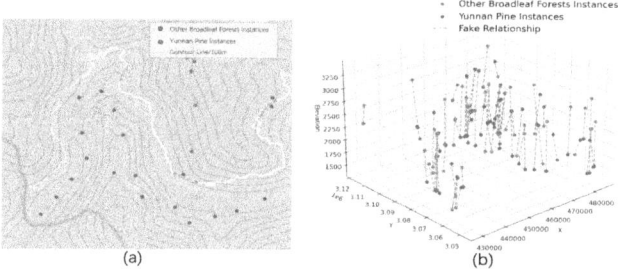

Fig. 7. (a) The distribution of Yunnan Pine and Other Broadleaf Forests in real space on a real dataset (part). (b) The fake adjacency relationship of the Yunnan Pine, Other Broadleaf Forests pattern.

4.3 Comparison of Time Consumption

This section delves into the efficiency of the ISCPM-ENM, Joinless, and the post-mining algorithm proposed in this study when processing different datasets.

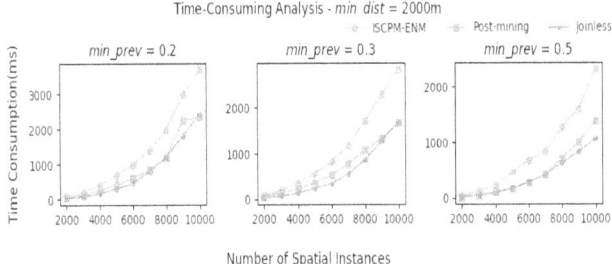

Fig. 8. Effect of the number of instances when $min\_prev = 0.2$, 0.3, and 0.5.

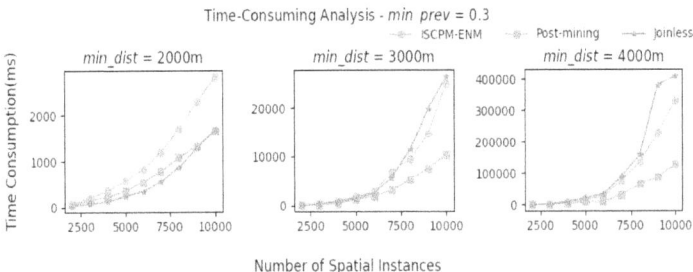

Fig. 9. Effect of the number of instances when $min\_dist = 2000$ m, 3000 m, and 4000 m.

As shown in Fig. 8, when calculating the neighbor relationship, ISCPM-ENM, by considering elevation, takes longer than Joinless for similar-sized datasets. However, when the data processing is conducted using the post-mining method proposed in this paper, its time efficiency is comparable to that of Joinless. The underlying reason for this phenomenon is that, after the initial mining phase, the post-mining method has effectively excluded spatial data instances irrelevant to the mining task. Therefore, when introducing the elevation attribute for subsequent mining in the dataset, there is no need to consider these excluded data instances again. Figure 9 shows that as the min_dist increases, the gap in time consumption between ISCPM-ENM and Joinless gradually decreases. At larger min_dist and on larger datasets, the time consumption of ISCPM-ENM is even lower than that of Joinless. This is because, as the min_dist increases, the number of table instances that need to be generated by Joinless increases exponentially, while the method using ISCPM-ENM, due to the added calculation of elevation, generates fewer table instances at the same min_dist. The post-mining method is faster than both ISCPM-ENM and Joinless. It shows that adding elevation to the mining process doesn't greatly increase time compared to Joinless, highlighting the secondary method's significant efficiency advantage over both.

4.4 Comparison of Memory Consumption

To comprehensively evaluate the performance of the algorithm, we also compared the peak memory usage during the execution of the three algorithm to measure the efficiency of memory usage.

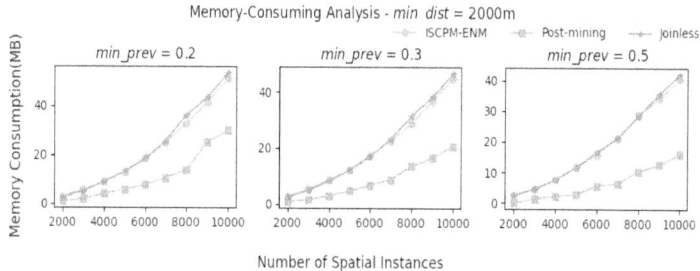

Fig. 10. Memory consumption of the number of instances when $min\_prev = 0.2, 0.3$, and 0.5.

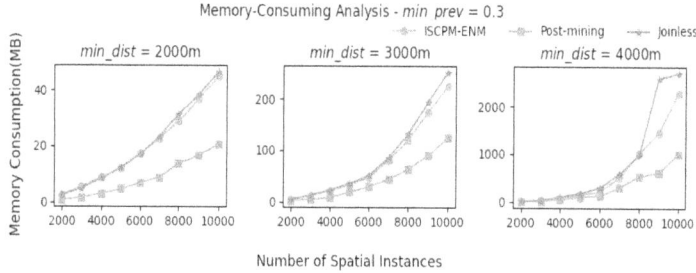

Fig. 11. Memory consumption of the number of instances when $min\_dist = 2000$ m, 3000 m, and 4000 m.

Figure 10 shows the memory consumption for Joinless, ISCPM-ENM, and the post-mining process as the dataset size increases, under the same min_dist but with different min_prev values. The results demonstrate that the memory usage of ISCPM-ENM is close to that of Joinless, even with the addition of computations for the elevation dimension. This similarity in memory consumption is attributed to the improved calculation of neighbor relationships in ISCPM-ENM, which does not result in more table instances during the candidate pattern generation phase than with Joinless. Since the main source of memory consumption in the algorithm is this phase, the memory consumption of ISCPM-ENM is equal to or less than that of Joinless. The memory required for the post-mining process is only half of that required by the first two methods. This reduction is due to the dataset being flagged before the post-mining; the process excludes spatial instances marked as false, significantly reducing memory overhead during candidate pattern generation.

Figure 11 illustrates the change in peak memory consumption of the three methods with the increase in dataset size, under the same min_prev values but with different min_dist. Consistent with the previous analysis, the memory consumption of the post-mining method remains superior to both ISCPM-ENM and Joinless.

5 Conclusion

This paper addresses a key issue in the field of co-location pattern mining: the traditional algorithms have not fully considered the impact of the elevation attribute of spatial instances on pattern mining. Through the introduction of an improved neighbor relationship calculation method, this paper successfully implements a co-location pattern mining method that improves the measurement of neighbor relationships between spatial instances. This method effectively mines subsets of spatial features significantly influenced by the elevation attribute, based on considering the elevation attribute of spatial instances. The experimental results show that the method proposed in this paper not only improves the interpretability of the results but also enhances their practicality. Furthermore, this paper proposes a post-mining method that enhances mining efficiency and spatio-temporal efficiency by adding the elevation attribute to existing data for post-mining. The introduction of this method not only provides a new perspective on the application of the elevation attribute in co-location pattern mining but also offers a new methodology for research in related fields. Future work could further explore the impact of other spatial attributes on co-location pattern mining and how the methods of this paper can be applied to a broader range of spatial data analysis scenarios to promote the development and application of spatial data mining technology.

Acknowledgments. This work was supported by the National Natural Science Foundation of China (62276227, 62266050, 62306266), the Yunnan Fundamental Research Projects (202201AS070015, 202401AT070450), and the Postgraduate Research and Innovation Foundation of Yunnan University (KC-23235527, TM-23236919).

References

1. Yao, X., Jiang, X., Wang, D., Yang, L., Peng, L., Chi, T.: Efficiently mining maximal co-locations in a spatial continuous field under directed road networks. Inf. Sci. **542**, 357–379 (2021)
2. Li, J., Wang, L., Yang, P., Zhou, L.: A novel algorithm for efficiently mining spatial multi-level co-location patterns. IEEE Trans. Knowl. Data Eng. (2024)
3. Wang, D., Wang, L., Jiang, X., Yang, P.: Rcpm_cfi: a regional core pattern mining method based on core feature influence. Inf. Sci. **658**, 119895 (2024)
4. Qiu, P., Gong, Y., Zhao, Y., Cao, L., Zhang, C., Dong, X.: An efficient method for modeling nonoccurring behaviors by negative sequential patterns with loose constraints. IEEE Trans. Neural Netw. Learn. Syst. **34**(4), 1864–1878 (2021)
5. Gao, X., Gong, Y., Xu, T., Lü, J., Zhao, Y., Dong, X.: Toward better structure and constraint to mine negative sequential patterns. IEEE Trans. Neural Netw. Learn. Syst. **34**(2), 571–585 (2020)
6. Shekhar, S., Huang, Y.: Discovering spatial co-location patterns: a summary of results. In: International Symposium on Spatial and Temporal Databases, pp. 236–256. Springer (2001)
7. Peng, H., Xia, H., Chen, H., Zhi, P., Xu, Z.: Spatial variation characteristics of vegetation phenology and its influencing factors in the subtropical monsoon climate region of southern china. PLoS ONE **16**(4), e0250825 (2021)
8. Wang, X., Lei, L., Wang, L., Yang, P., Chen, H.: Spatial colocation pattern discovery incorporating fuzzy theory. IEEE Trans. Fuzzy Syst. **30**(6), 2055–2072 (2022)
9. Hu, Z., Wang, L., Tran, V., Chen, H.: Efficiently mining spatial co-location patterns utilizing fuzzy grid cliques. Inf. Sci. **592**, 361–388 (2022)
10. Li, J., Wang, L., Chen, H.: dgridtopk-fcpm: a top-k spatial co-location pattern mining algorithm based on fuzzy theory and d-grids. J. Tsinghua Univ. (Sci. Technol.) **61**(9), 943–952 (2021)
11. Tran, V., Wang, L.: Delaunay triangulation-based spatial colocation pattern mining without distance thresholds. Stat. Anal. Data Min. ASA Data Sci. J. **13**(3), 282–304 (2020)
12. Barua, S., Sander, J.: Mining statistically sound co-location patterns at multiple distances. In: Proceedings of the 26th International Conference on Scientific and Statistical Database Management, pp. 1–12 (2014)
13. Yoo, J.S., Bow, M.: Mining spatial colocation patterns: a different framework. Data Min. Knowl. Disc. **24**, 159–194 (2012)
14. Qian, F., Chiew, K., He, Q., Huang, H.: Mining regional co-location patterns with k nng. J. Intell. Inf. Syst. **42**, 485–505 (2014)
15. Deng, M., Cai, J., Liu, Q., He, Z., Tang, J.: Multi-level method for discovery of regional co-location patterns. Int. J. Geogr. Inf. Sci. **31**(9), 1846–1870 (2017)

PFG: Generation of Paper-Style Handwritten Formulas for Enhancing Handwritten Mathematical Expression Recognition

Ze Liu, Kai Zhang[✉], Yanghai Zhang, Zhe Yang, Qi Liu, and Enhong Chen

State Key Laboratory of Cognitive Intelligence, University of Science and Technology of China, Hefei, China
kkzhang08@ustc.edu.cn

Abstract. As formula recognition models grow in complexity, the demand for handwritten formula data has surged. To address this, formula image generation models have been proposed to convert LaTeX printed formulas into handwritten form. However, the leading formula generation model, FormulaGAN, is limited to generating handwritten formulas without paper texture. This paper introduces a generative model called 'Paper-Style FormulaGAN' (PFG). By employing self-attention mechanisms and Vision Transformer, PFG successfully generates handwritten formulas with paper texture, expanding the dataset to enhance the performance of formula recognition models in practical scenarios. Experimental results demonstrate that formulas generated by PFG closely resemble real handwritten formulas compared to FormulaGAN. Moreover, recognition models trained with synthetic data from PFG outperform models trained with data from other methods across multiple evaluation metrics.

Keywords: generation model · formula recognition · data generation

1 Introduction

Recent studies by Sun et al. [28] have shown that increasing samples size significantly enhances deep neural network performance. Successful architectures like ResNet [11] and DenseNet [13] rely on large-scale labeled datasets like ImageNet [5], while newer models such as BERT [8], GPT [2], and LLaMA [31] require even larger datasets. Techniques such as weight decay [17] and batch normalization [14] help mitigate data scarcity but are not fundamental solutions. Transfer learning transfers knowledge from a data-rich source domain to a data-scarce target domain. For instance, Zhang et al. proposed IATN [41] and Eatn [40], which used transfer Learning for cross-domain sentiment transfer, effectively addressing data scarcity in the target domain.

Formula recognition systems require extensive labeled handwritten formula data, which is more challenging and costly to obtain compared to printed formula datasets. Existing datasets like CROHME [23] and HME-100K [18] offer handwritten data but suffer from variability in font and writing style. To address this, Matthias Springstein et al. proposed FormulaGAN [27], an attention-based GAN that converts printed formulas into handwritten form, effectively aiding in model training. However, FormulaGAN-generated formulas lack the paper texture (see Fig. 1) found in real handwritten formulas, posing a limitation for real-world applications.

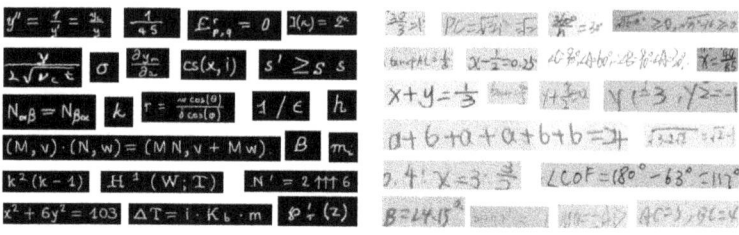

(a) Formulas without paper texture (b) Formulas with paper texture

Fig. 1. Handwritten Formulas without(a) and with(b) paper texture.

This paper introduces a generative model named 'Paper-Style FormulaGAN' (PFG), designed to produce handwritten formulas with paper texture, thereby improving the accuracy of formula recognition models through additional training data. PFG operates by first converting printed formulas into a handwritten style, followed by rendering these with paper texture (see Fig. 2).

$$p, q, r \in C(a, \infty)$$
$$A(x, y) \to B(y, z)$$
$$(a, 1) : x \mapsto a - x$$
$$x^2 + 76x + 1444$$
$$x = ay^2 + by + c$$

Task 1 →

$$p, q, r \in C(a, \infty)$$
$$A(x, y) \to B(y, z)$$
$$(a, 1) : x \mapsto a - x$$
$$x^2 + 76x + 1444$$
$$x = ay^2 + by + c$$

Task 2 →

$$p, q, r \in C(a, \infty)$$
$$A(x, y) \to B(y, z)$$
$$(a, 1) : x \mapsto a - x$$
$$x^2 + 76x + 1444$$
$$x = ay^2 + by + c$$

Fig. 2. Convert printed formulas into handwritten form (Task 1); Apply paper texture to formula images (Task 2).

Experiments reveal that formulas generated by PFG more closely resemble actual handwritten formulas than those by FormulaGAN (see Table 1), leading to superior performance of recognition models trained with PFG data across different metrics (see Table 2).

2 Related Work

2.1 Generative Adversarial Networks

Generative Adversarial Networks (GANs) [10] have been successful in several computer vision applications, including image generation [22,26], image translation [4,45], and face synthesis [16,25]. A GAN comprises a generator, producing samples, and a discriminator, distinguishing between real and fake samples. The effectiveness of GANs is primarily due to the 'Adversarial Loss' [10], which optimizes the generator and discriminator's performance.

Image-to-image translation using GANs transforms images from one domain to another. Pix2pix [15] relies on CGANs [22] and requires paired training data, which can be challenging to obtain. To address this issue, Zhu et al. introduced CycleGAN [45], which performs unpaired image translation using generators G and F and discriminators DX and DY. CycleGAN incorporates 'Cycle Consistency Loss' [45] to ensure that key attributes are preserved during style transfer.

Attention modules have been widely applied in computer vision [34]. To improve high-level semantic understanding in CycleGAN, researchers are integrating attention mechanisms [29]. Liang et al. introduced ContrastGAN [19], which uses object masks as additional inputs. Chen et al. [3] added an attention network to create attention maps that guide image translation. Hao Tang et al. [30] developed AttentionGAN, which generates multiple attention masks for both the foreground and background, effectively segmenting the input image into distinct layers.

2.2 Formula Recognition

Formula recognition methods primarily use an encoder-decoder architecture with attention mechanisms. The encoder extracts key features, while the decoder predicts token sequences from these features. Deng et al. [7] first proposed this approach, training models on 100,000 formulas extracted from scientific publications. They further enhanced it with RNNs to capture horizontal and vertical embeddings [6]. Zhang et al. [43] used multi-scale convolutional neural networks and incorporated two attention layers. Wang et al. [33] introduced a model combining CNNs with stacked LSTM decoders and positional embeddings. Furthermore, Yang et al. [35] proposed model GAP based on the Encoder-Decoder model, which significantly improves the recognition capability of multi-line formulas.

Zhang et al. [38] developed an attention-based model using DenseNet as the encoder and introduced a multi-scale attention mechanism to address the recognition challenges of mathematical symbols at different scales. Yuan et al. [36] integrated syntactic information into the recognition model by transforming LaTeX markup sequences into parsing trees using grammar rules, capturing the syntactic context effectively. Observing that traditional encoder-decoder models inadequately represent tree-like annotations, Zhang et al. [39] proposed a tree-structured decoder specifically for generating tree-like annotations.

Zhao et al. [44] introduced a Transformer-based model named 'Comer.' They developed an innovative Attention Refinement Module (ARM) that adjusts attention weights using historical alignment data while maintaining parallel processing capabilities. Additionally, they introduced self-coverage and cross-coverage concepts to leverage alignment information from both the current and previous layers.

2.3 Handwritten Formula Generation

Matthias Springstein et al. [27] introduced FormulaGAN, an attention-based GAN designed to convert printed formulas into handwritten ones. This system generates a vast dataset of synthetic samples, including tens of thousands of mathematical expressions. Experiments [27] demonstrate that these synthetic datasets are effective for pre-training and developing complex models.

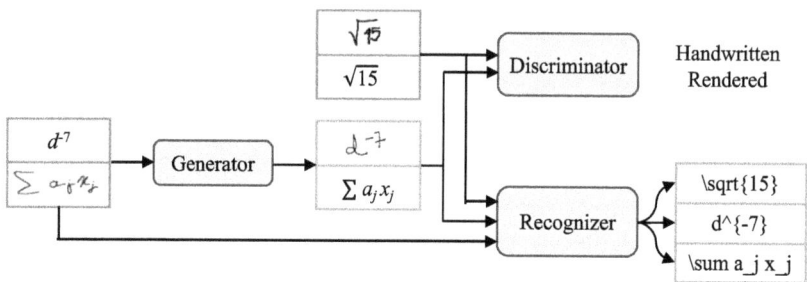

Fig. 3. The Structure of FormulaGAN.

FormulaGAN marks a significant breakthrough in formula recognition, particularly by addressing the scarcity of large-scale datasets for handwritten formulas. As the size of these synthetic datasets expands, new opportunities emerge for advancing research and applications in handwritten mathematical formula recognition.

3 Paper-Style FormulaGAN (PFG)

Directly converting printed formulas into handwritten forms with paper texture in a single step is highly challenging and often unsuccessful. Therefore, this task is divided into two sub-tasks (see Fig. 2). For Task 1, FormulaGAN [27] is utilized, as illustrated in Fig. 3. FormulaGAN comprises a Generator, a Discriminator, and a Recognizer. The Generator transforms the input LaTeX image into its corresponding handwritten form. The Discriminator evaluates the similarity of the generated output to real handwritten formulas, while the Recognizer ensures the accuracy of the converted content by comparing it to the original LaTeX.

Task 2 aims to integrate paper texture onto the data generated by FormulaGAN. The upcoming section will utilize CycleGAN as the framework for developing suitable generators and discriminators. The detailed design process for both components will be outlined below.

3.1 Generator Construction

To achieve a paper texture on formula images, the model must consider multiple layers. Treating the input image as a single layer can cause confusion between the background and the formula (see Fig. 4). Inspired by the self-attention mechanism [29] and AttentionGAN [30], the Generators in Task 2 treat each channel of the feature map as a token. Below, we detail the structure of Generator G (see Fig. 5), noting that Generators G and F share the same design.

Fig. 4. The model struggled to distinguish between the background and the formula when treating the input image as a single layer.

As shown in Fig. 5, the generator operates in the following manner: Initially, the input image x is processed by an Encoder using a ResNet architecture, producing a feature map F. The content decoder extracts essential information from F, combining it with x to form a content matrix C. Simultaneously, F is inputted into the attention mask decoder to compute attention weights, resulting in an attention mask A. A modulates C via element-wise multiplication, generating the masked content MC. Finally, MC is summed to yield the output of generator, $G(x)$, which represents the transformed image with a paper texture.

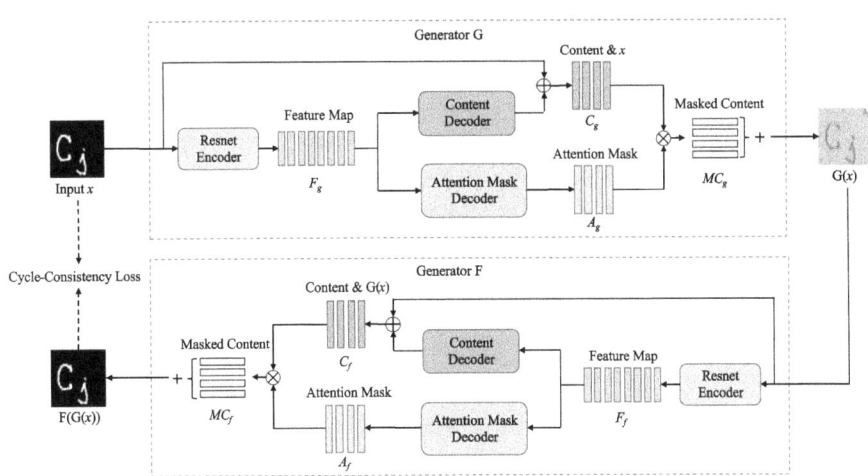

Fig. 5. The structure of the Generator.

3.2 Discriminator Construction

Formula images contain not only background details but also intricate stroke sequences (see Fig. 1). PatchGAN Discriminators (PGD) [15], commonly used in traditional CycleGAN framework, excel at recognizing local image features (see Fig. 6), but they are less effective at capturing the sequential information present in formula images.

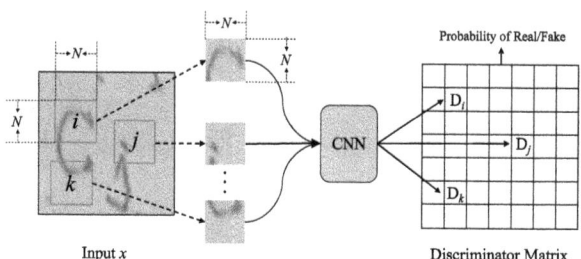

Fig. 6. The structure of PGD.

We introduced a Vit [9] based discriminator, depicted in Fig. 7. The performance of models can be significantly improved by processing data in chunks. For instance, Zhang et al. proposed DR-BERT [42], which enhances model performance by understanding semantics in chunks. The discriminator design proceeds as follows: the input formula image x is segmented into smaller tokens, each representing a distinct segment. These tokens undergo processing by a Transformer encoder, followed by a MLP that computes the likelihood of each segment being real or fake. The output of discriminator provides the probability of each segment being Real/Fake, akin to the output format of a PGD.

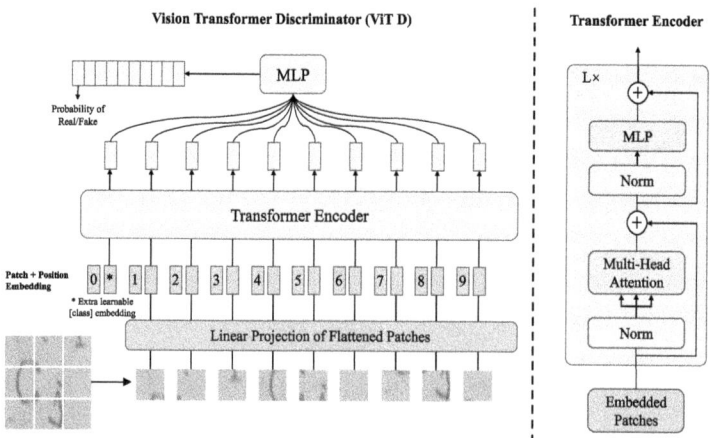

Fig. 7. The structure of the Vision Transformer Discriminator.

The discriminator is designed to output a set of vectors rather than a single value to assess the authenticity of various subregions. To enhance overall image recognition capability, we add a learnable class token (token 0 in Fig. 7) to the discriminator. Notably, the built-in self-attention mechanism in Vit [9] improves our ability of discriminator to recognize stroke sequences within formula images.

3.3 Loss Function

PFG comprises a pair of generators G and F alongside their respective adversarial discriminators DX and DY. Consider two distinct image domains, X and Y, with training images $x \in X$ and $y \in Y$. The generator G translates images from the source domain X to the target domain Y, aiming to deceive the discriminator DY, which, in turn, enhances its ability to differentiate between generated images $G(x)$ and real images y. Similarly, the generator F and discriminator DX engage in the same adversarial process.

Adversarial loss [10] is applied to both generators G and F, as well as their corresponding discriminators DX and DY. The objective function is defined as Eq. 3.

$$L_{GANreal} = \mathbb{E}_{x \sim pdata(x)}[\log D_x(x)]] + \mathbb{E}_{y \sim pdata(y)}[\log D_y(y)] \quad (1)$$

$$L_{GANfake} = \mathbb{E}_{x \sim pdata(x)}[\log(1 - D_y(G(x))) + \mathbb{E}_{y \sim pdata(y)}[\log(1 - D_x(F(y)))] \quad (2)$$

$$\min_{G,F} \max_{D_x,D_y} L_{GAN} = L_{GANreal} + L_{GANfake} \quad (3)$$

Cycle Consistency Loss [45] ensures that the generator maintains image consistency across different domains. For an image x in domain X, the cycle should return it to its original form, $x \to G(x) \to F(G(x)) \approx x$. Similarly, for y in domain Y, $y \to F(y) \to G(F(y)) \approx y$. This loss, defined by Eq. 4, ensures that the reconstructed image $F(G(x))$ closely matches the input x.

$$\min_{G,F} L_{cycle} = \mathbb{E}_{x \sim pdata(x)}[\|F(G(x)) - x\|_1] + \mathbb{E}_{y \sim pdata(y)}[\|G(F(y)) - y\|_1] \quad (4)$$

The Identity Preserving Loss [46] ensures that when the model processes an image from the target domain, the output remains consistent with the original input. Defined by Eq. 5, this loss function enforces that the generated output $G(y)$ ($y \in Y$) closely resembles y itself, maintaining the identity of the input image.

$$\min_{G,F} L_{identity} = \mathbb{E}_{x \sim pdata(x)}[\|F(x) - x\|_1] + \mathbb{E}_{y \sim pdata(y)}[\|G(y) - y\|_1] \quad (5)$$

The overall loss function is defined in Eq. 6, where L_{GAN}, L_{cycle}, and $L_{identity}$ correspond to Adversarial Loss, Cycle Consistency Loss and Identity Preserving Loss, respectively. The parameters λ_{cycle} and $\lambda_{identity}$ control the relative importance of each loss term. During training, λ_{cycle} is set to 10 and $\lambda_{identity}$ is set to 0.5.

$$\min_{G,F} \max_{D_x,D_y} L_{Total} = L_{GANreal} + L_{GANfake} + \lambda_{cycle}L_{cycle} + \lambda_{identity}L_{identity} \quad (6)$$

4 Experiments

4.1 Training Detail

To train PFG, two datasets labeled as X and Y are utilized (see Fig. 1). Dataset X is obtained from NTCIR-12 MathIR [37] and converted into handwritten formulas using FormulaGAN. Dataset Y is sourced from HME-100K [18]. To ensure a balanced training process, the length distribution of formulas in both datasets X and Y is made consistent. Ultimately, both datasets consist of 80,000 images.

4.2 Generated Image Quality Assessment

To assess the data quality of PFG, we randomly chose 10,000 formulas from the HME100K dataset [18]. We also selected 10,000 formulas from the FormulaGAN-processed dataset. By calculating the Fréchet Inception Distance (FID) [12] between these two sets of images, we obtained a measure of their dissimilarity. Then, we employed the PFG model to generate 10,000 handwritten formulas and recalculated the FID using the original 10,000 real formulas. Additionally, we divided the dataset into 5-length intervals, randomly selected 10,000 images from each subset, and performed the same calculations as before.

Table 1. Table comparing FID values of PFG and FormulaGAN generated images.

Data	Length					
	all	1~5	6~10	11~15	16~20	21~25
FormulaGAN	**106.66**	**121.49**	**106.12**	**111.47**	**118.02**	**117.95**
PFG	273.58	308.52	285.73	278.16	284.89	283.99
Data	Length					
	26~30	31~35	36~40	41~45	46~50	≥ 51
FormulaGAN	**125.60**	**125.92**	**133.03**	**145.54**	**155.19**	**163.14**
PFG	294.13	293.10	298.13	314.92	324.29	321.38

The results (see Table 1) indicate that the FID between PFG-generated images and real formulas is significantly smaller than FormulaGAN, suggesting that the formulas generated by PFG are closer to real data.

4.3 Evaluation of PFG in Data Augmentation

To evaluate the data augmentation capability of PFG, we randomly extracted 12,000 images from the HME100K dataset [18] to construct the "Baseline" training set. Additionally, 4,000 images were selected respectively as the validation and test sets. We combined the data generated by FormulaGAN with the Baseline training set to form the mixed dataset 'FormulaGAN'. Similarly, we applied

PFG and PFG-PGD (where the discriminator in the PFG model is replaced by a variant of PGD [15]) to generate two other augmented datasets, namely 'PFG' and 'PFG-PGD'.

Ensuring consistency in model parameters, we trained WYGIWYS [7] and DenseWAP [38], and evaluated their performance using metrics BLEU [24], METEOR [1], ROUGE [20] and CIDEr [32]. During the training process, the maximum number of epochs was set to 100, and the model parameters that performed best on the validation set were selected. The initial learning rate was set to 1×10^{-3}, and a Cosine Learning Rate Scheduler [21] was employed to dynamically adjust the learning rate.

Table 2. Table presenting the evaluation results of various data augmentation strategies on the performance of the WYGIWYS and DenseWAP.

Model	Training Set	BLEU	METEOR	ROUGE	CIDEr
WYGIWYS	Baseline	0.8021	0.5426	0.8975	7.2789
	FormulaGAN	0.8071	0.5456	0.8971	7.3165
	PFG-PGD	0.8071	0.5440	0.8971	7.3102
	PFG	**0.8103**	**0.5481**	**0.9013**	**7.3584**
DenseWAP	Baseline	0.7552	0.5164	0.8716	6.8321
	FormulaGAN	0.7203	0.4934	0.8468	6.4546
	PFG-PGD	0.7389	0.5057	0.8634	6.6946
	PFG	**0.7907**	**0.5376**	**0.8920**	**7.2310**

The results (see Table 2) demonstrate that data augmentation with PFG achieves the best performance. In contrast, other methods provide minimal augmentation benefits, resulting in negligible improvements in recognition accuracy.

5 Conclusion

In this paper, we introduce 'Paper-Style FormulaGAN' (PFG), a novel model that converts printed formulas into handwritten form with a paper texture. PFG aims to enhance the recognition of handwritten formulas by augmenting the training dataset. Specifically, PFG operates in two stages: first, it transforms LaTeX formulas into handwritten style; second, it applies a paper texture effect to the handwritten outputs.

Experimental results show that compared to FormulaGAN, images generated by PFG have a lower Fréchet Inception Distance (FID) with real images, indicating superior image quality. Moreover, models trained with data generated by PFG also achieve higher scores across various evaluation metrics.

Acknowledgments. This research was partially supported by the National Natural Science Foundation of China (No. 62337001), Anhui Provincial Natural Science Foundation (No. 2308085QF229), and the Fundamental Research Funds for the Central Universities (No. WK2150110034).

References

1. Banerjee, S., Lavie, A.: Meteor: An automatic metric for MT evaluation with improved correlation with human judgments. In: Proceedings of the ACL Workshop on Intrinsic and Extrinsic Evaluation Measures for Machine Translation and/or Summarization, pp. 65–72 (2005)
2. Brown, T., et al.: Language models are few-shot learners. Adv. Neural. Inf. Process. Syst. **33**, 1877–1901 (2020)
3. Chen, X., Xu, C., Yang, X., Tao, D.: Attention-gan for object transfiguration in wild images. In: Proceedings of the European Conference on Computer Vision (ECCV), pp. 164–180 (2018)
4. Choi, Y., Choi, M., Kim, M., Ha, J.W., Kim, S., Choo, J.: Stargan: unified generative adversarial networks for multi-domain image-to-image translation. In: Proceedings of the IEEE Conference on Computer Vision and Pattern Recognition, pp. 8789–8797 (2018)
5. Deng, J., Dong, W., Socher, R., Li, L.J., Li, K., Fei-Fei, L.: ImageNet: a large-scale hierarchical image database. In: 2009 IEEE Conference on Computer Vision and Pattern Recognition, pp. 248–255. IEEE (2009)
6. Deng, Y., Yu, Y., Yao, J., Sun, C.: An attention based image to latex markup decoder. In: 2017 Chinese Automation Congress (CAC), pp. 7199–7203. IEEE (2017)
7. Deng, Y., Kanervisto, A., Ling, J., Rush, A.M.: Image-to-markup generation with coarse-to-fine attention. In: International Conference on Machine Learning, pp. 980–989. PMLR (2017)
8. Devlin, J., Chang, M.W., Lee, K., Toutanova, K.: Bert: pre-training of deep bidirectional transformers for language understanding. arXiv preprint arXiv:1810.04805 (2018)
9. Dosovitskiy, A., et al.: An image is worth 16x16 words: transformers for image recognition at scale. arXiv preprint arXiv:2010.11929 (2020)
10. Goodfellow, I., et al.: Generative adversarial nets. In: Advances in Neural Information Processing Systems vol. 27 (2014)
11. He, K., Zhang, X., Ren, S., Sun, J.: Deep residual learning for image recognition. In: Proceedings of the IEEE Conference on Computer Vision and Pattern Recognition, pp. 770–778 (2016)
12. Heusel, M., Ramsauer, H., Unterthiner, T., Nessler, B., Hochreiter, S.: Gans trained by a two time-scale update rule converge to a local nash equilibrium. In: Advances in Neural Information Processing Systems, vol. 30 (2017)
13. Huang, G., Liu, Z., Van Der Maaten, L., Weinberger, K.Q.: Densely connected convolutional networks. In: Proceedings of the IEEE Conference on Computer Vision and Pattern Recognition, pp. 4700–4708 (2017)
14. Ioffe, S., Szegedy, C.: Batch normalization: accelerating deep network training by reducing internal covariate shift. In: International Conference on Machine Learning, pp. 448–456. pmlr (2015)

15. Isola, P., Zhu, J.Y., Zhou, T., Efros, A.A.: Image-to-image translation with conditional adversarial networks. In: Proceedings of the IEEE Conference on Computer Vision and Pattern Recognition, pp. 1125–1134 (2017)
16. Karnewar, A., Wang, O.: Msg-gan: multi-scale gradients for generative adversarial networks. In: Proceedings of the IEEE/CVF Conference on Computer Vision and Pattern Recognition, pp. 7799–7808 (2020)
17. Krogh, A., Hertz, J.: A simple weight decay can improve generalization. In: Advances in Neural Information Processing Systems, vol. 4 (1991)
18. Li, B., et al.: When counting meets hmer: counting-aware network for handwritten mathematical expression recognition. In: European Conference on Computer Vision, pp. 197–214. Springer (2022)
19. Liang, X., Zhang, H., Xing, E.P.: Generative semantic manipulation with contrasting gan. arXiv preprint arXiv:1708.00315 (2017)
20. Lin, C.Y.: Rouge: a package for automatic evaluation of summaries. In: Text Summarization Branches Out, pp. 74–81 (2004)
21. Loshchilov, I., Hutter, F.: Sgdr: stochastic gradient descent with warm restarts. arXiv preprint arXiv:1608.03983 (2016)
22. Mirza, M., Osindero, S.: Conditional generative adversarial nets. arXiv preprint arXiv:1411.1784 (2014)
23. Mouchère, H., Viard-Gaudin, C., Zanibbi, R., Garain, U.: Icfhr2016 crohme: competition on recognition of online handwritten mathematical expressions. In: 2016 15th International Conference on Frontiers in Handwriting Recognition (ICFHR), pp. 607–612. IEEE (2016)
24. Papineni, K., Roukos, S., Ward, T., Zhu, W.J.: Bleu: a method for automatic evaluation of machine translation. In: Proceedings of the 40th Annual Meeting of the Association for Computational Linguistics, pp. 311–318 (2002)
25. Pidhorskyi, S., Adjeroh, D.A., Doretto, G.: Adversarial latent autoencoders. In: Proceedings of the IEEE/CVF Conference on Computer Vision and Pattern Recognition, pp. 14104–14113 (2020)
26. Radford, A., Metz, L., Chintala, S.: Unsupervised representation learning with deep convolutional generative adversarial networks. arXiv preprint arXiv:1511.06434 (2015)
27. Springstein, M., Müller-Budack, E., Ewerth, R.: Unsupervised training data generation of handwritten formulas using generative adversarial networks with self-attention. In: Proceedings of the 2021 Workshop on Multi-Modal Pre-Training for Multimedia Understanding, pp. 46–54 (2021)
28. Sun, C., Shrivastava, A., Singh, S., Gupta, A.: Revisiting unreasonable effectiveness of data in deep learning era. In: Proceedings of the IEEE International Conference on Computer Vision, pp. 843–852 (2017)
29. Sutskever, I., Vinyals, O., Le, Q.V.: Sequence to sequence learning with neural networks. In: Advances in Neural Information Processing Systems, vol. 27 (2014)
30. Tang, H., Liu, H., Xu, D., Torr, P.H., Sebe, N.: Attentiongan: unpaired image-to-image translation using attention-guided generative adversarial networks. IEEE Trans. Neural Netw. Learn. Syst. **34**(4), 1972–1987 (2021)
31. Touvron, H., et al.: Llama: open and efficient foundation language models. arXiv preprint arXiv:2302.13971 (2023)
32. Vedantam, R., Lawrence Zitnick, C., Parikh, D.: Cider: Consensus-based image description evaluation. In: Proceedings of the IEEE Conference on Computer Vision and Pattern Recognition, pp. 4566–4575 (2015)

33. Wang, Z., Liu, J.C.: Translating math formula images to latex sequences using deep neural networks with sequence-level training. Int. J. Doc. Anal. Recogn. (IJDAR) **24**(1), 63–75 (2021)
34. Xu, D., Wang, W., Tang, H., Liu, H., Sebe, N., Ricci, E.: Structured attention guided convolutional neural fields for monocular depth estimation. In: Proceedings of the IEEE Conference on Computer Vision and Pattern Recognition, pp. 3917–3925 (2018)
35. Yang, Z., Liu, Q., Zhang, K., Tong, S., Chen, E.: Gap: a grammar and position-aware framework for efficient recognition of multi-line mathematical formulas. In: Proceedings of the 17th ACM International Conference on Web Search and Data Mining, pp. 901–910 (2024)
36. Yuan, Y., Liu, X., Dikubab, W., Liu, H., Ji, Z., Wu, Z., Bai, X.: Syntax-aware network for handwritten mathematical expression recognition. In: Proceedings of the IEEE/CVF Conference on Computer Vision and Pattern Recognition, pp. 4553–4562 (2022)
37. Zanibbi, R., Aizawa, A., Kohlhase, M., Ounis, I., Topic, G., Davila, K.: Ntcir-12 mathir task overview. In: NTCIR (2016)
38. Zhang, J., Du, J., Dai, L.: Multi-scale attention with dense encoder for handwritten mathematical expression recognition. In: 2018 24th International Conference on Pattern Recognition (ICPR), pp. 2245–2250. IEEE (2018)
39. Zhang, J., Du, J., Yang, Y., Song, Y.Z., Wei, S., Dai, L.: A tree-structured decoder for image-to-markup generation. In: International Conference on Machine Learning, pp. 11076–11085. PMLR (2020)
40. Zhang, K., et al.: Eatn: an efficient adaptive transfer network for aspect-level sentiment analysis. IEEE Trans. Knowl. Data Eng. **35**(1), 377–389 (2021)
41. Zhang, K., Zhang, H., Liu, Q., Zhao, H., Zhu, H., Chen, E.: Interactive attention transfer network for cross-domain sentiment classification. In: Proceedings of the AAAI Conference on Artificial Intelligence, vol. 33, pp. 5773–5780 (2019)
42. Zhang, K., et al.: Incorporating dynamic semantics into pre-trained language model for aspect-based sentiment analysis. arXiv preprint arXiv:2203.16369 (2022)
43. Zhang, W., Bai, Z., Zhu, Y.: An improved approach based on cnn-rnns for mathematical expression recognition. In: Proceedings of the 2019 4th International Conference on Multimedia Systems and Signal Processing, pp. 57–61 (2019)
44. Zhao, W., Gao, L.: Comer: Modeling coverage for transformer-based handwritten mathematical expression recognition. In: European Conference on Computer Vision, pp. 392–408. Springer (2022)
45. Zhu, J.Y., Park, T., Isola, P., Efros, A.A.: Unpaired image-to-image translation using cycle-consistent adversarial networks. In: Proceedings of the IEEE International Conference on Computer Vision, pp. 2223–2232 (2017)
46. Zhu, J.Y., et al.: Toward multimodal image-to-image translation. In: Advances in Neural Information Processing Systems, vol. 30 (2017)

Author Index

A
Ai, Yuang 47
Ao, Xiang 1, 189, 232

C
Cai, Shengze 147
Cao, Jie 47
Chen, Enhong 361
Chen, Haonan 15
Chen, Hongmei 345
Chen, Jiaen 121
Chen, Kailiang 121
Chen, Shufen 121
Chen, Wenbin 121
Chen, Xingyan 275
Chen, Yanjiang 275
Chen, Yanping 260
Chen, Yutong 275
Chu, Xuesen 147

D
Ding, Yanhui 62
Dou, Zhicheng 15
Du, Hongyun 62
Du, Junping 90

F
Fan, Jiping 90
Feng, Jing 174
Fu, Jiaojiao 290

G
Gan, Aoran 102
Gao, Huachao 78
Gao, Yuanjun 202
Ge, Hao 78
Gu, Bin 159
Guan, Zeli 90
Guo, Fanglin 247

Guo, Wei 147
Guo, Yi 290
Guo, Yuyao 189

H
Han, Tongle 174
He, Qing 1, 232
He, Ran 47
Hu, Wentao 78
Huang, Huaibo 47
Huang, Ruizhang 260

J
Jia, Shengjie 62
Jin, Hai 202
Jin, Haoxin 159

L
Li, Ang 90
Li, Chongshou 134
Li, Chunxiao 78
Li, Dazhong 78
Li, Tianrui 134
Li, Xinwei 319, 333
Li, Xuqiang 275
Li, Yang 78
Li, Zhigang 121
Liang, Meiyu 90
Liu, Qi 102, 361
Liu, Wuying 218
Liu, Yang 1
Liu, Yuchen 305
Liu, Ze 361
Liu, Zhaofeng 102

M
Mao, Jiaxin 15
Miao, Kuan 134

N
Niu, Renzhong 121
Niu, Xiaolong 78

P
Pan, Jibao 134
Peng, Lilan 134

Q
Qi, Jianpeng 305
Qin, Yongbin 260
Qu, Yunhan 174

R
Ren, Yawei 78

S
Shi, Xuanhua 202
Song, Wenbin 319, 333
Song, Yulun 78

T
Tong, Shiwei 102
Tran, Vanha 345

W
Wang, Jitian 260
Wang, Lin 218
Wang, Lizhen 345
Wang, Shuai 319, 333
Wang, Xin 290
Wu, Feng 159
Wu, Tao 47
Wu, Tianqi 147
Wu, Xiujie 62

X
Xia, Ziqi 174
Xiao, Weidong 34
Xu, Liang 345
Xu, Weizhi 62
Xu, Yanyan 78
Xu, Zhao-Dong 319, 333
Xue, Zhe 90

Y
Yan, Long 78
Yang, Bowen 290
Yang, Guangwen 147
Yang, Yikun 121
Yang, Zhe 361
Yi, Zhongchao 275
Yin, Baijian 319
Yu, Guoxin 232
Yu, Hao 102
Yu, Hui 62
Yu, Yanwei 305

Z
Zhang, Kai 102, 361
Zhang, Shiqi 34
Zhang, Wuyang 134
Zhang, Xinyue 189
Zhang, Xuehua 174
Zhang, Yanghai 361
Zhang, Yiming 232
Zhao, Long 159
Zhao, Peiyi 333
Zhao, Wenbin 159
Zhao, Xiaofei 247
Zhao, Yangwu 1
Zheng, Zixuan 159
Zhou, Yangfan 290
Zhu, Jiran 62
Zhu, Quntao 202
Zou, Anqi 260

www.ingramcontent.com/pod-product-compliance
Ingram Content Group UK Ltd.
Pitfield, Milton Keynes, MK11 3LW, UK
UKHW022328130225
455076UK00007B/180